S0-AKY-067

BIOMIMETIC POLYMERS

BIOMIMETIC POLYMERS

Edited by
Charles G. Gebelein
Youngstown State University
Youngstown, Ohio

PLENUM PRESS • NEW YORK AND LONDON

Library of Congress Cataloging-in-Publication Data

Biomimetic polymers / edited by Charles G. Gebelein.
p. cm.
Based on the proceedings of the American Chemical Society Symposium on Enzyme Mimetic and Related Polymers, held at the Third Chemical Congress of North America, July 5-8, 1988, in Toronto, Ontario, Canada.
Includes bibliographical references and index.
ISBN 0-306-43708-2
1. Biomimetic polymers--Congresses. I. Gebelein, Charles G.
QD382.B47B56 1990
547.7--dc20 90-46930
CIP

Based on the proceedings of the American Chemical Society Symposium on Enzyme Mimetic and Related Polymers, held at the Third Chemical Congress of North America, July 5–8, 1988, in Toronto, Ontario, Canada

ISBN 0-306-43708-2

Printed in the United States of America

PREFACE

The term biomimetic is comparatively new on the chemical scene, but the concept has been utilized by chemists for many years. Furthermore, the basic idea of making a synthetic material that can imitate the functions of natural materials probably could be traced back into antiquity. From the dawn of creation, people have probably attempted to duplicate or modify the activities of the natural world. (One can even find allusions to these attempts in the Bible; e.g., Genesis 30.)

The term "mimetic" means to imitate or mimic. The word "mimic" means to copy closely, or to imitate accurately. Biomimetic, which has not yet entered most dictionaries, means to imitate or mimic some specific biological function. Usually, the objective of biomimetics is to form some useful material without the need of utilizing living systems. In a similar manner, the term biomimetic polymers means creating synthetic polymers which imitate the activity of natural bioactive polymers. This is a major advance in polymer chemistry because the natural bioactive polymers are the basis of life itself. Thus, biomimetic polymers imitate the life process in many ways. This present volume delineates some of the recent progress being made in this vast field of biomimetic polymers.

Chemists have been making biomimetic polymers for more than fifty years, although this term wasn't used in the early investigations. The pioneering work of Overberger and others helped open new vistas of bioactive polymer research in the early 1950s. This was followed by intensified research in the 1960s and 1970s, lead by Donaruma, Samour, Vogl, Takemoto, Goodman, Ringsdorf, Levy and others. By the mid-1970s, the scholarly domain of biomimetic polymers was well established. What remained was to translate this research into practical applications. This effort continues today, but some biomimetic polymers are already showing promise in several biomedical operations.

In this book, we'll consider a few of the major classes of biomimetic polymers. The natural bioactive polymers are divided into several classes and mimics of most are represented herein. Enzymes are possibly the best know natural bioactive polymer, and the first seven chapters consider polymers which emulate the catalytic activity of these important polypeptides. The first two papers describe some organic synthesis reactions catalyzed by polymeric materials. In the first paper, Wolff describes polymers with chiral cavities and their use in organic synthesis. Burdick and Schaeffer then describe the use of thin films of some biocatalysts, similar to photographic films, to synthesize some important biochemicals.

The next paper (Mathias, et al.) describes enzyme-like activity for some simple polymers based on 4-diallylaminopyridine. These polymeric catalysts were useful in hydrolysis and esterification reactions. Carraher, et al., describe the synthesis of some potentially bioactive poly-

(amino acids) containing platinum or titanium atoms. Zeolites, inorganic polymers, are then considered as potential biomimetics with various sized cavities (Herron).

The sixth and seventh papers consider natural polymers whose properties have been modified to give enzyme-like behavior. In the first, Hilvert, the bioactivity of catalytic antibodies is described; this is an especially promising biomimetic realm, with dozens of new papers each year. The second paper (Keyes and Albert) discusses the process of "re-educating" ordinary protein molecules to teach them "new tricks" and make them act like a regular enzyme.

Heparin, an anionic polysaccharide, is the central polymer in the next three papers. In the first of these papers, Linhardt and Loganathan review the chemistry and biological activities of heparin and the heparinoids. The extracorporeal removal of heparin from the blood is then considered by Yang and Teng. The final paper in this group, Sharma, considers the blood compatibility of some potential dialysis-membranes, derived from poly(vinyl alcohol), and containing heparin-like polymers.

The membrane theme continues in the paper by Imanishi and Kimura, who consider synthetic polypeptides as membrane compartment polymers for the enkephalins. Penczek and Klosinski then describe synthetic versions of two important classes of bioactive polymers, the teichoic acids (found in the cell walls of many bacteria) and the nucleic acids. These analogs are based on phosphates and related polymeric systems. The synthetic teichoic acids permit the separation of different ions from mixtures. The synthetic nucleic acids are amenable to large scale synthesis, unlike the oligonucleotides. Takemoto, et al., also describes nucleic acid analogs, but their methodology is based on polypeptides or poly(ethyleneimines) with pendant nucleic base units. The final synthetic nucleic acid paper, Gebelein, discusses some potential medical applications for the nucleic acid analogs. In the last paper, Wang describes a sustained release system which uses insulin to increase the growth rate in rats.

Ten of the sixteen papers in this book were presented at a symposium on "Enzyme Mimetic and Related Polymers," which was organized, by the Editor, for the Third Chemical Congress of North America in Toronto, Canada, June, 1988. The sponsoring groups were the American Chemical Society Division of Polymeric Materials:Science and Engineering and the Biotechnology Secretariat. Their support is gratefully acknowledged. All of the manuscripts were centrally word-processed by CG ENTERPRISES.

Charles G. Gebelein

CONTENTS

BIOMIMETIC REACTIONS USING ORGANIZED POLYMERIC SUPPORTS

G. Wulff

Institute of Organic Chemistry & Macromolecular Chemistry
University of Düsseldorf, Universitätsstr. 1
D-400 Düsseldorf, F.R.G.

For the preparation of polymeric supports possessing organized structures, an imprinting procedure on the basis of template approach was used in crosslinked polymers as well as in silicas. In this manner, chiral cavities possessing definite shapes and specific arrangements of the functional groups (corresponding to the templates) were obtained within the polymer. These polymers could be used for separation of racemates. Using an analogous procedure, functional groups with defined cooperativity could be obtained on the surfaces of silica. Highly reactive 1.3.2-dioxaborole moieties were attached to polymers and were used as reagents for an aldol type reaction with aldehydes. In this manner, α,β-dihydroxy aldehydes, particularly carbohydrates are thus obtained.

INTRODUCTION

Several attempts have been made during recent years towards designing catalysts and reagents functioning in a manner similar to enzymes. The most important problem associated with this approach is to achieve the right stereochemical arrangement at the active site of the catalyst. One of the unique characteristics of enzyme catalysis is the binding of the reacting substrate in a perfectly fitting cavity or cleft of the enzyme that contains functional groups in the correct stereochemistry for binding, catalysis, and group transfer. This particular feature has been very difficult to mimic while designing synthetic catalysts.

During the last few years remarkable progress has been achieved in the design of organic molecules based on molecular recognition. The winners of the last year's Nobel Prize, D. J. Cram[1] and J.-M. Lehn,[2] used low molecular weight crown ethers or cryptates as the molecular hosts with cavities for specific binding. It was also possible to fix catalytically active functional groups in the right vicinity to the reacting groups of the bound substrate. Remarkable enhancement in rate and selectivity were observed though turn-over numbers of such reactions were rather poor. Other hosts have been used in the form of cyclodextrins by Bender[3] and Breslow[4] or in the form of certain concave molecules by

Biomimetic Polymers
Edited by C. G. Gebelein
Plenum Press, New York, 1990

Rebek[5] and others for performing such type of selective chemical operations.

Use of polymers as the carrier for the active site of the catalyst offers another possibility for designing such enzyme-like catalysts. Basically, the use of polymers makes the system further complicated compared to its low molecular weight counterparts, since the support needs to be made organized. On the other hand the use of polymeric substrates would offer certain advantages by taking the macromolecular nature of the enzymes into consideration. In fact many of the unique features of enzymes are directly related to their polymeric nature. This is particularly true for the high cooperativity of the functional groups and the dynamic effects such as the induced fit, the allosteric effect, and the steric strain as exhibited by the enzymes.

For obtaining polymeric catalysts, during the early years, catalytically active groups and binding groups were mostly introduced into polymers by copolymerization of the appropriate monomers bearing the desired functionalities. By this method one obtains a polymer with randomly distributed functional groups (Figure 1A). Another possibility involves the grafting of side chains containing the desired arrangement of functional groups onto the parent polymer (Figure 1B). A third possibility is the polymerization of monomers with desired arrangement of functional groups. In this case the groups are localized in the main chain one after another, as in some hormone receptors (Figure 1C).

On the contrary, in the case of natural enzymes, the functional groups responsible for the specificity are located at quite distant points from each other along the peptide chain and are brought into spatial relationship as a result of specific folding of the chain. In this case, both the functional group sequence in the chain and the peptide's tertiary structure, i.e., its topochemistry, are decisive (Figure

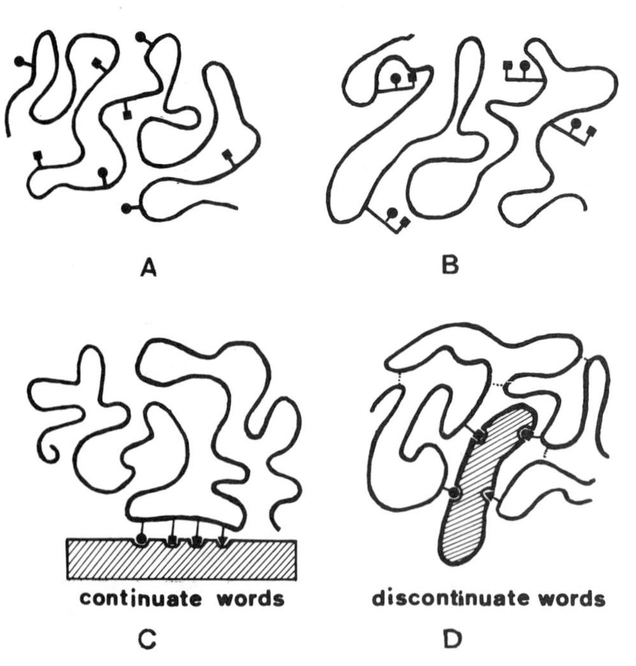

Figure 1. Possible arrangements of functional groups in synthetic and natural polymers.[8]

1D). This type of arrangement has been termed as the "discontinuate word" by R. Schwyzer.[6] In this case complex, three-dimensional steric arrangements of the functional groups can be obtained.

This review on biomimetic reactions using organized polymers consists of two parts. The first part of this article deals with the method developed to prepare defined cavities in polymers by an imprinting procedure using a template approach. With this method it becomes possible to introduce functional groups into the polymer in a discontinuate word arrangement.[7-10] Furthermore, the use of the template technique to introduce functional groups fixed in a defined distance on the surface of a solid support is described.[11,12] The second part of this article deals with the use of a new polymeric reagent for performing a biomimetic type of reaction.[13,14]

MOLECULAR RECOGNITION IN POLYMERS PREPARED BY IMPRINTING WITH TEMPLATES

In order to prepare a polymeric synthetic model of the binding site of an enzyme, we used a new approach.[7-10] The functional groups to be introduced were bound in the form of polymerizable vinyl derivatives to a suitable template molecule. This monomer was subsequently copolymerized under appropriate conditions to produce highly crosslinked polymers having chains in a fixed arrangement. After removal of the template, free cavities of the type shown in Figure 2 were formed with a shape and a three dimensional arrangement of functional groups corresponding to that of the template. The functional groups in this polymer are located at quite different points along the polymer chain and they are held in spatial relationship with one another by the crosslinking. This approach to prepare a cavity differs from those of Cram,[1] Lehn,[2] and others, who used crown type compounds to provide a low molecular weight moiety carrying the desired stereochemical information. On the other hand, this method has some similarity with that of Dickey[15] who precipitated silicic acid in presence of certain templates to produce silicas with affinity for the concerned templates.

For the optimization of the above technique we have chosen the monomer [1] as one of the template monomer.[9,16,17] The template is phenyl-α-D-mannopyranoside to which two molecules of 4-vinylbenzeneboronic acids are bound by diester linkages. The boronic acid groups act as the binding sites. Since a chiral template was chosen, the accuracy of the steric arrangement of the binding sites in the cavity could be tested by the

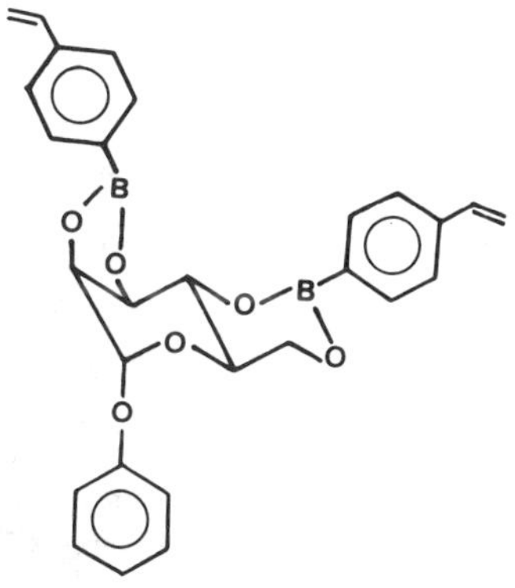

Structure of monomer [1].

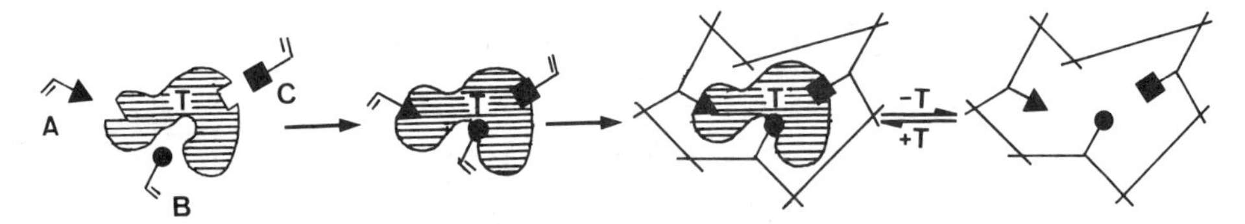

Figure 2. The preparation of defined functional cavities by an imprinting approach.[8]

ability of the polymer to resolve the racemate of the template after cleavage of the original template. The monomer was copolymerized by free radical initiation in the presence of an inert solvent (porogenic agent) with large amount of bifunctional crosslinking agent. Under these conditions macroporous polymers possessing a permanent pore structure and a high inner surface area were obtained. This would result in a good accessibility and a low swelling ability and hence a limited mobility of the polymer chain.

From this type of polymer, the templates could be split off to the extent of 60-90% using water or alcohol (see Figure 3a and 3b). When this polymer is treated with the racemate of the template in a batch procedure under equilibrium conditions, the enantiomer, which has been used as the template for the preparation of the polymer, is taken up preferentially. If the specificity is expressed by the separation factor α, which is the ratio of the distribution coefficients of L and D-form between solution and polymer, values ranging from 1.20 to 4.76 were obtained depending on the equilibration condition and the polymer structure.[9,16,17] The highest α-value obtained so far is 4.76. In this case, in the simple batch procedure a maximal enrichment of the D-form at the polymer to the extent of 70-80% was observed.

The enantiomer selectivity of these polymers is strongly dependent on the type and amount of crosslinking agent used during the polymerization.[16,17] With ethylene glycol dimethacrylate as the crosslinking agent, it was observed that for polymers containing <10% crosslinking (Figure 4) virtually no specificity was observed. Up to 50% crosslinking, the α value increases linearly to 1.50. From 50-66.7%, a dramatic increase in α value from 1.50 to 3.04 was observed, thus implying a fourfold increase in selectivity in this range. With further increase in crosslinking up to 95% the specificity rises to α = 3.66. On the other hand, the use of butanediol dimethacrylate and especially p-divinylbenzene as crosslinking agents showed a much lower specificity as a function of crosslinking percentage.

Polymers obtained with ethylene glycol dimethacrylate as crosslinker retained their specificity for a longer period. Even under high pressure in a high performance liquid chromatography (HPLC) column, the activity remained for months. This was true even when used at 70-80°C. On the contrary polymers crosslinked with divinylbenzene gradually lost their specificity at higher temperature.[10,18,19] Interestingly, at 80°C the α value for racemic resolution was further increased to 4.56.[20]

In addition to the requirement of rigidity for cavity stability, the polymers should, at the same time, possess some degree of flexibility. This is necessary to enable a fast reversible binding of the substrates within the cavities. Cavities of accurate shape but without any flexibility present kinetic hindrance to reversible binding.

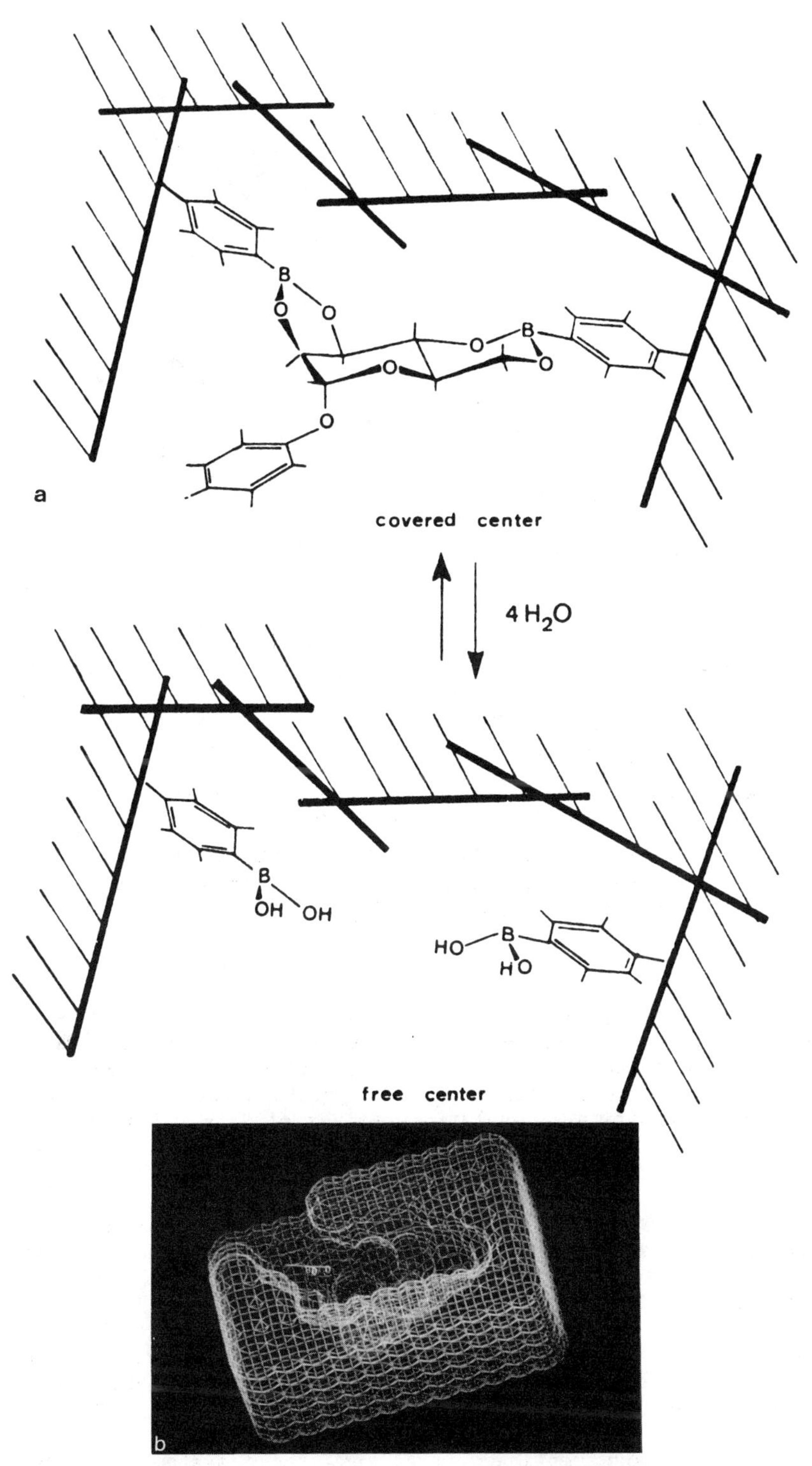

Figure 3. (a) Removal and uptake of the template from an imprinted polymer.[9] (b) Computer graphical representation of the imprinted cavity carrying phenyl-α-D-mannopyranoside as the template.

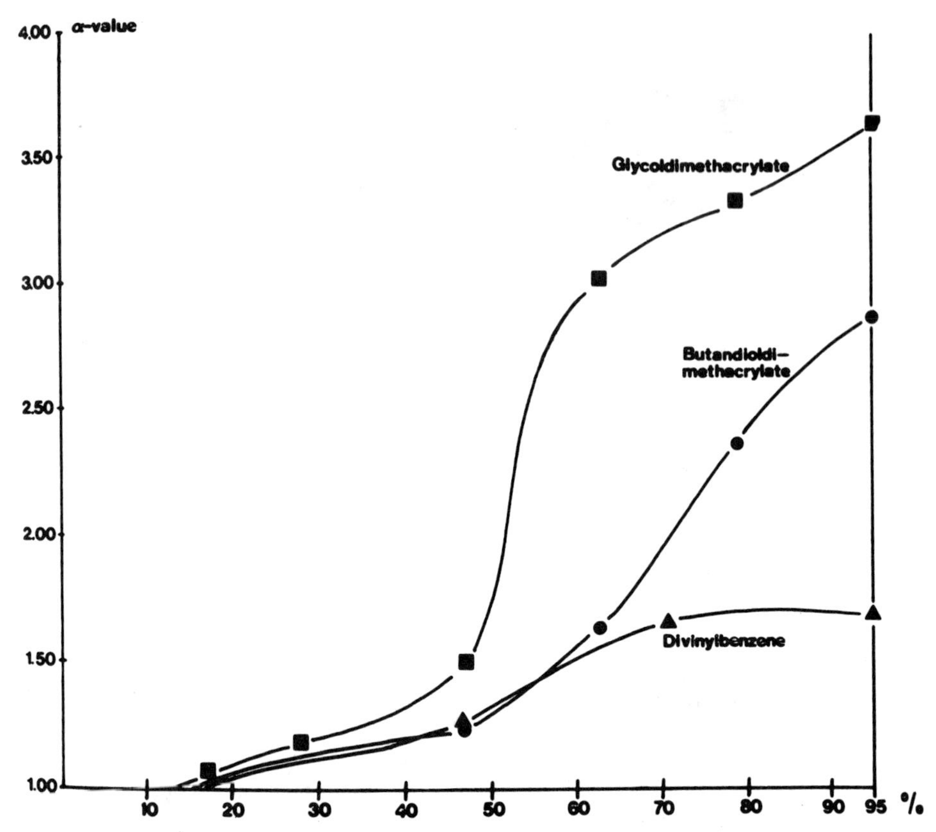

Figure 4. Dependence of the specificity of the polymers for racemate resolution upon the kind and amount of crosslinking agent.[16,17]

Polymers carrying chiral cavities obtained by polymerization in the presence of suitable templates can be used for the chromatographic separation of the racemates of the template molecules.[19-21] The selectivity of the separation process is fairly high (separation factors up to $\alpha = 4.56$) and at higher temperature resolutions values of $Rs = 2$ have been obtained, as well as satisfactory base-line separations have been obtained (see Figure 5).

These sorbents can be prepared conveniently and possess excellent thermomechanical stability. Even when used at 80°C under high pressure for a long time, no leakage of the stationary phase during chromatography is observed.

In the mean time, many examples for imprinting with templates are known from our group and from other groups,[10] like those of Shea,[22] Neckers,[23] and Mosbach.[24]

Besides the above mentioned novelties, these systems are associated with some shortcomings. One of the shortcomings of this method is the slow kinetics of mass transfer. This might arise due to the binding reactions employed. The formation of the boron diesters within the cavity could be sterically hindered. But, it was observed by Mosbach and his coworkers,[24] that even with non-covalent interactions slow mass transfer

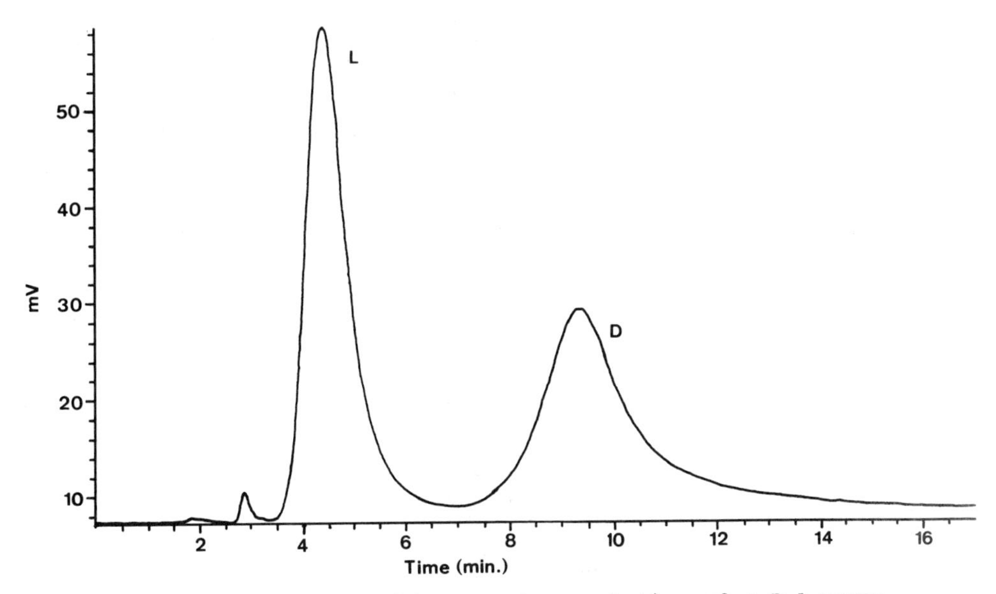

Figure 5. Chromatographic racemic resolution of α-D,L-mannopyranoside on imprinted polymers (see Figure 3). Chromatography at 70°C; flow rate 1 mL/min; eluent as gradient acetonitrile containing 5% aqueous NH_3 (25%) and water containing 5% NH_3 (25%) ranging from 9:1 to 5:5. (M. Minarik, G. Wulff, unpublished data).

was encountered. Therefore, it appears that a two-point binding within a cavity is in general a kinetically restricted process.[25]

Another drawback is the high selectivity of the separation. While other less selective sorbents possess a broader applicability, with respect to substrates our sorbents can be employed only for the resolution of racemates of the template or for racemates possessing a similar structure to the templates. For substrates of different chemical structure, therefore, a new sorbent must be prepared. However, these highly selective and durable polymers can be used advantageously for routine analysis and, in particular, for preparative separations.

EXACT PLACEMENT OF FUNCTIONAL GROUPS ON THE SURFACES OF RIGID MATRICES

The selectivity observed in the type of racemic resolution discussed above is an outcome of the combination of an accurate cavity-shape fitting and the exactness of the arrangement of the functional groups. It was of prime interest to examine whether the arrangement of the functional groups alone can give rise to selectivity due to the distinct distance between two functional groups. In other words, the question is whether a two-dimensional instead of a three-dimensional information transfer will be adequate to bring about selectivity.

It was attempted to introduce two amino groups arranged at a distinct distance into crosslinked polymers with the aid of a template (see Figure 6a). Furthermore, a similar imprinting procedure on the surface of silica was performed through the formation of siloxane bonds. With this method it was possible to locate two functional groups on a more or less plane

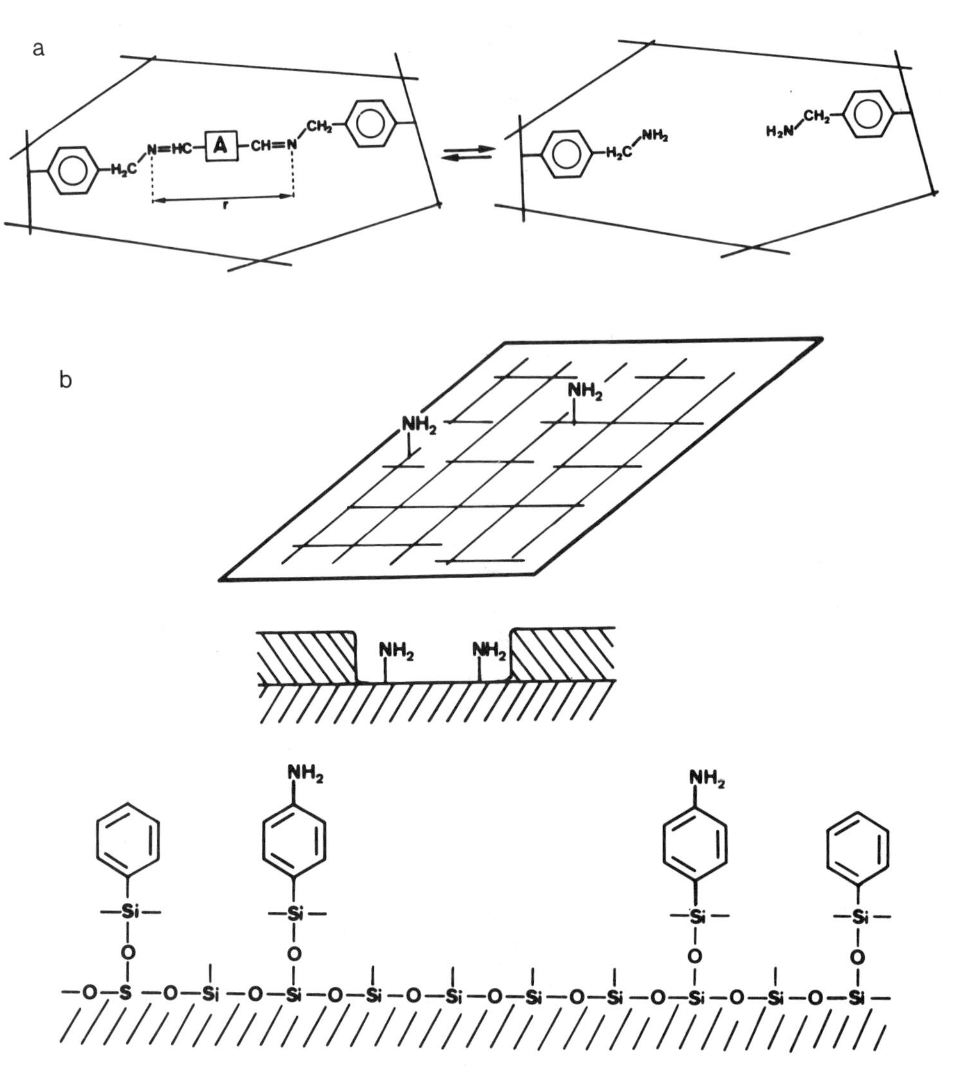

Figure 6. (a) Introduction of two amino groups in a distinct distance apart into a crosslinked polymer.[11,12]
(b) Introduction of two amino groups in a distinct distance apart on a plane surface of silica.[11,12]

surface of silica (see Figure 6b) thus enabling to investigate the selectivity caused by distance accuracy alone.[11,12]

For imprinting purposes, the bis(azomethines) [2] and [3] were used as template monomers. They were radically copolymerized with ethylene glycol dimethacrylate in the presence of inert solvent to produce macroreticular polymers. The dialdehyde templates could be split off under mild conditions. The mean distance between the remaining amino groups should be 0.72 nm in the case of [2] and 1.56 nm in the case of [3] as template monomer. For comparison purposes a polymer with randomly distributed amino groups was prepared from [4].

In order to elucidate the role of distance accuracy for selectivity both the polymers were equilibrated with an equimolar mixture of the three dialdehydes [5], [6] and [7]. For a two-point binding a distance of

0.72, 1.14, and 1.56 nm, respectively between the amino groups in the polymer are required. With other distances only a one-point binding is expected.

The apparent binding constants obtained under the conditions used are presented in Table 1. Polymers [A-1] and [A-2] showed the highest uptake for the dialdehyde having the same distance of the functional groups as the template. The selectivity was calculated as the selectivity factor α and corrected for the different nonspecific adsorption to the polymer with statistically distributed amino groups; using Equation 1, where K_5 = apparent binding constant for substance [5] on Polymers [A-1] or [A-2] and K_{5stat} = the apparent binding constant for substance [5] on Polymer [A-3].

$$\alpha'_{5,6} = (K_5/K_{5stat})/(K_6/K_{6stat}) \qquad \text{(Equation 1)}$$

The observed selectivity is surprisingly high in the case of [A-1], with $\alpha'_{5,7} = 5.37$. Even a difference in distance of only 0.42 nm (4.2 A) shows a selectivity of $\alpha'_{5,6} = 4.60$. Similar work on the distance accuracy of groups in crosslinked polymer was performed by the group of Shea.[26]

Besides a selective two-point binding, the foregoing examples might contain some shape selectivity as well. In order to eliminate the latter effect, if any, two amino groups were attached to the surface of silica

Table 1. Selectivity of polymers with each of the two amino groups in a defined distance.[11,12]

Monomer 2

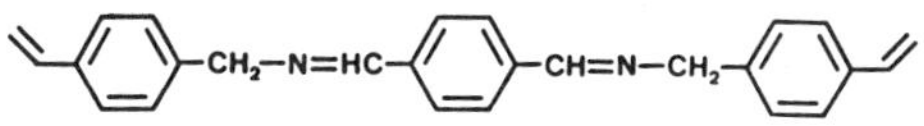

Monomer 3

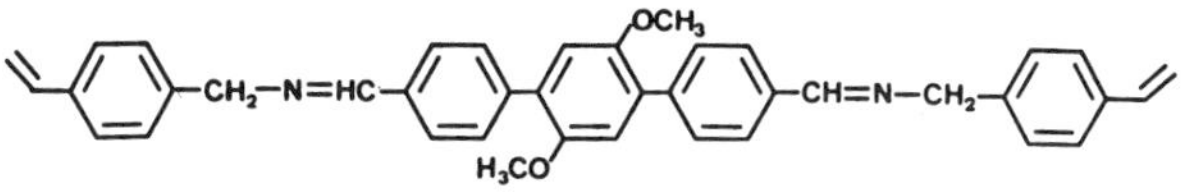

	Splitting Percentage	Distance r of groups (nm)	Apparent Binding Constants of 5	6	7	Selectivity
Polymer A-1 from 12	84%	0.72	1.38	0.20	0.20	$\alpha_{5,6} = 4.60$ $\alpha_{5,7} = 5.37$
Polymer A-2 from 3	91%	1.56	0.42	0.30	0.59	$\alpha_{7,6} = 1.69$ $\alpha_{7,5} = 1.81$
Polymer A-3 from 4	~90%	—	0.18	0.12	0.14	—

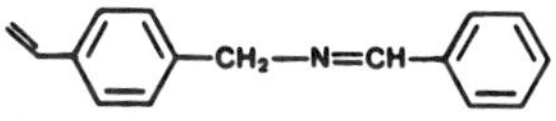

4 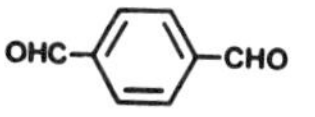 5 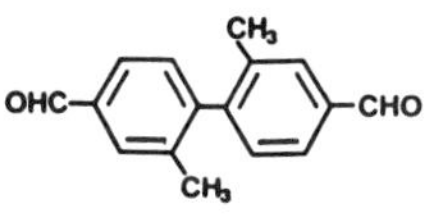 6 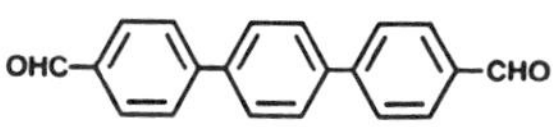 7

The distance of the groups was determined from molecular models, assuming the most probable conformation. Equilibration in acetonitrile containing 2.5% water and 10 µmol/L toluene sulfonic acid. The equilibration was performed with a mixture of the dialdehydes [5], [6] and [7]. Selectivity was corrected for non-specific binding to polymer [A-3].

at a distance from one another using the template monomers [8] and [9]. In this case the attachment to the surface is achieved by condensation through the formation of siloxane bonds between the methoxy silane group of [8] and [9] and the silanol groups on the surface of the silica. Most of the remaining silanol groups were afterwards capped by reaction with hexamethyldisilazane to avoid nonspecific adsorption. The templates could be split off over 95% (see Figure 6b). Unlike polymers, in this case the position of the two amino groups should not be changed as a result of chain mobility, swelling, or shrinking. The distance can only be altered by conformational changes within the functional group part.

The selectivity was determined by equilibration with an equimolar mixture of the two template dialdehydes [5] and [11] (see Table 2). Both the silicas showed a significant difference in binding preferring their own templates with α-values of 1.74 and 1.67. This clearly suggests that by the distance selectivity alone and with differences of only 0.33 nm (between [5] and [11]), substrate selectivity can be observed.

Imprinting on the surface of silica is thus a further extension of the original imprinting method. The method appears to be applicable to several other conceivable examples.

A NEW POLYMERIC REAGENT

Polymers bearing cavities of defined shape and with arrangements of functional groups in predetermined stereochemistry appear to serve as enzyme models. Many enzymes contain a cofactor, called coenzyme, which reacts with the bound substrate. Towards this end, we are looking for suitable reagent groups acting in a similar manner. To begin with, these investigations are performed with low molecular weight compounds transferred to polymers. The large range objectives has been subsequently directed towards incorporating these species within organized polymeric

Table 2. Selectivity of modified silicas with each of the two amino groups in a defined distance.[11,12] For more details, see the legend for Table 1.

Monomer 8: $H_3CO-Si(CH_3)_2-C_6H_4-N=CH-C_6H_4-CH=N-C_6H_4-Si(CH_3)_2-OCH_3$

Monomer 9: $H_3CO-Si(CH_3)_2-C_6H_4-N=CH-C_6H_4-CH_2-C_6H_4-CH=N-C_6H_4-Si(CH_3)_2-OCH_3$

	Splitting Percentage	Distance r of groups (nm)	Apparent Binding Constants of $OHC-C_6H_4-CHO$ 5	$OHC-C_6H_4-CH_2-C_6H_4-CHO$ 11	Selectivity α'
Silica modified with 8	>95%	0.72	4.91	2.58	1.74
Silica modified with 9	>95%	1.05	9.07	13.77	1.67
Silica modified with 10 (at random)	—	—	2.26	2.05	—

Monomer 10 $H_2N-(CH_2)_3-Si(OCH_3)_3$

media with the anticipation of mimicing the reactions carried out by enzymes.

In order to synthesize certain sugars, the glycolaldehyde anion $[CHOH-CHO]^-$ appears to play an important role as a synthon since it can be exploited strategically via an aldol reaction for the synthesis of α,β-dihydroxyaldehydes, carbohydrates in particular.

Cyclic enediolates should be suitable as synthetic equivalents for the above mentioned synthon. We have been interested in exploring the potentiality of the previously unknown 1,3,2-dioxaboroles [12], having substituents only in the 2-position, as building blocks for the synthesis of carbohydrates.[13,14] These compounds [12] were synthesized by reaction of the boronic acid with the diol [15], which was obtained from vinylene carbonate [13] and anthracene [14] by the Diels-Alder reaction and subsequent hydrolysis. These esters [16] were sublimated at 160-220°C/ 10^{-2} mbar and could be converted almost quantitatively by thermolysis in the gas phase at 550°C/10^{-2} mbar into analytically pure 1,3,2-dioxaboroles [12]. The total yields with respect to [13] were over 80% (see Figure 7). Another possibility for obtaining [12] involves a three step synthesis starting with ethylene glycol as can be seen from Figure 8.[27]

The 1,3,2-dioxaborole reacts with aldehydes in an aldol fashion. In order to restrict the reaction at a single addition to aldehydes, [12] was immobilized on a crosslinked polymer. This strategy for site separation inhibits any further reaction of the primary addition products. Polymer bound [12] (= [22]) could be obtained quite easily by triethylamine-catalyzed, rapid transesterification (see Figure 9).

As an example of the applications of [22], the synthesis of L-ribose is described in Figure 10. 2,3-O-cyclohexylidene-L-glyceraldehyde (S)-[23] was allowed to react with [22] (CH_2Cl_2, room temperature, 12-36 hr.). The addition products could be isolated from the polymer using methanol/water. Removal of the protecting group (by ion exchange) furnished the aldopentoses of the L-series in 70-95% yield with respect to the 1,3,2-dioxaborole units, or 75% with respect to (S)-[23].

Enantiomerically pure L-ribose was obtained by liquid chromatographic separation (LiCHrosorb-NH_2, acetonitrile/water 85:15; $[\alpha]^{20}_D = +19.7$ {c = 0.22, H_2O}).

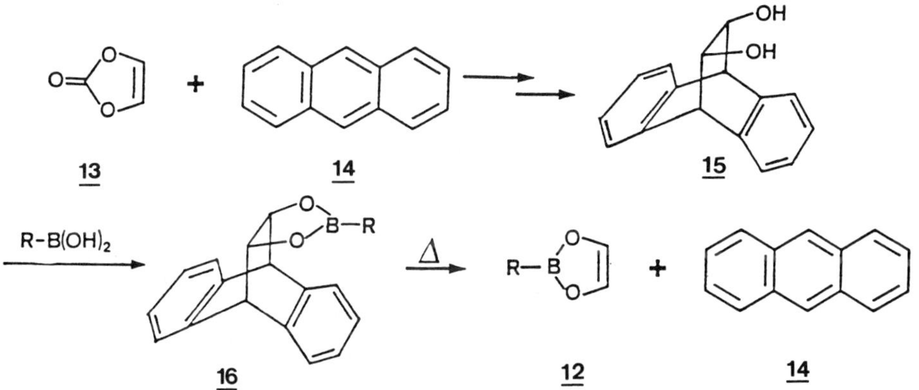

Figure 7. Preparation of phenyl-1,3,2-dioxaborole via the retro-Diels-Alder reaction.[13,14]

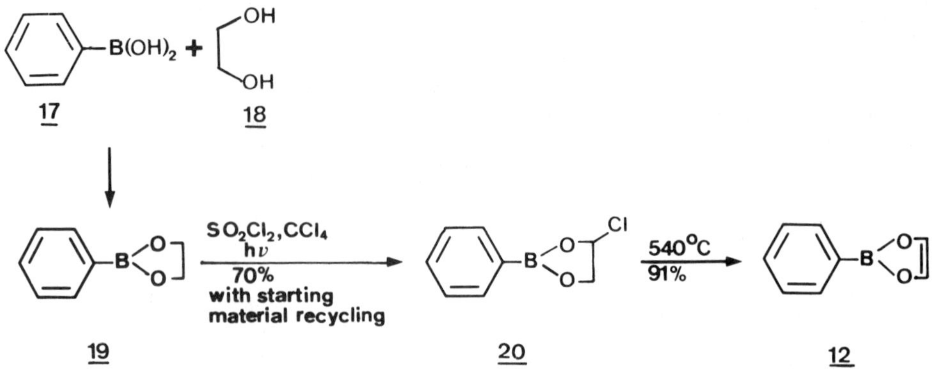

Figure 8. Preparation of phenyl-1,3,2-dioxaborole via chlorination of the 1,3,2-dioxaborinane.[27]

The nucleophilic attack of the enediolate [**22**] at the aldehyde group of [**23**] is controlled *ul*-selectively (1,2-induction *ca.* 9:1) by the diastereofacial influence of the cyclohexylidene-dioxy residue according to the Felkin-Anh model.[28] An enantiofacial control (1.3-induction) was seldom observed. The threo-erythro ratio at C-2 and C-3 was 1:2, similar to that found with achiral aldehydes.

The synthetic procedure enables the lengthening of sugar chains by two C atoms and thus offers a novel possibility for the synthesis of rare sugars. Another possibility for the synthesis of ribose has recently been described by Masamune and Sharpless and their coworkers.[29] In an elegant work they started with (R)-[23] as in our case, and obtained D-ribose in 42% yield through a six-step synthetic procedure. The same yield is obtained as in our case where only two steps are needed.

OUTLOOK

Polymers and silicas with molecular recognition obtained by imprinting with templates can find broad application in highly selective separations both in the batch procedure and in chromatography. Furthermore within the cavities stereo-selective reactions can be performed[30] or reactive groups inside the cavity can be used as selective polymeric reagents.[31]

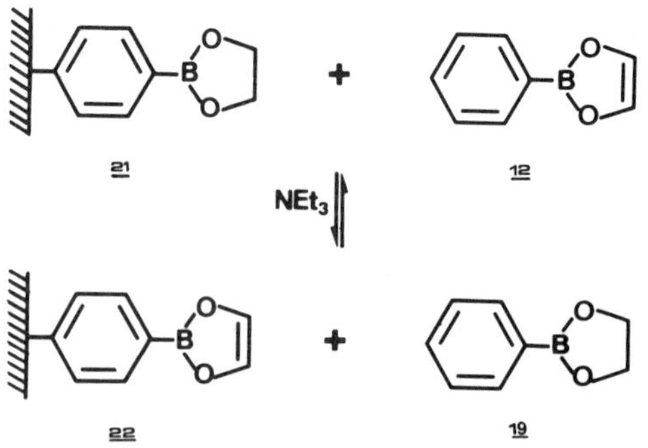

Figure 9. The preparation of polymer-bound 1,3,2-dioxaboroles.[13,14]

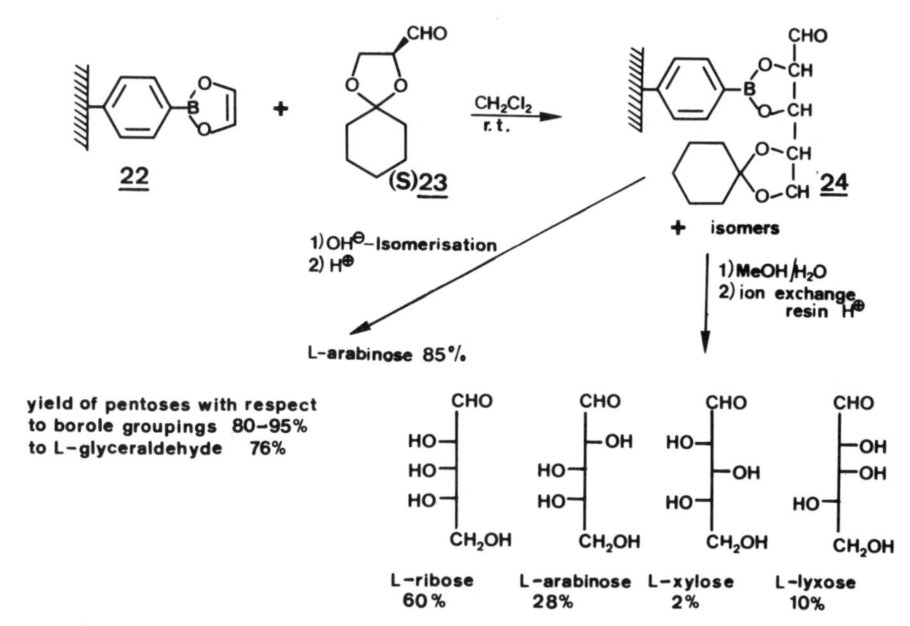

Figure 10. Preparation of L-ribose via polymer-bound 1,3,2-dioxaboroles.[13,14]

The final aim in the application of these polymers is the use as catalysts working in a fashion similar to enzymes.[10] The first steps in this direction have been reached. It is now possible to prepare cavities of a specific shape. Furthermore, there is a method to introduce into the cavity functional groups in a defined topology and some binding sites are available which meet the requirements for substrate binding.[10,32] Also, some new highly reactive reagent groups for use at the polymer have been synthesized. The next step has to be the development of a suitable catalytic environment. It is important to have a cooperative action of several catalytically active groups. To obtain a fast turn-over in catalysis, the binding of the substrate should be much tighter compared to the product. To speed up the turn-over it could be advantageous to use as the template a substrate resembling the stereochemistry of the transition state of the reaction instead of that of the substrate, or the product.

On the basis of the results obtained until now, it seems that the way to polymeric enzymes models is more difficult than to those with low molecular weight models, but the application of polymeric supports shows several advantages. They can mimic the natural enzymes more perfectly and they can be handled and processed more easily. So increased efforts in this direction might be worthwhile.

ACKNOWLEDGMENTS

Thanks are due to "Fonds der Chemischen Industrie", the "Minister für Wissenschaft und Forschung des Landes Nordrhein-Westfalen" and to "Deutsche Forschungsgemeinschaft" for financial support.

REFERENCES

1. D. J. Cram, Science, **240**, 760 (1988).
2. J.-M. Lehn, Angew. Chem. Int. Ed. Eng., **27**, 89 (1988).
3. M. L. Bender & M. Komiyama, "*Cyclodextrin Chemistry*," Springer Verlag, Berlin, 1978.
4. R. Breslow, Science, **218**, 532 (1982).
5. J. Rebek, Jr., Science, **235**, 1478 (1987).
6. R. Schwyzer, Proc. Fourth Inter. Congress on Pharmacology, 5, 196 (1970).
7. G. Wulff & A. Sarhan, Angew. Chem., **84**, 364 (1972); Angew. Chem. Inter Ed. Engl., **11**, 3411 (1972).
8. G. Wulff, A. Sarhan & K. Zabrocki, Tetrahedron Lett., **1973**, 4329.
9. G. Wulff, W. Vesper, R. Grobe-Einsler & A Sarhan, Makromol. Chem., **178**, 2799 (1977).
10. For a comprehensive review see: G. Wulff in: "*Polymeric Reagents and Catalysts*," W. T. Ford, Ed., ACS Symposium Series 308, Washington, 1986, p. 186.
11. G. Wulff, B. Heide & G. Helfmeier, J. Am. Chem. Soc., **108**, 1089 (1986).
12. G. Wulff, B. Heide & G. Helfmeier, Reactive Polymers, **6**, 299 (1987).
13. G. Wulff & A. Hansen, Angew. Chem., **98**, 552 (1986); Angew. Chem. Inter. Ed. Engl., **25**, 560 (1986).
14. G. Wulff & A. Hansen, Carbohydr. Res., **164**, 123 (1987).
15. F. H. Dickey, Proc. Natl. Acad. Sci. U.S., **35**, 227 (1949).
16. G. Wulff, R. Kemmerer, J. Vietmeier & H.-G. Poll, Nouv. J. Chim., **6**, 681 (1982).
17. G. Wulff, J. Vietmeier & H.-G. Poll, Makromol. Chem., **188**, 731 (1987).
18. G. Wulff & W. Vesper, J. Chromatog., **167**, 171 (1978).
19. G. Wulff & M. Minarik in: "*Chromatographic Chiral Separations*," N. Zief & L. J. Crane, Eds., Marcel Dekker, New York, 1988, p. 15.
20. G. Wulff & M. Minarik, J. High Resolut. Chromatog. Chromatog. Commun., **9**, 607 (1986).
21. G. Wulff, H.-G. Poll & M. Minarik, J. Liqu. Chrom., **9**, 385 (1986).
22. K. J. Shea & E. A. Thompson, J. Org. Chem., **43**, 4253 (1978).
23. J. Damen & D. C. Neckers, J. Am. Chem. Soc., **102**, 3265 (1980).
24. R. Ashady & K. Mosbach, Makromol. Chem., **182**, 687 (1981); L. Anderson, B. Sellergren & K. Mosbach, Tetrahedron Letters, **25**, 5211 (1984); B. Sellergren, B. Ekberg & K. Mosbach, J. Chromatog., **137**, 1 (1985).
25. G. Wulff, D. Oberkobusch & M. Minarik in: "*Proceedings of the XVIIIth Solvay Conference*," G. V. Binst, Ed., Springer, Berlin, 1986.
26. K. J. Shea & T. K. Dougherty, J. Am. Chem. Soc., **108**, 1091 (1986).
27. G. Wulff, P. Birnbrich & A. Hansen, Angew. Chem., **100**, 1197 (1988); Angew. Chem. Inter. Ed. Engl., **27**,1158 (1988).
28. N. T. Anh, Topics Current Chem., **88**, 145 (1980).
29. S. Y. Ko, W. M. Lee, S. Masamune, L. A. Reed, K. B. Sharpless & F. J. Walker, Science **220**, 949 (1983).
30. G. Wulff & J. Vietmeier, Makromol. Chem., in press.
31. G. Wulff & T. Röhnisch, unpublished results.
32. G. Wulff, Pure Appl. Chem., **54**, 2093 (1982).

APPLICATION OF THIN-FILM BIOCATALYSTS TO ORGANIC SYNTHESIS

Brent A. Burdick* and James R. Schaeffer
Life Sciences Research Laboratories
Eastman Kodak Company
Rochester, New York 14650

Coating methods developed for use in the preparation of photographic and clinical analytical materials have been adapted for use in the construction of immobilized thin-film biocatalysts. Syntheses of gluconic acid, pyruvic acid, aspartic acid, alanine and ribavirin were catalyzed with thin-film biocatalysts. Several of the synthetic processes utilize bioreactors composed of spiral wound immobilized biocatalysts. The advantages of the method are discussed.

INTRODUCTION

Although most approaches to immobilization of enzymes and cells involve the preparation of particles,[1-5] a number of approaches have been described which involve thin-film formation.[6,7] Photographic, clinical, and synthetic processes are described which involve chemical reactions in thin-films.

Monochrome photography has for its objective the recording and preservation of brightness differences. Color photography extends monochrome photography by adding to the brightness dimension the attribute of color. This achievement is brought about through the use of complex chemical reactions. In practice this is done in a thin-film.[8]

When the appropriate dye-forming couplers are incorporated into the blue, green, and red sensitized emulsions, the three layers may be developed after exposure in a color developer to produce a negative in colors complimentary to those of the exposed object. Printing this color negative onto a similar material leaves, after processing, a positive image (Figure 1). An example of a color forming chemical reaction of the type found in color photographic materials is described in Figure 2.

Just as quantitative chemical reactions take place within the various layers of a complex multilayered photographic film, to record the amount of red, blue, and green light impinging upon the film to produce an

* Present address: Meiogenics, 9160 Red Branch Road, Columbia, MD, 21045

Biomimetic Polymers
Edited by C. G. Gebelein
Plenum Press, New York, 1990

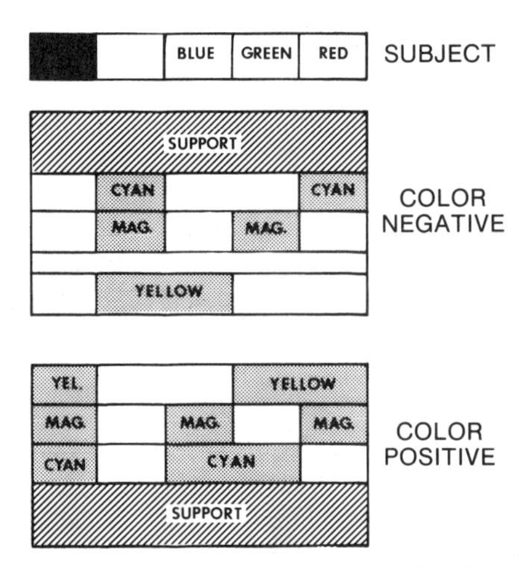

Figure 1. Subtractive color system employing dyes that subtract one-third of the visible spectrum and transmit the other two-thirds.

image, so can quantitative analytical systems be designed for clinical diagnostic use. Here knowledge gained in the construction of photographic films in relation to the coating of multilayer systems consisting of many different and varied chemical compounds, hydrophobic solvents, polymeric mordants, etc., can be transferred and extended. Coated analytical systems for clinical diagnostic application will, however, require the successful manipulation and survival of sensitive biologically active reagents (e.g., enzymes, antibodies, cofactors) which presents a particular challenge.

A schematic representation of a coated, multilayer film for clinical diagnostic use is presented in Figure 3. The film consists of or comprises three distinct regions or zones:

(1) A spreading zone for the application and spreading of the clinical sample (blood, serum, plasma, urine)
(2) A reagent zone which may consist of one or many layers in which unique reactions take place
(3) A support zone which is essentially a transparent polymeric support upon which the above are coated.

The advantages inherent in such an assembly of layers are many. Several physical separation steps can proceed within the film to allow for the removal of interfering substances in the applied sample, thus offer-

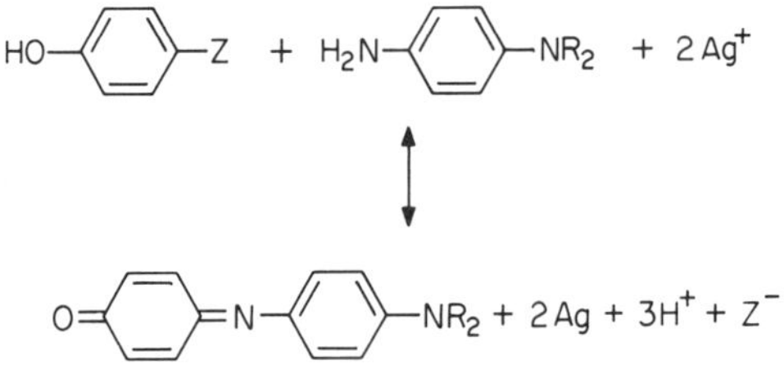

$$HO{-}C_6H_4{-}Z + H_2N{-}C_6H_4{-}NR_2 + 2Ag^+ \updownarrow O{=}C_6H_4{=}N{-}C_6H_4{-}NR_2 + 2Ag + 3H^+ + Z^-$$

Figure 2. Dye formation of the type found in color photographic processes.

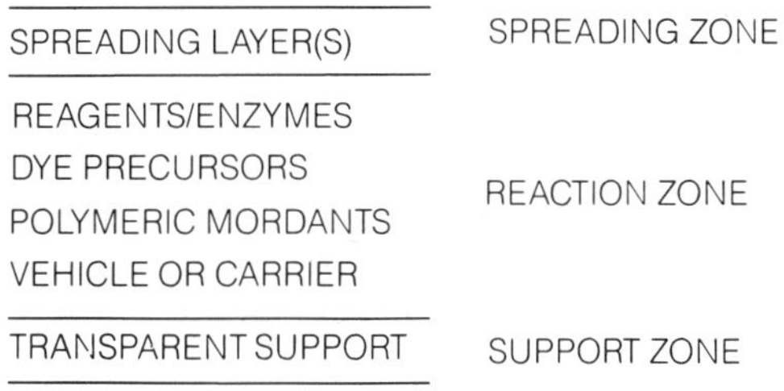

SPREADING LAYER(S)	SPREADING ZONE
REAGENTS/ENZYMES DYE PRECURSORS POLYMERIC MORDANTS VEHICLE OR CARRIER	REACTION ZONE
TRANSPARENT SUPPORT	SUPPORT ZONE

Figure 3. Multilayer film for clinical diagnostic use exhibiting three major reaction zones.

ing improved clinical sensitivity and specificity. Sequential chemical reactions can occur within the separate layers comprising the spreading and reagent zones, allowing several steps or conversions to occur without operator intervention. Furthermore, as may be obvious, all the necessary and required reagents for the process are coated within the various layers, which allows the operator considerable ease of use as well as reliability.

In order to illustrate the construction and operation of a representative film for clinical diagnostic use, let us focus on the system devised for analysis of serum triglyceride levels which represents a challenging and complex problem.[9]

A cross-sectional view of the triglyceride film is illustrated in Figure 4. Triglycerides are strongly associated with lipoproteins in serum and dissociation must occur prior to analysis. The spreading layer contains a surfactant (TX-100) to dissociate the lipoprotein complex to free triglycerides which diffuse to the next lower layer or the hydrolysis layer. This layer contains an immobilized lipase enzyme reagent which converts the released triglyceride to glycerol and fatty acids by virtue of enzymatic activity and specificity. Glycerol then diffuses downward through the spacer layer into the glycerol detection layer, where glycerol is converted through subsequent enzyme activity to hydrogen peroxide, which can then be detected visually through coupling with horseradish peroxidase which catalyzes triarylimidazole leuco dye oxidation. The entire sequence of events is represented in Figure 5.

The triglyceride clinical diagnostic film is a particularly good example of the ability to separate or confine given reactions to given layers within the film to produce accurate clinical data. Here, as is the case with many of the Kodak Ektachem films for other clinical analytes, the immobilization of specific enzymes to perform specific functions in a sequential manner within the film is the key to success.

The ability to immobilize enzymes in separate layers also allows the film builder to incorporate scavenging enzymes in a layer such that in-

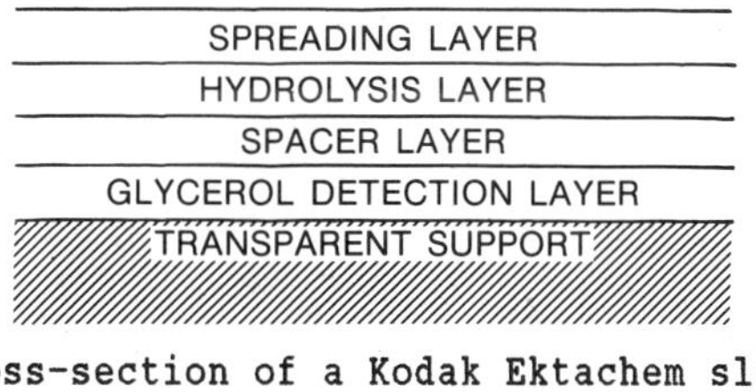

SPREADING LAYER
HYDROLYSIS LAYER
SPACER LAYER
GLYCEROL DETECTION LAYER
TRANSPARENT SUPPORT

Figure 4. Cross-section of a Kodak Ektachem slide for measuring triglycerides.

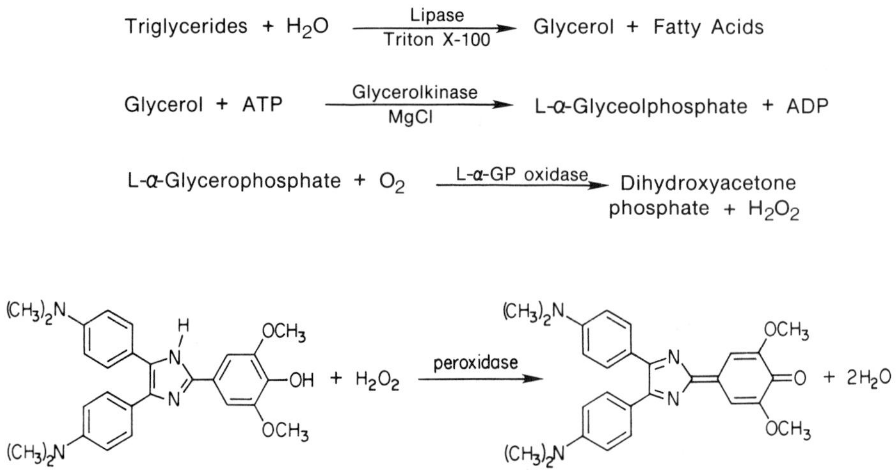

$$\text{Triglycerides} + H_2O \xrightarrow[\text{Triton X-100}]{\text{Lipase}} \text{Glycerol} + \text{Fatty Acids}$$

$$\text{Glycerol} + \text{ATP} \xrightarrow[\text{MgCl}]{\text{Glycerolkinase}} \text{L-}\alpha\text{-Glyceolphosphate} + \text{ADP}$$

$$\text{L-}\alpha\text{-Glycerophosphate} + O_2 \xrightarrow{\text{L-}\alpha\text{-GP oxidase}} \text{Dihydroxyacetone phosphate} + H_2O_2$$

$$\text{dye} + H_2O_2 \xrightarrow{\text{peroxidase}} \text{dye (oxidized)} + 2H_2O$$

Figure 5. Reaction sequence for triglyceride quantitation.

terfering substances introduced within the sample can be converted into innocuous by-products. An example would be the incorporation of a layer containing immobilized ascorbate oxidase, which will oxidize ascorbate to a form which no longer interferes with subsequent dye chemistry development.

As factors and requirements dictate, additional specialized layers may be added to allow other processes to occur that would not be possible within the film design depicted in Figure 3. Whole blood samples cannot be directly processed on such a design as cells and cell debris would rapidly clog the available pores within the spreading layer (a composite TiO_2 or $BaSO_4$-cellulose acetate polymer layer). To extend the use of the clinical diagnostics films to include whole blood samples, an additional layer consisting of spherical polymer beads can be placed on top of the serum or plasma spreading layer. This allows for the filtration and separation of cellular material in blood within the film and has proven to be useful for clinical analytes of interest.[10,11]

With the knowledge and experience gained in coating viable, immobilized enzyme reaction layers, it then becomes possible and useful to define other applications. With the removal of the uppermost spreading layer from the clinical diagnostic film, as depicted in Figure 6, underlying layers containing immobilized enzymes may be exploited for their biosynthetic properties and powers. We would now like to turn to a discussion

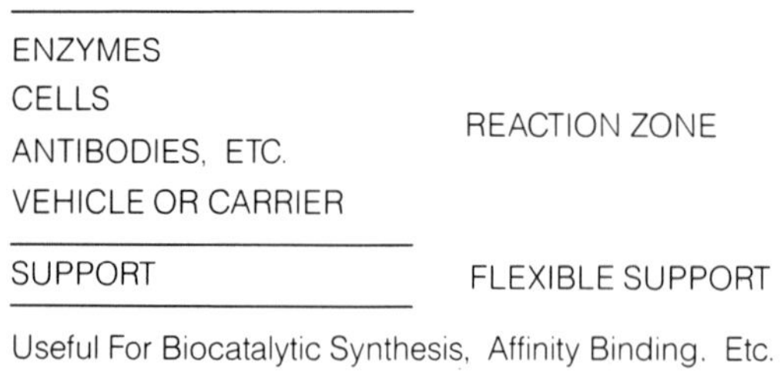

ENZYMES CELLS ANTIBODIES, ETC. VEHICLE OR CARRIER	REACTION ZONE
SUPPORT	FLEXIBLE SUPPORT

Useful For Biocatalytic Synthesis, Affinity Binding. Etc.

Figure 6. A Kodak Ektachem slide film with spreading layer removed.

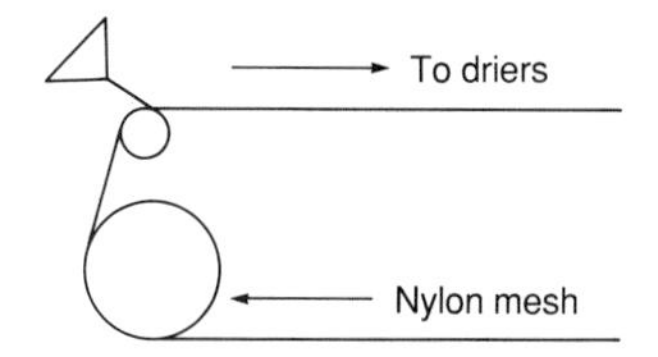

Figure 7. Hopper-coating assembly used for large scale preparation of thin-film biocatalysts.

of our use of these systems in biocatalytic schemes to produce specialty chemicals of interest in a potentially large scale operation.

The successful large scale preparation of thin-film immobilized biocatalysts depends in part on the choice of a machine coating method. Among the possible methods available are hopper, dip and bar coating. Although dip coating is briefly described, most of this paper is devoted to a description of hopper coating methods for preparation of thin-film biocatalysts.

The hopper coating assembly (Figure 7) allows for large scale preparation of biocatalytic membranes.[12,13] The process involves delivering an enzyme solution or cell suspension in the appropriate vehicle (gelatin or synthetic polymer) to the surface of the support in a continuous manner. The enzyme or cell suspension in the appropriate vehicle is applied to the surface through an extrusion hopper at the required coverage. The coated support is then dried by passage through drying ovens. Crosslinking of the enzymes or cells may be accomplished either by addition of the crosslinking agent at the point of application or by a second pass of crosslinking agent.

The dip coating assembly (Figure 8) also allows for large scale preparation of biocatalytic membranes. This method differs from the hopper coating method in that the support is continuously passed through a coating bath. The support which is coated on both sides is dried by passage through drying ovens. Crosslinking of the enzymes or cells may be accomplished by passing the coated support through a second bath containing a crosslinking agent.

Biocatalyst coatings have been used for preparation of several materials, such as pyruvic acid, gluconic acid, aspartic acid, alanine, and ribavirin.

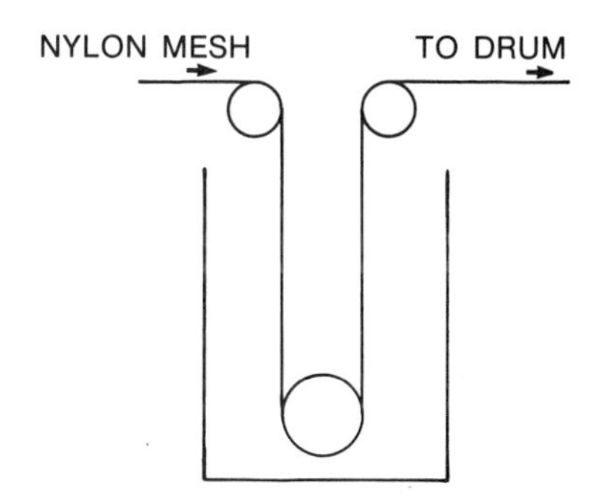

Figure 8. Dip-coating assembly used for large scale preparation of thin-film biocatalysts.

EXPERIMENTAL

Materials

The nylon mesh fabric (Scan-Fablok fabric, obtained from Fablok Mills, Union City, NJ), backed with a thin layer of removable pigmented polyethylene, consisted of nylon threads approximately 1 mm wide and 0.5 mm thick woven into crosshatch structures with interstices approximately 2 x 2 mm. As received, the nylon mesh had been previously treated with an acrylate sizing material to enhance rigidity. Polyethylene backing was applied by extrusion coating. The material was slit into 0.127-meter-wide sections to allow for biocatalyst coating. Ninon, a woven polyester, was obtained from Applied Technology Division. Pellon FS-2108, a non-woven polyolefin, was purchased from Pellon Corp., Chelmsford, MA. Glucose oxidase (28.6 U/mg) was obtained from Sigma Chemical Co. Catalase (3650 U/mg) was purchased from Worthington Corp., Freehold, NJ. L-Lactic acid was obtained from Aldrich Chemical Co., Milwaukee, WI. DuPont Zonyl FSN fluorosurfactant also was used. All other organic chemicals and solvents were Kodak products.

Cells of *Escherichia coli* (ATCC11303), *Pseudomonas dacunhae* (ATCC-21191, ATCC19121, IAM1152) and *Brevibacterium acetylicum* (ATCC21665) were produced using standard microbiological techniques.[14-16]

Synthetic polymers were prepared by known free radical polymerization methods.[17,18]

Analytical Methods

Organic Acid Assay. The concentrations of fumarate and malate, present during the course of the bioconversion, were determined by gas chromatographic analysis of samples drawn from the reactor reservoir.[19]

Amino Acid Assay. The amount of aspartic acid or alanine produced was determined by HPLC.[20,21]

Ribavirin Assay. Ribavirin content of the reaction mixture was assayed by HPLC.[22]

Immobilization of Enzymes and Cells by Hopper Coating Methods

Gelatin melts, containing enzymes, were mixed with a crosslinker and then applied onto the surface of the nylon mesh fabric at a rate corresponding to a (dry) laydown of 2.7-32.3 g/m^2 gelatin by extrusion coating. This range of gel laydown was designed to just coat the surface of the fiber at the lowest laydown and to coat the fiber as well as fill the cells formed by the woven fibers with gel/biocatalyst at the higher gel coverages.

Crosslinking agents were either bisvinylsulfone methyl ether (BVSME), glutaraldehyde, or oxidized dextran (MW = 40,000). The coated mesh was then chill set and passed through a series of dryers. After drying, coatings were stored in the refrigerator until evaluation. In cases where it was desirable to remove the polyethylene backing for evaluation, this was accomplished by careful stripping by hand or by applying and removing tape attached to the poly backing.

Immobilization of Glucose Oxidase-Catalase in Gelatin

Six variations on enzyme coverage/gelatin coverage in the coated biocatalysts were explored to optimize coverages relative to surface area (Table 1).

Gluconic Acid Synthesis Utilizing Glucose Oxidase-Catalase Coatings

In 400 mL of distilled water, 3.2 g (1.8×10^{-2} mole) of α-D-glucose was dissolved. To the stirred solution, pieces of stripped, glutaraldehyde-treated coating (coating 1, Table 1, 0.093 m^2) were added. This suspension was stirred at room temperature for 24 hr. The pH of the suspension was maintained between 6 and 8 by addition of 1 N sodium hydroxide solution. The conversion of glucose to gluconic acid was greater than 90% based on the consumption of sodium hydroxide. The coating was removed by vacuum filtration, washed with distilled water (100 mL), and then resuspended in distilled water. The combined filtrate and wash was freeze-dried, yielding 4.0 g of product (100%). An infrared spectrum obtained on the product was identical with an infrared spectrum of an authentic sample of sodium gluconate.

Immobilization of Lactate Oxidase-Catalase in Gelatin on Polyethylene-Backed Nylon Mesh.[12]

Lactate oxidase (2 g, 2600 U) was dissolved in 5 mL of distilled water and mixed in 100 g of stirred 20% type IV gelatin at 40°C. After the enzyme had been thoroughly mixed, dry catalase (0.2 g, 3650 U/mg) was

Table 1. Immobilization of biocatalysts on nylon mesh, as g/m^2.[12]

Enzyme or Cells	Gelatin Type IV	Buffer K2HPO4, pH 7	Surfactant FSN	Hardener[a]
0.32 (GOD)[b] 0.054 (catalase)[c]	10.7	0.32	0.21	0.31 [B]
0.16 (GOD)[b] 0.026 (catalase)[c]	5.4	0.16	0.11	0.16 [B]
0.32 (GOD)[b] 0.054 (catalase)[c]	10.7	0.32	0.21	0.31 [G]
0.16 (GOD)[b] 0.026 (catalase)[c]	5.4	0.32	0.21	0.31 [G]
0.64 (GOD)[b] 0.11 (catalase)[c]	21.4	0.64	0.42	0.51 [B]
0.96 (GOD)[b] 0.16 (catalase)[c]	32.1	0.96	0.64	0.83 [B]

(a) BVSME [B] or glutaraldehyde [G]
(b) 30.6 U/mg
(c) 3650 U/mg

added, followed by 1.0 g of dipotassium phosphate and 0.4 g of DuPont Zonyl FSN fluorosurfactant. The total weight of the melt was brought to 400 g by addition of distilled water. The pH was adjusted to one of three levels (6, 7, and 8) prior to final dilution. The coating was done as described above.

Conversion of Lactic Acid to Pyruvic Acid with Immobilized Lactate Oxidase-Catalase

In 100 mL of an aqueous solution containing 0.45 g (4.9×10^{-3} mole) of L-lactic acid and 1.4 g (1×10^{-2} mole) of mono-potassium phosphate adjusted to pH 7, was placed one 0.3-meter section of coating cut into pieces 1 cm on end. The mixture was shaken for 48 hr at 100 rpm and 30°C, then passed through a sintered glass funnel. To the filtrate, 2,4-dinitrophenylhydrazine reagent (1.6 g 2,4-DNP, 6 mL sulfuric acid, 8 mL water, and 40 mL alcohol) was added. After standing 15 min, the product was collected, washed with water, and air dried; it weighed 0.52 g (38.5%). An infrared spectrum obtained on the product was identical with the spectrum of an authentic sample (Table 2).

Immobilization of *Escherichia coli* Cells in Gelatin on Nylon Mesh Fabric

Thawed *Escherichia coli* cells (32% solids, 433 g dry weight) were blended for 2-4 min in a Waring blender. During the blending enough water was added to bring the total weight to 2600 g. The blended cells were dispersed in 1 kg of 20% type IV deionized gelatin, 10 g of potassium phosphate was added, and the pH of the mixture was adjusted to 7.5 (5% HCl). Fluorocarbon surfactant, Zonyl FSN (3.0 g) was added as a coating aid and the total weight of the melt was brought to 4 kg with distilled water.

The melt was coated at 214 g/m² onto 12.54 cm wide Fablok backed with pigmented polyethylene at a rate of 6 m/min using a conventional 10.2 cm wide 0.02 cm slit extrusion hopper. Hardener (BVSME, 2% by weight of the

Table 2. Oxidation of L-lactic acid to pyruvic acid with lactate oxidase-catalase immobilized in gelatin.[12]

Coating pH	Crosslinker[a]	Lactic Acid	% Pyruvic Acid[b]
7	Glutaraldehyde (1)	L	39
7	Glutaraldehyde (2)	L	none
7	Glutaraldehyde (1)	D,L	42
6[c]	Glutaraldehyde (1)	L	47
8[d]	Glutaraldehyde (1)	L	29
7	Oxidized dextran (1)	L	39
6	Oxidized dextran (1)	L	none
8	Oxidized dextran (1)	L	none

(a) Number of treatments.
(b) Recovery of 2,4-dinitrophenylhydrazone derivative from an authentic sample of pyruvic acid was 78%.
(c) Run 2, pyruvic acid produced in 36% yield.
(d) Run 2, pyruvic acid produced in 21% yield.

gelatin) was mix melted at the hopper with the gelatin-*E. coli* mixture. The coating was then chill set and dried at 85-100°F, 10% relative humidity.

A second hardening with glutaraldehyde was carried out (after one week) by passing the coating through the machine, and an aqueous solution of glutaraldehyde (0.3%) was coated onto the pre-formed membrane at 96.3 g/m². The coating was dried and allowed to harden at room temperature for 24 hr prior to refrigeration. After standing a total of 48 hr at room temperature, a portion of the film was wound into three tightly packed cylinders and loaded into a 10.2 cm x 45.8 cm stainless steel column (5.1 m² of film).

An additional film was coated utilizing *E. coli* in gelatin at 2.3 g cells per gram gelatin. Hardening was done with a mix melt of glutaraldehyde and a wash coating of glutaraldehyde after the coating stood one week. This film was wound and packed into a 10.2 cm (diam) x 45.8 cm stainless steel column (5.1 m² of film).

Conversion of Ammonium Fumarate to Aspartic Acid with Immobilized *Escherichia coli*

Twelve liters of aspartic acid production medium [116 g (1.0 mole) fumaric acid, 0.2 g magnesium chloride in 1 L distilled water adjusted to pH 8.5 with ammonium hydroxide] was recirculated through the biocatalyst column (BVSME, glutaraldehyde hardening) at 600 mL/min at 37°C for 72 hr. The reaction mixture was drained from the column and the column was washed with 1 L of distilled water (total volume of reaction mixture 13).

The product was isolated from 3.5 L of reaction mixture by acidification to pH 2.8 at 90°C with 60% sulfuric acid. The aspartic acid crystallized before the solution cooled to room temperature. The solid product was collected (286 g) and the filtrate was placed in the refrigerator for 18 hr. An additional 85 g of product was collected. Infrared spectra obtained on both samples were identical with the spectrum obtained on an authentic sample of aspartic acid. The yield of aspartic acid from 13 L was 1058 g (86.1%), [α]D 22 = +24.7 (C = 2.002, 6 N HCl).[13]

Analysis: Calcd. for $C_4H_7NO_4$: C, 36.1; H, 5.3; N, 10.5.
Found: C, 36.3; H, 5.3; N, 10.4.

Preparation of Immobilized *Escherichia coli* in Poly[N-(3-aminopropyl)-methylacrylamide hydrochloride] on Polyethylene-Backed Nylon Mesh

A suspension of cells in the synthetic polymer was prepared by combining blended *E. coli* cells (40 g dry weight, 125 g wet weight at 32% solids) with 635 g of a 3% aqueous solution of poly[N-(3-aminopropyl)-methacrylamide hydrochloride]. Potassium phosphate (2 g monobasic salt) was added and the pH of the suspension was adjusted to 7.5 with stirring through the addition of 5% hydrochloric acid. Zonyl FSN (0.8 g) was added and the total weight of the suspension was made up to 4 kg. The suspension was coated on the surface of the nylon mesh at a delivery rate sufficient to the cell-coverage specified 10.8 g/m² dry weight cells/m². Conventional extrusion hopper coating (0.02 cm hopper opening) application was used. A 3.2% (wt/vol) solution of crosslinking polymer, oxidized dextran, was added at the point of application. The nylon mesh was dried at 85-100°F and 10% RH. After a week of curing at room temperature and 50% RH, the coating was ready for evaluation (Table 3).

Table 3. Synthesis of aspartic acid with immobilized *E. coli* prepared using synthetic polymers.

Polymer[a]	Polymeric Hardener	Percent Yield Aspartic Acid[b]			
		1	2	3	4
1	7	62	87	52	52
1	6	66	84	68	80
1	5	64	82	83	88
2	7	50	85		
3	7	64	82	61	68
4	7	58	76	55	74
4	7	68	80	67	73

(a) The polymer structures are shown below. A control coating was made which contained gelatin crosslinked with polymer 7. Yields of aspartic acid in runs 1-4 were 59%, 47%, 40% and 48%, respectively.
(b) Runs made over a 48 hr period.

1) $-CH_2CH)_{80}-(CH_2CH)_3-(CH_2C)_5-(CH_2CH)_{10}-$
CH_3
C=O NH CH CH_3 CH_3 | C=O NH_2 | C=O NH $(CH_2)_3$ NH_2 HCl | C=O NH CH_2 NH C=O $-(CHCH_2)-$

2) $(CH_2CH)_{90}-(CH_2C)_3-(CH_2C)_3-$
CH_3 CH_3
C=O NH_2 | C=O NH $(CH_2)_3$ $N(CH_3)_3^{\oplus}$ $Cl^{\ominus}$ | C=O NH $(CH_2)_3-NH_2{\bullet}HCl$

3) $-(CH_2CH)_{80}-(CH_2CH)_{20}-$
C=O NH_2 | C=O NH $(CH_2)_3$ $NH_2{\bullet}HCl$

4) $-(CH_2C)-$
CH_3
C=O NH $(CH_2)_3$ $NH_2{\bullet}HCl$

5) $-(CH_2CH)_{90}-(CH_2CH)_{10}-$
C=O NH_2 | CHO

6) $-(CH_2CH)_{90}-(CH_2CH)_{10}-$
C=O NH_2 | CHO

7) CH_2 O OH CH CH O O | O CH_2 O OH OH OH O

Conversion of Ammonium Fumarate to L-Aspartic Acid with Immobilized *Escherichia coli* Cells

In 100 mL of aspartic acid production medium [116 g (1 M) of fumaric acid and 0.2 g magnesium chloride dissolved in 1 L of distilled water adjusted to pH 8.5 with ammonium hydroxide] three-tenths of a meter of coating was suspended after removal of the polyethylene backing and cutting into pieces 1 cm on end. The suspension was shaken at 37°C and 150 rpm for 48 hr. The immobilized biocatalyst was removed from the reaction medium by means of vacuum filtration and suspended in fresh aspartic acid production medium (100 mL). The filtrate was heated to 90°C, acidified to pH 2.8 with 60% sulfuric acid, and then allowed to cool to room temperature. After standing 18 hr under refrigeration, the solid product (10.6 g, 80% yield, Table 4) was collected with vacuum assist and air dried. An infrared spectrum obtained on the product was identical with the infrared spectrum of an authentic sample of aspartic acid. [α]D 22 = +24.5 (C = 2.002, 6 N HCl).

Analysis: Calcd. for $C_4H_7NO_4$: C, 36.1; H, 5.3; N, 10.5.
Found: C, 35.9; H, 5.3; N, 10.5.

Immobilization of *Brevibacterium acetylicum* Cells in Gelatin on Polyethylene-Backed Nylon

Gelatin melts containing cells of *Brevibacterium acetylicum* were applied onto the surface of the nylon mesh fabric at a rate corresponding to a dry laydown of 10.7 g/m^2 gelatin, 10.7 g/m^2 cells, 214 mg/m^2 FSN, and 535 mg/m^2 potassium phosphate all at pH 7. A dual melt (cells and hardener) was applied with a conventional slit extrusion hopper at a coating speed of 61 dm/min. The cell melt was coated at 214 g/m^2 and the glutaraldehyde at 36.4 g/m^2 (25 mg glutaraldehyde in 3.4 mL). The coated fabric was chill set at 40°F and passed through a series of dryers (85-100°F). After drying the web was allowed to stand for a week at room temperature and then stored under refrigeration.

Table 4. Batch production of ribavirin with immobilized *Brevibacterium acetylicum*.

Polymer	Cell Content (g dry wt/L)	Catalyst	Temperature (°C)	Time (hr)	Yield (g/L)
Gelatin	3.3	coating	30	24	0
		(nylon mesh)	30[a]	24	0
Gelatin	3.3	coating	60	24	0.8
		(nylon mesh)	60[a]	24	4.5
Polymer[b]	3.3	coating (nylon mesh)	60	48	8.0
Gelatin (high salt)	5.7	coating (nylon mesh)	60	48	5.1
Gelatin (high salt)	5.7	coating (Kodak Estar base)	60	48	4.7

(a) A second run was made with the catalyst.
(b) Poly[acrylamide-co-N-(3-aminopropyl)-methacrylamide hydrochloride].

Production of Ribavirin with *Brevibacterium acetylicum*-Gelatin Coating in a Column Bioreactor

A 6 m length of coating was stripped of its backing and cut into two equal lengths. Each length was machine rolled on a plastic core to a diameter of 5 cm. After trimming to fit the column, the two rolls were placed in a column reactor 0.3 meter in length. The total square meters of catalyst coating in the reactor was 0.61. The void volume of the reactor was 180 mL.

Immobilization of *Brevibacterium acetylicum* Cells in Poly[acrylamide-co-N-(3-aminopropyl)-methacrylamide Hydrochloride] on a Polyethylene Backed Nylon Mesh Fabric by Machine Coating

Melts containing poly[acrylamide-co-N-(3-aminopropyl)-methacrylamide hydrochloride] (80/20) and cells were applied to the surface of the fabric at a rate corresponding to a dry laydown of 10.7 g poly[acrylamide-co-N-(3-aminopropyl)-methacrylamide hydrochloride]/m^2, 10.7 g cells/m^2, 214 mg FSN/m^2, and 535 mg potassium phosphate/m^2 at pH 7. A dual melt was applied with a conventional slit extrusion hopper. The coating speed was adjusted to 61 dm/min. The cell melt was coated at 214 g/m^2 and the glutaraldehyde at 36.4 g/m^2 (25 mg glutaraldehyde in 3.4 mL of water). The coating was dried at 85-100°F. The coating was allowed to stand a week at room temperature and then stored under refrigeration.

Immobilization of *Brevibacterium acetylicum* Cells in High Salt-Gelatin Films by Hand Coating

High salt content gelatin was prepared by dissolving 400 g of type IV gelatin and 200 g of sodium sulfate in 1500 g of hot distilled water. In 20 g of the above solution stirred at 37°C, 1 g dry weight of cells was dispersed. The melt was coated on nylon mesh on a conventional coating block with an 8 mil knife at 40°C. The coating was chill set and allowed to dry. The melt was coated at 364 to 428 g/m^2. The film was crosslinked by dipping into 1% glutaraldehyde solution for 5 min, then washed with distilled water and allowed to dry. The coating was kept 72 hr at room temperature and then stored under refrigeration.

Evaluation of Immobilized Biocatalysts for Ribavirin Production in Shake Flasks

The biocatalysts were shaken in 100 mL of ribavirin production medium [5.6 g 1,2,4-triazole-3-carboxamide, 13.6 g inosine and 10.5 g monopotassium phosphate in 100 mL of distilled water adjusted to pH 7 with 45% potassium hydroxide] at 30 and 60°C. The reaction was done on a New Brunswick shaker at 100 rpm for 24 to 48 hr (Tables 5 and 6).

Production of Ribavirin with *Brevibacterium acetylicum*-Poly[acrylamide-co-N-(3-aminopropyl)-methacrylamide Hydrochloride] Coating in a Column Bioreactor

A 6 m length of coating was stripped of its backing and cut into two

Table 5. Specific productivities for the production of ribavirin with immobilized *Brevibacterium acetylicum* cells.

Catalyst Type	Fermentation Batch	Process	Time (hr)	Yield (g/L)	SP[a]
Gelatin particle	5	shake	24	0.8	0.019
Carrageenan particle	5	shake	24	0.4	0.008
Gelatin coating (nylon)	5	shake	48	2.3	0.016
Polymer coating[b] (nylon)	16	shake	48	8.0	0.055[c]
Polymer coating[b] (nylon)	7	shake	48	1.1	0.007
Polymer coating[b] (nylon)	9	column	2[d]	2.2	0.03
Gelatin coating[b] (nylon) (high salt)	21	shake	48	5.1	0.019
Gelatin coating (Kodak Estar base) (high salt)	21	shake	48	4.7	0.017

(a) Suspended cell data indicate specific productivity range between 0.023 and 0.039 (specific productivity may be defined as grams of product per gram of dry weight cells per hr).
(b) Poly[acrylamide-co-N-(3-aminopropyl)-methacrylamide hydrochloride].
(c) Since this experiment was run only one time, the value 0.03 will be used as the upper range limit for specific productivities for immobilized cells.
(d) Column residence time 2 hr.

Table 6. Production of ribavirin in a continuous bioreactor at 60°C.

	Fraction	Flow Rate (mL/min)	Yield (g/L)
[A]	1	1.2	0.2
	2	2.0	0.4
	3	2.0	0.4
[B]	1	1.5	2.2
	2	1.5	2.0
	3	3.0	1.3
	4	3.0	1.1
	5	3.0	1.5
	6	3.0	1.0
	7	3.0	1.0
[C]	2	0.7	4.3
	3	0.7	4.3

[A] = Gelatin-nylon mesh biocatalyst.
[B] = Poly[acrylamide-co-N-(3-aminopropyl)-methacrylamide hydrochloride]-nylon mesh biocatalyst.
[C] = Poly[acrylamide-co-N-(3-aminopropyl)-methacrylamide hydrochloride]-nylon mesh biocatalyst. (Duplicate preparation of biocatalyst.)

equal lengths. Each length was machine rolled on a plastic core to a diameter of 5 cm. After trimming to fit the column, the two rolls were placed in a 0.3-meter reactor column. The total area of the catalyst was 0.61 m^2. Ribavirin production medium was pumped through the reactor at three different flow rates: 0.7, 1.5 and 3.0 mL/min. The void volume of the column was found to be 180 mL. The residence times were 200, 120, and 60 min. The temperature of the reagents and the reactor was maintained at 60°C. Fractions of reactor output were automatically collected every 1.5 hr (Table 6).

Preparation of Immobilized *Pseudomonas dacunhae* in Gelatin on Polyethylene-Backed Nylon Mesh

Gelatin melts containing cells were coated onto the surface of nylon mesh fabric at a dry laydown rate of 10.7 g/m^2 gelatin, 10.7 g/m^2 cells, 214 mg/m^2 FSN and 535 mg/m^2 potassium phosphate at pH 7. A dual melt of cells and hardener was applied with a standard X-hopper at a coating speed of 61 dm/min. The cell melt was coated at 214 g/m^2 and glutaraldehyde (hardener) at 37.5 g/m^2 (25 mg glutaraldehyde/3.5 mL). The coating was chill set at 40°F and dried at 85-100°F. After drying, the film stood at room temperature for one week before storing under refrigeration.

Evaluation of Alanine Production with *Pseudonomas dacunhae*-Gelatin Coatings in Shake Flasks

A 0.3-meter coating was stripped of its backing, cut into 2.5 cm squares and shaken in 100 mL of alanine production medium [133 g L-aspartic acid (1 M) adjusted to pH 5.5 with ammonium hydroxide and 2.5 mg pyridoxal-5-phosphate (0.1 mM)] at 37°C and 200 rpm. A control flask containing 1.0 g wet weight of free cells was prepared in like manner. The reactions were carried out in a bench-top shaker for two consecutive runs with freshly prepared medium for periods 76 and 168 hr. Both the immobilized cells and the free cells produced the same amount of alanine for each run (Table 7).

Preparation of Immobilized *Pseudomonas dacunhae* in Poly[acrylamide-co-N-(3-aminopropyl)-methacrylamide Hydrochloride] on Polyethylene-Backed Nylon Mesh

Poly[acrylamide-co-N-(3-aminopropyl)-methacrylamide hydrochloride] (80/20) cell melts were applied on the surface of nylon mesh fabric at a

Table 7. Alanine production in shake flasks for free and immobilized *Pseudomonas dachunae* (ATCC21192) cells.

Method	Time (hr)	Alanine Produced (g/L)[a]
Free cells	72	14.3
Free cells	168	13.3
Immobilized cells	72	14.3
Immobilized cells	168	13.2

(a) Alanine concentration determined by HPLC.

dry laydown rate of 10.8 g/m^2 each of polymer and cells, 216 mg/m^2 FSN and 540 mg/m^2 potassium phosphate at pH 7. A dual melt of cells and hardener was applied by machine. The cell melt was applied at 216 g/m^2 and glutaraldehyde hardener at 36.7 g/m^2 (25 mg glutaraldehyde in 3.4 mL). After drying at room temperature for one week, the coating was then stored under refrigeration.

Production of Alanine in a Column Bioreactor with *Pseudomonas dacunhae* in Gelatin on Nylon Mesh

A 6.1 m section of coating was stripped of its polyethylene backing and cut into two equal lengths. Each piece was machine rolled onto a plastic core to a 5.1 cm diameter. Both rolls were trimmed to fit a 30.5 x 5.1 cm reactor column giving a total area of 0.56 m^2 of biocatalyst. Alanine production medium was initially pumped through the reactor at a 1.2 mL/min rate with fractions collected every 90 min over a three-day period. The temperature was set at 37°C. The void volume of the column was 180 mL. Alanine content of the fractions was determined by HPLC (Table 8).

Production of Alanine in a Column Bioreactor Containing Squares of Nylon Mesh Coated with *Pseudomonas dacunhae*-Poly[acrylamide-co-N-(3-aminopropyl)-methacrylamide Hydrochloride]

A 0.46 m^2 section of *Pseudomonas dacunhae*-poly[acrylamide-co-N-(3-aminopropyl)-methacrylamide hydrochloride] coated on nylon mesh was cut into pieces 2.5 cm on end and placed in a column reactor 30.5 cm in length and 5.1 cm in diameter. The bioreactor was run as described in the previous experiment and fraction assayed for alanine (Table 9).

Immobilization of Enzymes and Cells by Dip Coating Methods

Fablok 909 and Ninon were corona discharge treated on both sides of the 25 cm wide webs prior to dip-coating. The coating solution contained 4.0 g of glucose oxidase, 0.8 g of catalase and 80 g of type IV gelatin

Table 8. Alanine production in a spiral wound bioreactor containing *Pseudomonas dachunae* on nylon mesh in gelatin.

Fraction	Flow (mL/min)	Temp. (°C)	Alanine (g/L)[a]
5	1.2	25	10.7
8	1.2	25	13.9
13	1.2	25	20.0
18	1.2	37	17.1
19	1.2	37	17.5
21	1.2	37	19.0
23	1.2	37	15.4
25	1.2	37	14.1

(a) Alanine concentration determined by HPLC.

Table 9. Production of alanine in a column bioreactor containing squares of nylon mesh coated with *Pseudomonas dacunhae*-poly[acrylamide-co-N-(3-aminopropyl)-methacrylamide hydrochloride]

Fraction	Flow (mL/min)	Temp. (°C)	Alanine (g/L)
13	1.5	33	10.3
14	1.5	33	9.5
15	1.5	33	12.3
16	1.5	33	11.9
2	3.0	32	4.5
4	3.0	32	6.3
6	3.0	32	7.2
8	3.0	32	8.7
10	3.0	32	8.3

in 4 L of distilled water. The coating was done on a Dixon coating machine at 40°C and then air dried at 60-70°C (surface temperature of coating <50°C). The enzyme coated films were then dip-coated at room temperature in a 1% glutaraldehyde solution in the manner described above.

Determination of biocatalytic activity. A measured amount of film (0.06-0.3 m) was cut into small pieces and placed in a defined amount (50-400 mL) of a 0.1 M α-D-glucose solution at pH 7. The solution was stirred at 25°C and titrated with 1 N sodium hydroxide solution until no further consumption of base was noted (Table 10).

Immobilization of Aspartic Acid Producing *Escherichia coli* by Dip Coating

Fablok 909 and Pellon FS21087 were corona discharge treated on both sides of the 25 cm wide films prior to dip-coating. The coating solution (8.5 L) contained 300 g of type IV gelatin and 300 g (dry weight) of *E. coli* (ATCC11303). The coating was done as described above for immobilization of glucose oxidase-catalase. The webs were then dip coated in 0.1% glutaraldehyde solution.

Table 10. Determination of catalytic activity of dip-coated fabrics.

Fabric	Biological	Area (m^2)	Time (hr)	Temp. (°C)	Gluconic Acid[a] (%)	Aspartic Acid[b] (%)
Fablok	GOD/cat.	1	6	25	97	
Ninon	GOD/cat.	1	6	25	93	
Fablok[c]	*E. coli*	0.043	48	37		48.4
Pellon[c]	*E. coli*	0.063	48	37		72.0

(a) Reaction medium composed of 400 mL of 0.1 M α-D-glucose.
(b) Reaction medium composed of 150 mL of 1 M ammonium fumarate.
(c) Samples dried at 90°C for 1 hr were as active as those dried at 60 to 70°C.

Determination of biocatalytic activity. A measured amount of membrane (0.04-0.06 m) was cut into small pieces and placed in a defined amount (50 to 400 mL) of aspartic acid production medium [116 g (1.0 mole) fumaric acid, 0.2 g magnesium chloride in 1 L distilled water adjusted to pH 8.5 with concentrated ammonium hydroxide]. The mixture was shaken (100 rpm) on a New Brunswick incubator shaker at 37°C for 48 hr (Table 10).

RESULTS AND DISCUSSION

Two methods of coating, hopper and dip coating, were evaluated in order to develop satisfactory methods for the preparation of thin-film biocatalysts.

Whole cells and enzymes were immobilized by entrapment in gelatin on a nylon mesh fabric by hopper-coating techniques. It was possible to prepare these films by this method because of a novel approach which places an easily removed polyethylene backing on the nylon mesh. The backing procedure was carried out by an extrusion coating technique. Molten polyethylene was applied as a sheet to the Fablok in such a manner that would allow for its easy removal after the film had been formed.

Synthesis of Gluconic Acid

A number of coatings containing a coupled enzyme system (glucose oxidase/catalase) in gelatin on the porous nylon mesh fabric were prepared. This immobilized enzyme system was designed for the production of gluconic acid.[12]

Variations included several gel/biocatalyst coverages in an attempt to determine optimal coverage/surface area (Table 1), and in one case bis-vinylsulfone methyl ether was replaced by glutaraldehyde as the crosslinking agent.

Evaluation of the immobilized biocatalyst was done by titration of a stirred suspension of the film in a glucose solution. In all of the coatings tested, glucose was converted to gluconic acid in greater than 90% yield (Figure 9). It was also found that glutaraldehyde could be substituted for bis vinylsulfone methyl ether as a crosslinking agent without loss of synthetic capability.

Films containing immobilized glucose oxidase and catalase in a gelatin carrier have been prepared by dip-coating fabrics such as Fablok (nylon mesh) and ninon (woven polyesters) into an aqueous gelatin solution containing the enzymes. The gelatin-enzyme coatings were then crosslinked by dip-coating into a second solution containing 1% aqueous glutaraldehyde.

Evaluation of the dip-coated immobilized biocatalyst was done by titration of a stirred suspension of the film in a glucose solution. Glucose was converted to gluconic acid in greater than 90% yield (Table 10). The dip-coating procedure was more difficult to control than the hopper-coating procedure in terms of reproducible amounts of catalyst per unit area of coating.

Synthesis of Pyruvic Acid

Machine coatings were prepared containing lactate oxidase (1444 U/m^2)

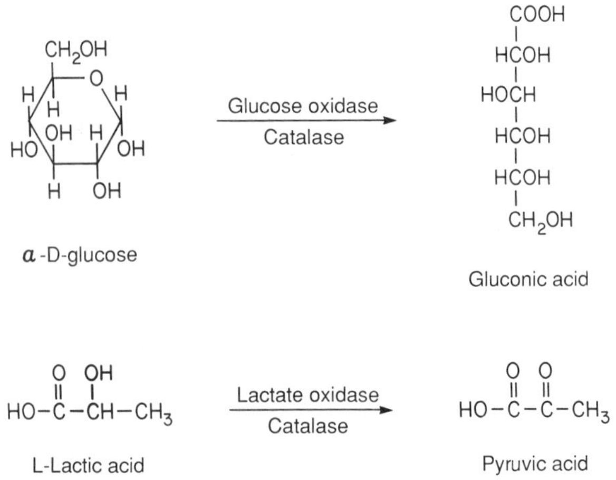

α-D-glucose $\xrightarrow[\text{Catalase}]{\text{Glucose oxidase}}$ Gluconic acid

L-Lactic acid $\xrightarrow[\text{Catalase}]{\text{Lactate oxidase}}$ Pyruvic acid

$HO-C(=O)-CH(OH)-CH_3 \rightarrow HO-C(=O)-C(=O)-CH_3$

Figure 9. Synthesis of gluconic and pyruvic acids catalyzed with immobilized enzymes.

and catalase (340,550 U/m²) in gelatin (10.7 g/m²). The gelatin was hardened with either glutaraldehyde or oxidized dextran (MW 40,000).[12]

The coatings were evaluated by cutting 0.3 m of film into pieces ~1 cm on end after removal of the polyethylene backing. A mixture containing the biocatalyst, lactic acid and buffer was shaken at 30°C and 100 rpm for 48 hr. Pyruvic acid formation (Figure 9) was demonstrated by treatment of the reaction mixture with 2,4-dinitrophenylhydrazine reagent, after removal of the biocatalyst.

L-Lactic acid is converted to pyruvic acid in 39% yield under these conditions (Table 2). Replacement of L-lactic acid with double the amount of D,L-lactic acid produced pyruvic acid in 42% yield.

The biocatalyst was inactivated by a second treatment with glutaraldehyde (Table 2). Coatings were made at three different pH levels (6, 7 and 8). Coatings made at pH 6 appeared to yield the greatest amounts of pyruvic acid.

When oxidized dextran was used in place of glutaraldehyde, in order to harden the gelatin, an active coating was obtained at pH 7. Coatings prepared at pH 6 and 8 were inactive.

The biocatalyst preparations described have not been optimized to obtain high catalytic activity or to preserve catalytic activity for long operation times.

We believe the results of these investigations indicate the immobilization of coupled enzyme systems by the described processes to be a viable method for preparing biocatalytic thin-films. With these films, it should be possible to assemble bioreactors for the continuous production of specialty chemicals, as demonstrated in the following examples.

Synthesis of Aspartic Acid

Thin-films were prepared containing cells of *E. coli* in the ratio of 24.6 g of cells to 10.7 g of gelatin per square meter of film.[13]

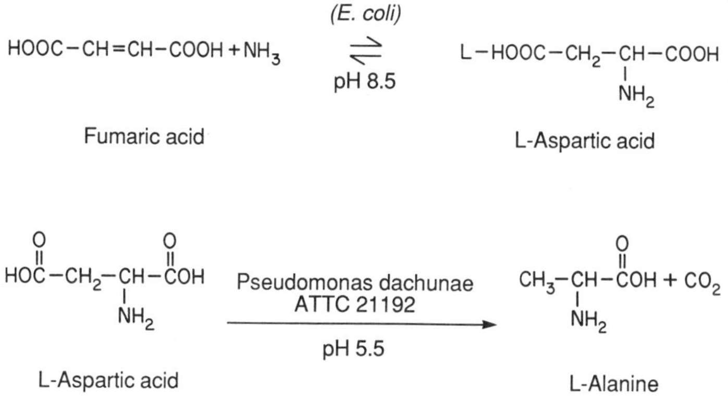

$$HOOC-CH=CH-COOH + NH_3 \underset{pH\ 8.5}{\overset{(E.\ coli)}{\rightleftharpoons}} L-HOOC-CH_2-\underset{NH_2}{\underset{|}{CH}}-COOH$$

Fumaric acid — L-Aspartic acid

$$HO\overset{O}{\overset{\|}{C}}-CH_2-\underset{NH_2}{\underset{|}{CH}}-\overset{O}{\overset{\|}{C}}OH \xrightarrow[pH\ 5.5]{\text{Pseudomonas dachunae ATTC 21192}} CH_3-\underset{NH_2}{\underset{|}{CH}}-\overset{O}{\overset{\|}{C}}OH + CO_2$$

L-Aspartic acid — L-Alanine

Figure 10. Synthesis of aspartic acid and alanine catalyzed with immobilized whole cells.

Initial coatings were made with bis-vinylsulfone methyl ether as the hardening agent; however, shake flask examination of the films prepared in this manner indicated that a substantial amount of gelatin and cells were lost when the coating was cut into pieces and shaken at 100 rpm in aspartic acid production medium. Therefore, a second wash coating of glutaraldehyde was used. The films thus produced did not lose appreciable amounts of solid when tested in the above manner.

Since glutaraldehyde was found to improve the integrity of the coatings previously hardened with bis-vinylsulfone methyl ether, it was substituted for the vinylsulfone. It was necessary, however, to use two glutaraldehyde treatments. The films were tightly wound on a polyolefin tubing core into 10.2 cm diameter cylinders (5.1 m^2 of film) and placed into 45.8 cm stainless steel columns.

Recirculation of aspartic acid production medium through an 45.8 cm reactor column (coating containing 2.3 g *E. coli*/g gelatin-bis-vinylsulfone methyl ether-glutaraldehyde hardening, at 600 mL/min and 37°C) produced aspartic acid in 86% yield (Figure 10). Comparable yields of aspartic acid were obtained with the glutaraldehyde hardened coating.

Analysis of samples of reaction mixture by gas chromatography taken over different reaction times indicated that an activation period was required before the columns produced a maximum yield for *E. coli* (Figures 11 and 12). A reduction in operating temperature of the column from 37° to 20°C led to a reduction in product yield from 84% to 72%.

Single pass runs were made with the reactor column containing immobilized *E. coli* film. It would appear that a greater than 20-min residence time may be needed to obtain 90% yields of aspartic acid.

Vinyl polymers were prepared by standard free radical polymerization techniques. The polymers were stored as aqueous solutions. Partially oxidized dextran was produced by periodate oxidation of 40,000 molecular weight dextran. Analysis for aldehyde content indicated 6.5 milliequivalents per gram of solid.

The polymers were selected to provide positively charged and neutral carriers. Selection was also based on availability, ease of synthesis and potential for modification. The crosslinking polymers selected contain aldehyde groups that react with amino groups in the carrier polymers. The

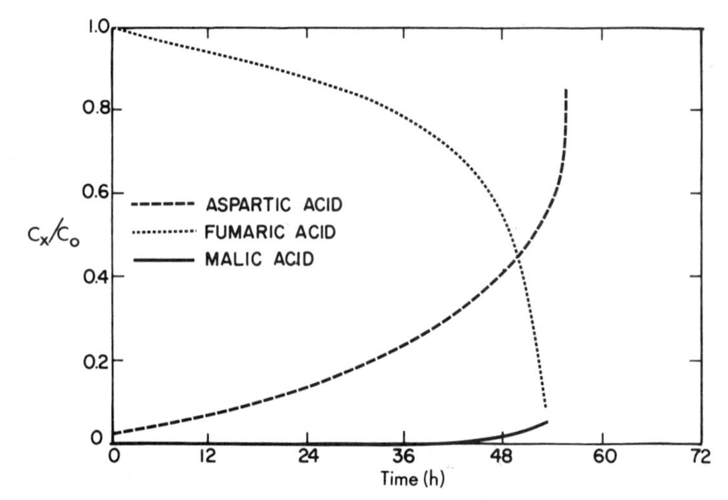

Figure 11. Preparation of aspartic acid with immobilized *E. coli* (run I). C_x/C_o represents the ratio of the compound concentration at time t to the concentration of the same compound at the initial starting point: (---) aspartic acid, (•••) fumaric acid, (——) malic acid.

structures of the synthetic polymers and crosslinking agents are shown in Table 3.

Four consecutive bioconversion runs were made with each aspartic acid biocatalyst coating (Table 3). In the one case, where decomposition of coatings were observed, fewer runs are reported.

Several of the polymers evaluated show excellent potential as replacements for gelatin as a carrier (Table 3). Coatings which appeared to yield the largest amounts of aspartic acid were prepared with polymers 1

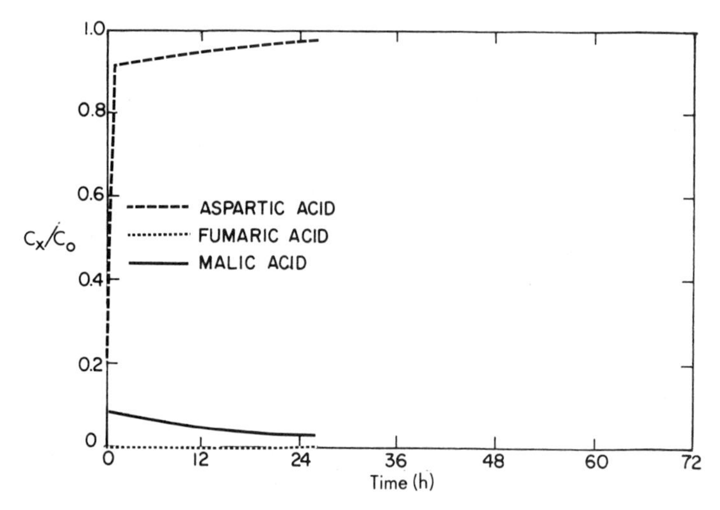

Figure 12. Preparation of aspartic acid with immobilized *E. coli* (run II): (---) aspartic acid, (•••) fumaric acid, (——) malic acid.

(oxidized dextran hardening and poly[acrylamide-co-m+p-vinylbenzaldehyde] hardening), 3 (oxidized dextran hardening), and 4 (oxidized dextran hardening). Polymer 2 did not appear satisfactory due to fragility of the coating which may have resulted from incomplete crosslinking.

The preparation of immobilized whole cell films by hopper coating shows potential for use in the manufacture of biocatalyst films that might be utilized in the manufacture of specialty chemicals such as amino acids.

Attempts to prepare immobilized *E. coli* by a dip-coating procedure proved even more difficult than the procedure used for enzyme immobilization. The major difficulty occurred in the drying operation as cells from the coating severely contaminated the drying ovens. Evaluation of the dip-coated films also indicated a lower yield of product (Table 10).

Synthesis of Ribavirin

Preliminary shake flask experiments with thin-film immobilized *Brevibacterium acetylicum* cells indicated that ribavirin (Figure 13) was produced from 1,2,4-triazole-3-carboxamide and inosine in 24 hr (Table 4). In column experiments yield data (Table 5) and visual observations of the films indicated that the gelatin coatings were physically less stable (more susceptible to erosion) than the poly[acrylamide-co-N-(3-aminopropyl)-methacrylamide hydrochloride] coatings. Specific productivities for the immobilized cells in the coated form ranged between 0.008 and 0.03 which compared favorably with those obtained with nonviable suspended cells, which ranged between 0.023 and 0.039 (Table 5).

At 30°C, ribavirin production ceased. Temperatures between 30° and 60°C were not examined.

The relationship between yield of ribavirin (g/L) and flow rate through the bioreactor was examined. A linear relationship between flow rate and concentration of ribavirin in the reactor output was found: 1.2 g/L at 3.0 mL/min, 2.1 g/L at 1.5 mL/min, and 4.3 g/L at 0.7 mL/min (Table 6).

Synthesis of Alanine

Initial shake flask experiments with various *Pseudomonas* species indicated that *Pseudomonas dacunhae* (ATCC21192) produced the largest amount of alanine in 16-18 hr (Figure 10).

Comparison between immobilized ATCC21192 and free cells produced from the identical fermentation lot indicated that both produced the same amount of alanine when exposed to alanine production medium under the same experimental conditions (Table 7). Attempts to obtain large quantities of this organism with the same high alanine-producing capability in fermentors were unsuccessful.

Although alanine-producing capability was substantially diminished, the cells were evaluated in a bioreactor for alanine production. The bioreactor experiments that were done with cells coated in either gelatin (Table 8) or poly[acrylamide-co-N-(3-aminopropyl)-methacrylamide hydrochloride] (Table 9) demonstrated that immobilization in either polymer did not diminish the alanine-producing capability of the cells.

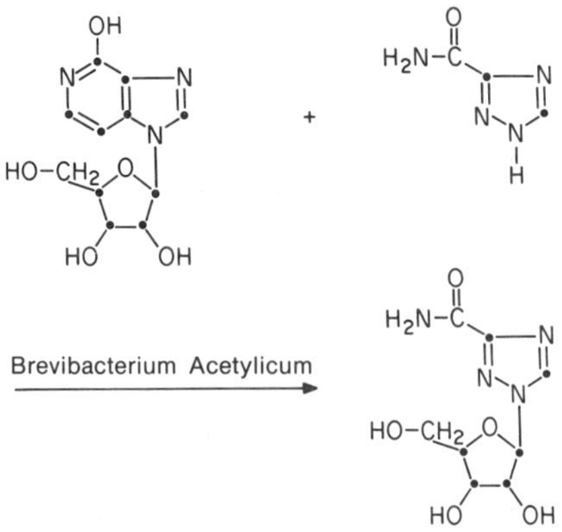

Brevibacterium Acetylicum

Figure 13. Synthesis of Ribavirin with immobilized cells of *Brevibacterium acetylicum.*

CONCLUSIONS

The use of thin-film biocatalysts in the synthesis of organic chemicals is just one aspect of the use of thin-films for a variety of purposes. In this paper we discuss methods by which biocatalysts, namely enzymes and cells, are successfully incorporated into biologically active thin-films. Two processes, hopper coating and dip coating, are discussed. The hopper-coating approach, although requiring more complex equipment, appears to be the process that is best controlled.

Thin-film immobilized biocatalysts were prepared and used for synthesis of gluconic acid, pyruvic acid, aspartic acid, alanine, and ribavirin. This approach to construction of immobilized biocatalysts offers answers to some of the problems encountered with the use of particles, such as the introduction of a rigid carrier in order to prevent column packing. Other problems that are addressed include the possibility of a modular approach to rapid replacement of spent catalyst and a more uniform distribution of the catalyst within the support.

Further, the development of more complex films would allow for layering of biocatalysts one upon the other in such a manner as to allow for the incorporation of multiple chemical reactions in a single film. Also, the films might be constructed in such a way as to permit the separation of reactants and products. This approach would require passage of the reaction mixture through the film instead of over the surface as is the case with the present configuration.

We believe that the thin-film approach to the development of biocatalysts offers great potential for the construction of complex and useful biocatalytic processes in the future.

ACKNOWLEDGMENTS

We wish to thank David Bellinger for development of the polyethylene-backed nylon mesh support, I. S. Ponticello, R. K. Hollister, and W. A. Bowman for samples of synthetic polymers, R. Snoke and E. Dohrn for batches of microorganisms, W. A. Napoli for bioreactor development, and J. C. Lievense for bioreactor synthesis data on ribavirin. We also wish to thank C. T. Abrams and J. Fyles for technical assistance.

REFERENCES

1. O. Zaborsky, "*Immobilized Enzymes,*" CRC Press, Cleveland, Ohio, 1973.
2. L. B. Wingard, Jr., Ed., "*Enzyme Engineering,*" Interscience Publishers, New York, N.Y., 1972.
3. R. A. Messing, Ed., "*Immobilized Enzymes for Industrial Reactors,*" Academic Press, New York, N.Y., 1975.
4. I. Chibata, Ed., "*Immobilized Enzymes - Research and Development,*" John Wiley and Sons, New York, N.Y., 1978.
5. I. Chibata and L. B. Wingard, Jr., Eds., "*Applied Biochemistry and Bioengineering, Immobilized Microbial Cells,*" Vol. 4, Academic Press, New York, N.Y., 1973.
6. W. R. Vieth, S. G. Gilbert, S. S. Wang, and R. Saini, U. S. Patent 3,758,396 (1973).
7. Japanese Patent 56131391 to Kansai Paint Co. Ltd. (1980).
8. A. Weissberger, "*The Theory of the Photographic Process,*" C. E. K. Mees and T. H. James, Eds., 3rd Ed., Macmillan Company, New York, N.Y., 1966, p. 382.
9. R. W. Spayd, B. Bruschi & B. A. Burdick, Clin. Chem., **24**, 1343 (1978).
10. B. A. Burdick, D. A. Hilborn & T. W. Wu, Clin. Chem., **32**, 1953 (1986).
11. B. A. Burdick, Clin. Chem., **33**, 310 (1987).
12. B. A. Burdick & J. R. Schaeffer, Biotech. Lett., **9**, 253-258 (1987).
13. B. A. Burdick & J. R. Schaeffer, Biotechnol. Bioeng., **31**, 390-395 (1988).
14. T. Sato, et al., Biotechnol. Bioeng., **27**, 1797 (1975).
15. M. C. Fusee & J. E. Weber, Appl. Environ. Microbiol., **48**, 694 (1984).
16. T. Fujishima & Y. Yamamoto, European Patent G093401 (1983).
17. W. R. Sorenson & T. W. Campbell, "*Preparative Methods of Polymer Chemistry,*" Interscience Publishers, New York, N.Y., 1961, p. 149.
18. S. R. Sandler & W. Karo, "*Polymer Synthesis,*" Vol. 1, Academic Press, San Diego, CA, 1974.
19. D. T. Canvin, Can. J. Biochem., **43**, 1281 (1965); A. B. Littlewood, "*Gas Chromatography,*" Academic Press, Inc., New York, N.Y., 1970.
20. W. Lindner, et al., J. Chromatogr., **185**, 323 (1979).
21. R. L. Heinrikson & S. C. Meredith, Anal. Biochem., **136**, 65 (1984).
22. C. D. Carr, Anal. Chem., **46**, 743 (1974).

POLYMERIC BIOMIMETIC CATALYSTS BASED ON 4-DIALLYLAMINOPYRIDINE

Lon J. Mathias*, Rajeev A. Vaidya and Gustavo Cei

Department of Polymer Science
University of Southern Mississippi
Hattiesburg, Mississippi 39406-0076

A new class of polymeric catalysts based on 4-(diallylamino)pyridine (DAAP) is described. Cyclopolymerization of DAAP yields a polymer whose monomer units are identical to 4-(pyrrolidino)pyridine (PPY), one of the most effective catalysts for nucleophilic substitution reactions. The homopolymer of DAAP proved to be more effective than PPY and DMAP [4-(dimethylamino)pyridine] in the hydrolysis of esters like p-nitrophenyl caproate. Copolymerization of DAAP with vinyl monomers of cationic, anionic and neutral character was found to yield catalysts that are even more reactive than the homopolymer when the rate is referred to the number of catalytic units actually present. The catalysis of the esterification of linalool was also investigated. Most of the copolymers investigated displayed lower activity than DMAP. However, catalytic activity similar to DMAP was observed for certain compositions of DAAP-DMAM copolymers.

INTRODUCTION

It is well known that pyridine and the 4-substituted (dialkylamino)-pyridines are effective nucleophilic catalysts for a series of reactions. Among others, these compounds are useful in catalyzing esterolyses, esterifications, silylations and alkylation reactions of hindered substrates. Several reviews of the synthesis and applications of these catalysts have appeared[1-3]. The easy commercial synthesis has made 4-(dimethylamino)pyridine (DMAP) [1] the most popular member of this class of supernucleophiles, although the pyrrolidine analogue (PPY) [2] is the most active (Scheme 1).[1]

Several research groups attempted binding of (dialkylamino)pyridines to a polymeric backbone. The advantages of polymeric reagents compared to monomeric ones are multiple.[4,5] For instance, polymeric reagents can be isolated after reaction by filtration or precipitation and reused. Therefore, not only lower production costs but also higher purity of the product is achieved. Insoluble polymeric reagents are also suitable for continuous flow industrial operations and also for reactions that have to be carried out to a certain conversion.

Biomimetic Polymers
Edited by C. G. Gebelein
Plenum Press, New York, 1990

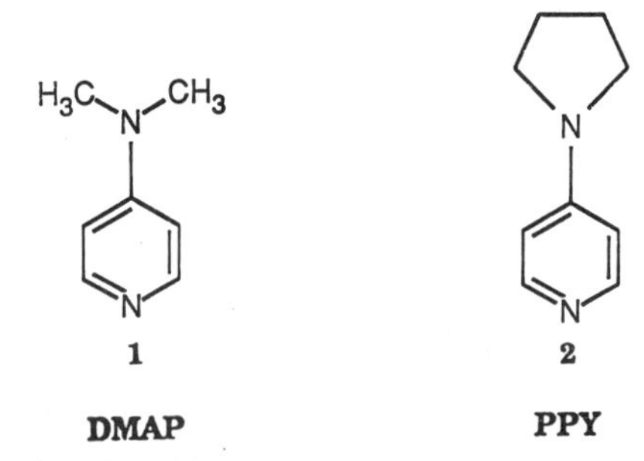

4-DIMETHYLAMINOPYRIDINE 4-PYRROLIDINOPYRIDINE

Scheme 1. **[1] and [2]. Catalysts** for nucleophilic **substitution.**

The first report of polymer bound (dialkylamino)pyridine dealt with the reaction of several carboxylic acid containing (dialkylamino)pyridines with commercial poly(ethyleneimine) (PEI) to give polymeric catalysts similar to [3] (Scheme 2).[6,7] These compounds were examined for catalytic behavior in the hydrolysis of p-nitrophenyl esters and found to be very reactive. Other research group incorporated (dialkylamino)pyridine into crosslinked chlorinated polystyrene [**4**], either by reaction on preformed polymer[8-11] or by synthesis of a monomer containing the desired functionality and subsequent polymerization.[12,13] More recently, polymers with structures similar to those obtained with our procedure have been obtained by reduction of 4-(N-maleimido)pyridine residues.[14] These polymers apparently react rapidly with benzyl halides, but insolubility greatly restricts characterization of the polymers and precludes their use as catalysts in homogeneous systems.

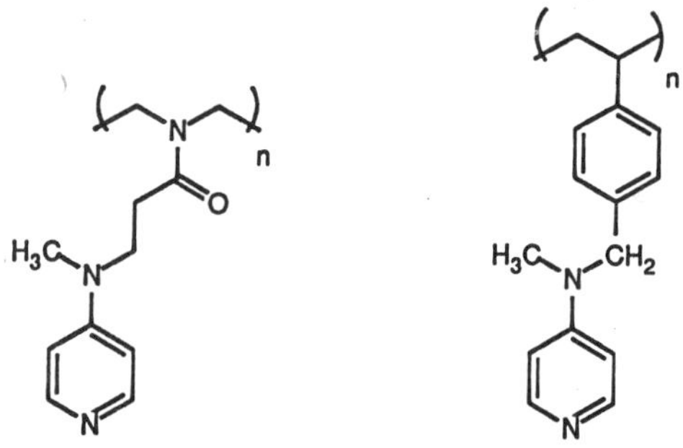

H_3C N n O 3

H_3C N CH_2 n 4

Scheme 2.

The polymeric catalysts synthesized in these reports have certain disadvantages related to the type of polymeric backbone used. Polymeric catalysts containing residual amine groups can react with the substrate in a way different to that expected from the supernucleophilic pyridine group. Due to the higher pKa of the amine group respective to pyridine, a cationic environment around the polymer chain is generated that can modify the activity of the catalyst.

Catalysts based on polystyrene supports are inherently **hydro**phobic and suffer from several drawbacks.[15] They are generally synthesized by reactions on preformed polymer. Such reactions are difficult to drive to completion. Only the catalysts obtained by polymerization of a supernucleophilic monomer result in formation of uniform and readily characterized polymers. Moreover, while PPY has the highest activity of known (dialkylamino)pyridines, almost all of the previous polymeric catalysts have used derivatives of DMAP, (benzylmethylamino)pyridine, or other pyridine derivatives which are 2 to 10 times less active than PPY.

The approach we chose to supernucleophilic polymeric catalysts is based on the previously described concept of generating the desired repeat unit structure during the polymerization process. We have used this concept in the synthesis of polymers containing crown ether units through a macrocyclopolymerization procedure.[16,17] In this paper we report synthesis and polymerization of DAAP [5] (Scheme 3), which leads to formation of the homopolymer [6] and a variety of copolymers possessing a range of solubilities and functional groups.

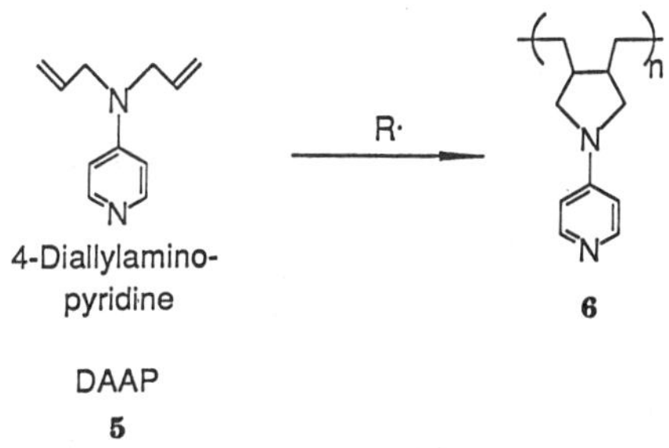

4-Diallylamino-
pyridine

DAAP
5

R·

6

Scheme 3. Synthesis of Poly(DAAP).

The cyclopolymerization reaction employed in the present investigation is based on the work of Butler et al., on polymerization of quaternary diallylamines.[18,19] Several literature reports dealing with enhanced catalytic activity due to cooperative effects between different units of the macromolecular catalyst, whether in natural enzymes[20] or synthetic catalysts,[21] led us to the consideration that the activity and selectivity of poly(DAAP) could be modified or even increased further by copolymerization with suitable monomers. We chose three comonomers, N,N-dimethylacrylamide (DMAM) which yields neutral repeat units, dimethyldiallylammonium chloride (DMDAACl) which gives positively charged ones, and acrylic acid (AA) which leads to carboxylate anions in the copolymers.

EXPERIMENTAL SECTION

1. Materials

Reagent grade solvents and deionized water were utilized for all synthetic procedures. Spectrophotometric grade acetonitrile and absolute methanol were used for the kinetic experiments. Diallylamine (Aldrich) was only 97% pure by GC. Distillation under normal pressure at 110°C gave a clear colorless liquid (purity >99% by GC). Acrylic acid (Aldrich) was freed from inhibitor by distillation under vacuum (bp 48°C at 2 KPa). Dimethyldiallylammonium chloride (DMDAACl) (Polysciences) was purchased as a 65% aqueous solution and used without further purification. Dimethylacrylamide (DMAM) (Polysciences) was distilled under vacuum (bp 75°C at 2 KPa).

4-Chloropyridine hydrochloride (Aldrich), 2,2-azobis(amidinopropane)-hydrochloride (V50) (Wako Chemicals), 4-(pyrrolidino)pyridine (Aldrich), 4-(dimethylamino)pyridine (Reilly Chemicals), linalool (Aldrich) and decane (Aldrich) were used without further purification. Acetic anhydride, triethylamine and toluene were distilled before use.

2. Synthesis of 4-(Diallylamino)pyridine (DAAP)

DAAP was synthesized by reaction of diallylamine and 4-chloropyridine. 4-Chloropyridine was obtained from 4-chloropyridine hydrochloride. 4-Chloropyridine hydrochloride (10 g) was dissolved in deionized water (20 mL) and excess 6N NaOH was added to the solution to reach a pH of 12. 4-Chloropyridine formed an organic layer above the aqueous phase. The mixture was extracted with four 50 mL portions of ether. The combined ether extracts were filtered over two layers of phase separation paper (Whatman PS) to remove water. The ether was evaporated and the remaining dark brown liquid was kept at high vacuum for 20 min. The overall yield based on 4-chloropyridine hydrochloride was 70-80%.

For the DAAP synthesis, 4-chloropyridine (13.24 g) was refluxed with excess diallylamine (28.04 g, 2.5-fold molar excess) under a nitrogen atmosphere for three days. The product mixture was neutralized with 1N NaOH and extracted with four portions of ether (100 mL each). The combined ether extracts were filtered over two layers of phase separation paper. The ether was evaporated to yield a dark reddish brown viscous liquid. This liquid was submitted to distillation at high vacuum with a short path distillation apparatus. A pale yellow liquid distilled at approximately 160°C. The distilled fraction reacts with oxygen (yellowing takes place if stored in air) and the distillation apparatus must be purged with N_2 to release the vacuum. This product proved to be DAAP (purity >99% by GC).

Homopolymerization of DAAP. DAAP (10 g) was dissolved in concentrated aqueous HCl (2.308 mL, 10% molar excess) at 0°C under a nitrogen atmosphere. Deionized water (8.3 mL) was added to form a 50% by weight solution of DAAP-HCl. The solution was degassed under high vacuum and the vial filled with N_2. The initiator (V-50) (0.156 g, 1 mol %) was added to the solution and the polymerization carried out at 65° C under a nitrogen atmosphere for 12 h. Two further doses of initiator (0.156 g each) were added to the system at 12 h intervals. The resulting polymer solution was dialyzed against 5% aqueous NaCl, and then water, to remove unreacted monomer and oligomers. The resulting solution was freeze-dried to obtain a dry product.

Copolymers of DAAP. All copolymerizations were carried out in Schlenk flasks at 60°C. The free radical initiator V50 was added before degassing. A 1.4 molar excess of HCl with respect to DAAP monomer was necessary to ensure polymerizability of the amine. Monomer solutions were submitted to three freeze-thaw cycles for degassing. After polymerization in a water bath, the flasks were chilled, the contents poured into dialysis bags (MW cutoff 3,000-6,000) and dialyzed against water for 4-5 days. The contents were then freeze-dried to give solid products for determination of conversion and characterization.

3. Characterization

Weight average molecular weight of poly(DAAP) was determined with low-angle laser light scattering using a Chromatix KMX-16 differential refractometer and a Chromatix KMX-6 spectrophotometer. The M_w value was calculated to be 8,000 by using the obtained dn/dc value of 0.20.

Solutions of polymer catalyst were prepared by diluting stock solutions used for M_w measurements. Concentrations of the stock solutions were verified by evaporating known weights of solutions to constant weights of polymer present; 10 values were obtained for each concentration to ensure accuracy. The calculated concentrations were confirmed by UV spectroscopy using a Beer's law relationship established for PPY.

NMR spectra were recorded with a Bruker MSL-200 spectrometer. UV spectra were recorded with a Perkin-Elmer 320 spectrophotometer. The pH of the buffers used for pKa determinations was checked with an Orion pH meter model 701-A, using an Orion combination pH probe 91-55. The calibration was performed with standard buffers from Curtin Matheson Scientific, Inc.

Viscometric measurements employed Cannon-Ubbelohde viscometers at 25°C. The solvent was a 0.1 M aqueous NaCl solution in order to reduce polyelectrolyte effects at high polymer dilutions.

4. Calculation of the Reactivity Ratios

Reactivity ratios in copolymerizations are usually evaluated with samples obtained at low conversions (< 5%).[22,23] If the conversion is accurately known, the Kelen-Tudos method[24] can be modified so that samples obtained at high conversion can also be used.[25] Copolymer compositions were determined by UV spectroscopy. The protonated DAAP unit shows an absorption maximum at 280 nm, The molar absorptivity of poly-(DAAP) in 10% aqueous HCl was determined to be 1.36×10^4 L/mol/cm. The copolymers with acrylic acid were only soluble in aqueous formic acid. Solutions were obtained by dissolving in concentrated formic acid followed by dilution with water. The molar absorptivity in 10% formic acid was 1.43×10^4 L/mol/cm.

5. Determination of pKa of the Copolymers

The pKa of organic acids and bases is usually calculated by titration of a solution of the electrolyte with strong base or acid while recording pH as a function of the amount of titrant. This procedure was not feasible in our case due to the weakness of the polybase and, in some cases, to the low solubility of the polymers. Instead, the titrations were carried out spectroscopically.[26,27] Protonated DAAP units absorb at 280 nm while unprotonated groups show an absorbance maximum at 260 nm. The molar fraction of protonated and free base may therefore be calculated with the following expression (Equation 1), where x(B) is the molar fraction of free DAAP units at the pH under investigation. The absorbencies at pH 1 and pH 13 at 260 and 280 nm were obtained from spectra of the corresponding polymer in 0.1 N HCl and 0.1 N NaOH, respectively. Several buffers were prepared with 0.1 M salt solutions[28] and the necessary amount of polymer solution was injected into the spectrometer cell containing the buffer. The amount added was small in comparison to the cell content to preclude significant changes of pH.

(Equation 1)

$$x(B) = \frac{\mathrm{Abs(280)/bs(260)]_{pH1}} - \mathrm{[Abs(280)/Abs(260)]_{pHp}}}{\mathrm{[Abs(280)/Abs(260)]_{pH1}} - \mathrm{[Abs(280)/Abs(260)]_{pH13}}}$$

6. Kinetic Measurements

Preliminary investigations on the esterolytic activity of poly(DAAP) were carried out in the following way: 3 mL of the aqueous buffer at the desired pH was pipetted into a cuvette. The substrates were dissolved in acetonitrile to make solutions at concentrations such that injection of a 1 µL of the solution into the buffer in the cuvette would result in a substrate concentration of 1 x 10^{-5} M. The polymer and PPY were dissolved in methanol to obtain solutions such that each microliter when injected into 3 mL of buffer yielded a catalyst concentration of 5 x 10^{-7} M. For catalyzed reactions the desired amount of catalyst was injected into the cuvette and the contents were shaken to ensure mixing. The cuvette was introduced into the UV spectrometer and the stirrer started. The data acquisition programs were loaded into the computer along with a data disk in the appropriate drive. The desired amount of substrate was injected and the stopwatch started. After exactly 1 min., data acquisition was initiated. Spectra were collected at time intervals of 3, 5 or 60 min. depending on the experiment. After the required number of spectra had been collected, the data analysis routine was initiated and the absorbance at 400 nm was determined as a function of time. Knowing the extinction coefficient for the p-nitrophenoxide ion at that pH and the initial substrate concentration, we calculated the conversion of substrate. Each kinetic run was repeated at least three times.

The body of the esterolytic experiments was performed as follows: the reaction was carried out in disposable polystyrene cuvettes (transparent at 400 nm) placed in a water bath at 30°C. Each experiment was repeated (up to five times) in order to ensure reproducibility. The absorption of p-nitrophenolate anion, at 400 nm (extinction coefficient of 15,600 L/mol/cm at pH 7.8)[27] was followed with a Bausch & Lomb Spectronic 710 instrument. The use of quartz cuvettes did not markedly affect the precision of the kinetic measurements.

The procedure involved filling the cuvette with buffer (3 mL) and injecting the necessary volume (usually 5 µL) of catalyst solution (in methanol or water). The final catalyst concentration (in mol of DAAP monomer units) was approximately 1 x 10^{-6} M. The exact catalyst concentrations are reported in the Tables. The cuvette was then capped, shaken and equilibrated in the water bath. The necessary amount of substrate (0.5-10 µL) was then added, and the stopwatch started. The cuvette was recapped, shaken and, after ensuring that no air bubbles were on the light path, the absorbance was measured. The initial substrate concentration was in the average 1-20 x 10^{-5} M. Absorbance measurements were repeated every ten minutes.

The pseudo first order rate constant (k_{obs}) was obtained as the slope of the plot of ln $([AB]_o/[AB]_t)$ against time, where $[AB]_o$ and $[AB]_t$ are the substrate concentrations at the onset of the reaction and at time t. The concentration of p-nitrophenyl caproate (PNPC), $[AB]_t$, was obtained by subtracting the p-nitrophenolate (PNP) anion concentration (calculated by means of Beer's law) from $[AB]_o$. In all cases, straight lines were obtained with satisfactory correlation coefficients.

However, differences between parallel experiments run under identical conditions were on the order of 20% in the worst cases. We could not reduce the scatter even using the indirect injection method suggested by Menger and Venkataram.[29] Reproducibility of the amounts delivered by the microsyringe was checked with a precision balance. Adsorption onto the cuvette walls was also possible; however, the results could not be significantly improved by replacing disposable cuvettes by quartz ones. A substrate concentration higher than the CMC (critical micellar concentration) can also lead to misleading results. According to Guthrie,[30] the possibility of micelle formation by the substrate p-nitrophenyl caproate can be excluded at the concentration of the study.

The esterification reaction was followed by GLC on a Hewlett Packard 5880A instrument with flame ionization detector and a J & W scientific fused silica megabore column, with 5% phenyl methyl polysiloxane. The temperature program consisted of a 3 min. period at 60°C, then heating (20°C/min) to 100°C and holding 4 min. at this temperature. Decane (GC standard) and linalool peak area ratios at different reaction times were compared with the initial ratio for determining percentage conversion. Duplicate experiments showed a measurement error of less than 2% in most cases. The reactor consisted of a 100 mL three neck flask fitted with a reflux condenser and magnetic stirrer. It was placed in an oil bath at 80°C and charged with toluene (20 mL), triethylamine (0.03 mol) and decane (0.5 g). The necessary amount of polymeric catalyst was dissolved in acetic anhydride (0.03 mol) and the resulting solution added to the reactor. The mixture was magnetically stirred for 30 min. under N_2 atmosphere, and linalool (0.02 mol) was added. After ensuring homogeneous mixing a 0.1 mL aliquot was taken and dissolved in 0.9 mL of $CHCl_3$. This solution was analyzed by GC. Other samples were taken over a 3 h period and were analyzed in the same way.

RESULTS

1. Synthesis and Characterization of the Copolymers

The goal of determining copolymer compositions with maximum catalytic activity can only be achieved through a quantitative evaluation of structure-activity relationships along with the kinetic determinations. In the system under study, catalytic activity depends not only on the molar fraction of DAAP in the polymer but also on the sequence length distribu tion, which can be determined if the reactivity ratios are known. However, in aqueous solutions, a certain fraction of the DAAP monomer units becomes protonated at a certain pH. The degree of protonation of the DAAP units needs to be known as well because only free DAAP monomer units are nucleophilic centers, and therefore active for catalysis. For weak bases like DAAP, the degree of protonation is related to the pH by the Henderson-Hasselbalch equation (Equation 2).

$$pH = pKa + \log [B]/[BH+] \qquad \text{(Equation 2)}$$

In this equation [B] and [BH+] are the concentrations of the free and protonated base. The pKa needs then to be calculated in order to determine the ratio of free to protonated base at the pH of the kinetic experiments. With this in mind, we carried out the analysis of the reactivity ratios and acid-base properties of the copolymers of DAAP used in the kinetic experiments.

<u>1A Poly(DAAP-co-DMAM).</u> All copolymers were synthesized as described in the experimental section. The dry polymers were brittle and yellow-colored, especially those with high DAAP content. They were soluble in methanol and aqueous HCl, less soluble in water and DMSO, and only swelled in $CHCl_3$, and were insoluble in acetone. Polymer compositions were determined by UV spectroscopy. Table 1 gives the data necessary for the calculation of the reactivity ratios. In all reactivity ratio symbols, monomer 1 (M_1) is DAAP, while M_2, M_3 and M_4 are DMAM, DMDAACl and

Table 1. Free radical copolymerization of DAAP with DMAM at 60°C in water.					
Sample	(a)	(b)	(c) x 10^3	(d)	(e)
A_1	0.099	0.029	4.7	20	0.05
A_2	0.531	0.121	10	240	0.1
A_3	0.698	0.240	15	1350	0.1
A_4	0.753	0.362	22	825	0.25
A_5	0.798	0.398	23	825	0.23
A_6	0.815	0.647	25	1350	0.1
A_7	0.850	0.458	24	825	0.17
A_8	0.902	0.554	25	825	0.15
A_9	0.947	0.799	26	825	0.11

(a) Molar fraction of DAAP in the feed
(b) Molar fraction of DAAP in the polymer
(c) Initiator concentration, mol/L
(d) Polymerization time (min.)
(e) Weight conversion

AA, respectively. Figure 1 gives the corresponding copolymerization diagram, clearly showing the low reactivity of DAAP in comparison with DMAM.

<u>1B Acid-base Behavior of Poly(DAAP-co-DMAM).</u> The Henderson-Hasselbalch equation, which describes the dependence of the degree of dissociation on pH, was modified by Katchalsky et al.,[31] in order to describe the titration of polyelectrolytes (Equation 3).

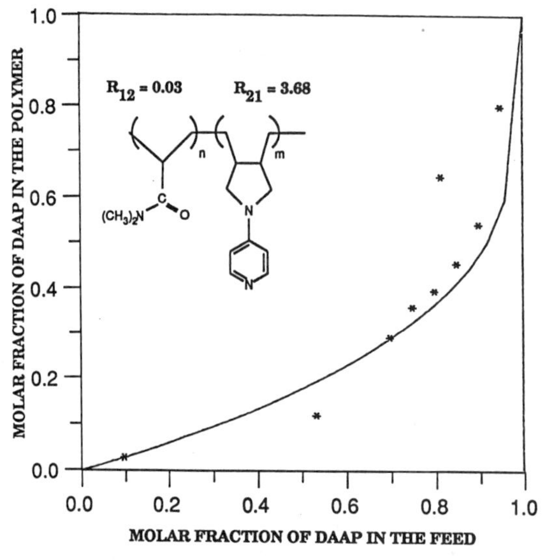

Figure 1. Copolymerization diagram for DAAP [1] with DMAM [2] at 60°C.

$$pH = pKa(app) + n \times \log([B]/[BH+]) \qquad \text{(Equation 3)}$$

The factor n takes into account the neighboring group effect, that is, the fact that electrostatic repulsion between vicinal protonated monomer units decreases the basicity of the polymeric base with respect to the low mol. weight analog.

This expression is an approximation in which pKa(app), the apparent pKa, is a function of the degree of ionization of the polyelectrolyte and of the ionic strength of the solution. With enough low molecular weight ions present, the electrostatic interactions between ionized groups along the polymer chain becomes negligible.[32] At high ionic strength, the pKa becomes a constant and is independent of the degree of ionization. Our results are evaluated according to Equation 3 in order to obtain pKa values that are valid at degrees of ionization close to 0.5 (pKa = pH) at the ionic strength of the buffers. These conditions are also present in the kinetic experiments.

The value of log([B]/[BH+]) in Equation 3 is calculated using Equation 1. pKa(app) is obtained by plotting log([B]/[BH+]) vs. pH. Figure 2 shows a typical titration curve (copolymer A_2) and Figure 3 the corresponding plot according to Eqation 3. Several repeat determinations were made in order to check the precision of these measurements. In all cases, the relative error in pKa was less than 2%. Figure 4 shows the dependence of pKa on copolymer composition for all the copolymers of this research. In DAAP-rich copolymers adjacent protonated monomer units repel each other leading to easier deprotonation. These copolymers are more acidic than DAAP-poor copolymers where adjacent DAAP units are less common. In fact, with increasing DMAM content, pKa values become closer to that of PPY (pKa = 10.5), whose structure is similar to that of the repeating units in poly(DAAP). The acid base behavior of the other copolymers is analyzed below.

1C. Poly(DAAP-co-DMDAACl). Copolymers of DAAP and DMDAACl were obtained as soft, yellow and hygroscopic products, especially those

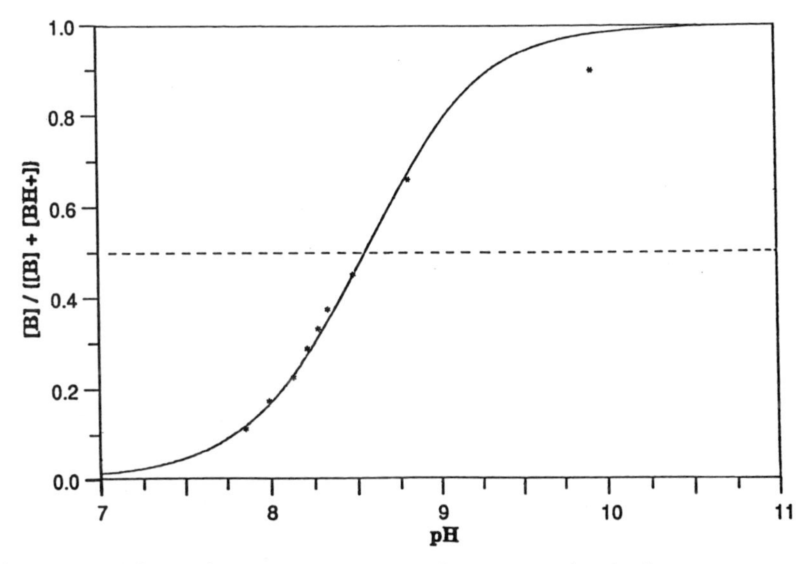

Figure 2. Titration curve of copolymer A_2 (poly(DAAP-co-DMAM)).

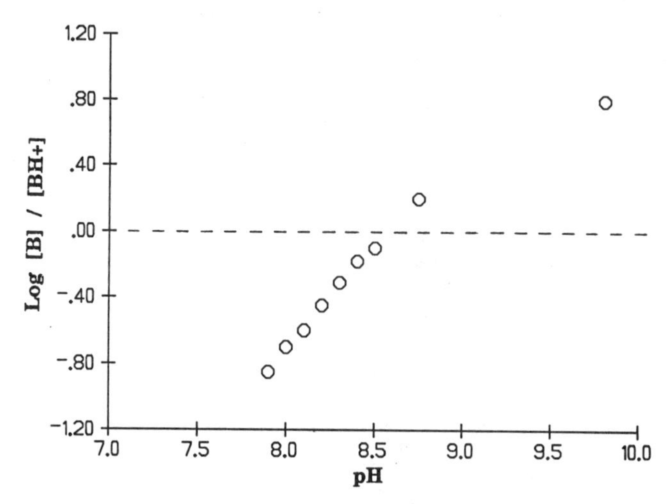

Figure 3. Determination of pKa of copolymer A2 (poly(DAAP-co-DMAM)) according to Equation 3.

containing larger ratios of DMDAACl units. These products were soluble in H_2O and methanol but insoluble in acetone, DMSO and $CHCl_3$. The calculation of the reactivity ratios was performed in the same way as for the DAAP-DMAM copolymers. Table 2 summarizes the results. The copolymerization diagram (Figure 5) shows the higher reactivity of DAAP in this system compared with the system DAAP-DMAM. Table 2 also includes the intrinsic viscosity values for several of the copolymers. We do not believe that the decrease in molecular weight with increasing DAAP content has a very strong effect on the pKa of the copolymers. Overberger and Okamoto showed that molecular weight differences have a very weak influence on the pKa of poly(4-vinylimidazole).[33] Differences in pKa(app) in this system, if present, result from different copolymer compositions.

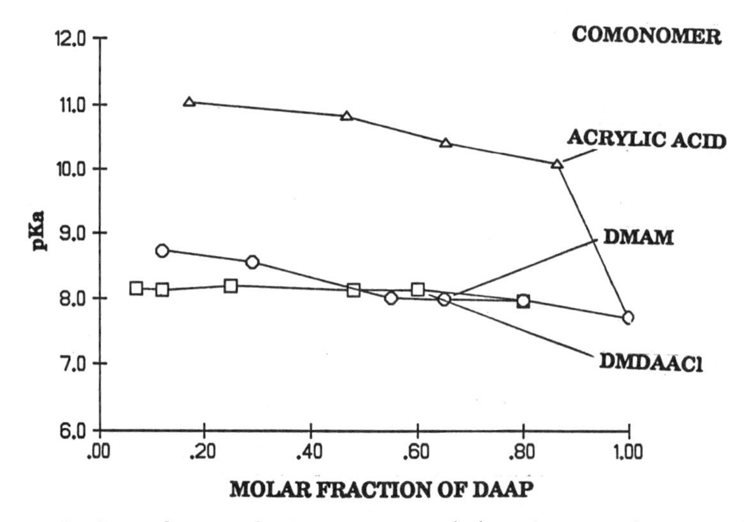

Figure 4. Dependence of pKa on composition for copolymers of DAAP with DMAM, DMDAACl and acrylic acid.

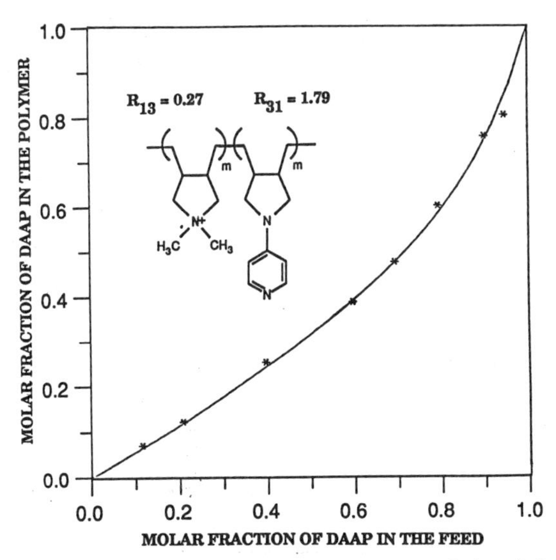

Figure 5. Copolymerization diagram for DAAP [1] with DMDAACl [3] at 60°C.

1D. Acid-base behavior of Poly(DAAP-co-DMDAACl). The titrations were carried out in the same way as for the system DAAP-DMAM. Figure 4 shows that there is no remarkable pKa change with composition. This result is rather surprising because coulombic interactions between charged DMDAACl units and pyridinium groups from DAAP units should facilitate their deprotonation and therefore lower the pKa(app). Probably coulombic interactions between vicinal DMDAAP and DAAP units are not so strong as between two protonated vicinal DAAP units. Not only the distance between

Table 2. Free radical copolymerization of DAAP with DMDAACl at 60°C.

Sample	(a)	(b)	(c) x 10^2	(d)	(e)	(f)
B_1	0.116	0.071	2.2	120	0.24	0.553
B_2	0.209	0.124	2.2	130	0.15	
B_3	0.402	0.254	2.3	130	0.07	0.168
B_4	0.502	0.422	1.9	280	0.08	
B_5	0.601	0.388	2.2	410	0.11	
B_6	0.697	0.477	2.4	280	0.06	
B_7	0.797	0.602	2.1	410	0.05	0.074
B_8	0.902	0.756	2.2	280	0.03	
B_9	0.947	0.802	2.1	410	0.03	0.045

(a) Molar fraction of DAAP in the feed
(b) Molar fraction of DAAP in the polymer
(c) Initiator concentration, mol/L
(d) Polymerization time (min.)
(e) Weight conversion
(f) Limiting viscosity number (dL/g)

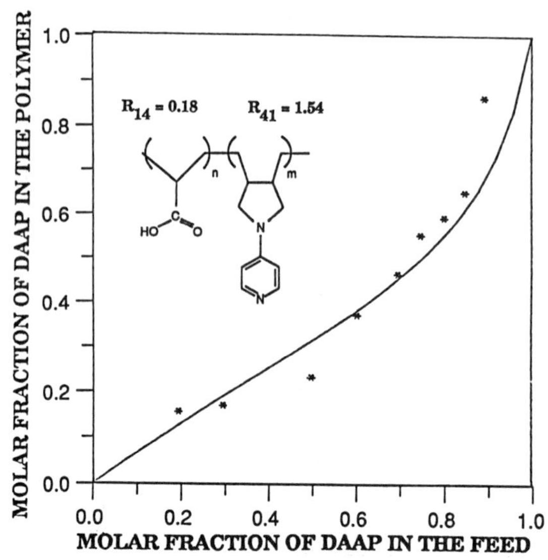

Figure 6. Copolymerization diagram for DAAP [1] with AA [4] at 60°C.

the two positively charged centers is on average larger, but also the presence of permanently charged DMDAAP units creates a counterion atmosphere around the polymer chain that may well compensate for the electrostatic repulsion between protonated DAAP and DMDAACl units.

1E. Poly(DAAP-co-AA). DAAP and acrylic acid were copolymerized under the conditions given in the experimental section. Table 3 gives the conditions for the individual copolymerizations. Glassy and hard products were obtained that displayed colors from clear to dark yellow depending on monomer composition. Copolymers with high acrylic acid content were

Table 3. Free radical copolymerization of DAAP with acrylic acid at 60°C.

Sample	(a)	(b)	(c)	(d)	(e)
C_1	0.200	0.157	0.006	20	0.320
C_2	0.303	0.171	0.013	20	0.374
C_3	0.496	0.234	0.019	45	0.234
C_4	0.597	0.373	0.021	60	0.299
C_5	0.693	0.466	0.022	130	0.280
C_6	0.745	0.554	0.023	190	0.275
C_7	0.800	0.594	0.023	130	0.194
C_8	0.848	0.652	0.025	190	0.147
C_9	0.893	0.863	0.025	190	0.090

(a) Molar fraction of DAAP in the feed.
(b) Molar fraction of DAAP in the polymer.
(c) Initiator concentration (mol/L).
(d) Polymerization time (min.).
(e) Weight conversion.

scarcely soluble in any solvent and gave turbid solutions or precipitated when dialyzed against water. Formic acid was found to be the only good solvent for this system. The polymers were dissolved first in concentrated formic acid and then diluted with water in order to obtain a 10% formic acid solution. No precipitation occurred then. The copolymerization diagram (Figure 6) shows that both DAAP and AA terminal radicals react preferably with an AA monomer.

1F. Acid-base Properties of Poly(DAAP-co-AA). Due to low solubility, we could not prepare directly aqueous solutions of these copolymers for UV measurements of pKa. The copolymers were first dissolved in formic acid, then diluted with distilled water, and neutralized with n-butylamine. No turbidity or precipitation was observed, even after addition of excess aqueous KOH. Figure 4 shows the values of pKa(app). At pH values close to the pKa of these copolymers, most of the acrylic acid units are deprotonated (pKa of poly(acrylic acid) = 6.17).[34] Increased electrostatic stabilization of the protonated pyridines in the vicinity of carboxylate anions is probably the reason for the change in pKa obtained for these copolymers compared to those with DMAM and DMDAACl.

2. Choice of Reaction and Substrates

Several factors have to be taken into account in the choice of reactions for studying the activity of a catalyst. A suitable reaction must proceed via known pathways to a set of unique products. No side reactions or reactive by-products should form. A facile way of measuring the extent of the reaction must exist. The substrates should be readily available and well characterized. Finally, data on the reaction catalyzed by other catalysts should exist in order to allow comparisons. The hydrolysis of p-nitrophenyl esters meets these criteria and will be discussed first.

Catalysis of bond forming reactions is more important from the synthetic point of view than that of bond-cleaving reactions. Since the initial work by Merrifield[52] on polymer assisted peptide synthesis, many research groups have investigated the application of polymeric reagents on synthesis.[4,5,53] Crosslinked polystyrene has been preferred as support, although poly(vinylpyridine)[54,55] and acrylic polymers[56] have also been investigated. Solubility and availability of reactive centers is of great importance. For instance, the amount of active centers in the reactor depends both on the composition and concentration of the copolymer under investigation and on its solubility behavior. This factor determines the availability of catalytic centers: the microenvironment may have a negative influence.[10,57] Polymeric catalysts have already been used for esterification reactions of tertiary alcohols like 1-methylcyclohexanol[58] and linalool[59] with acetic anhydride. In both cases polystyrene bound DMAP [1] was used as the catalyst. Some of the polymeric catalysts showed activity similar to DMAP. We decided to reinvestigate the reaction of linalool with acetic anhydride in toluene and compare the effectiveness of our polymeric catalysts with that of DMAP. A higher activity was expected for DAAP copolymers for two reasons:

(a) They were better catalysts than the low molecular weight catalysts for esterolyses.

(b) The monomer unit consists of PPY [2], which is a better catalyst than DMAP [1], which is attached to the polystyrene backbone in the catalysts used so far.[58,59]

3. Esterolysis of p-Nitrophenyl Caproate (PNPC)

The catalyzed reaction (Scheme 4) proceeds cleanly to yield only the phenoxide and carboxylate anions via attack on the acylpyridinium phenoxide. The nitrophenoxide anion is a very weak base and does not interfere with the reaction. p-Nitrophenyl esters of a series of straight-chain carboxylic acids with chain length varying from 1 to 18 carbons are commercially available.[35] Catalysis of the esterolysis reaction by polymer bound imidazole has been studied.[36] The reaction was also used to evaluate the activity of supernucleophiles bound to poly(ethyleneimine).[6,7] Mechanistic studies of the uncatalyzed reaction and the reaction aided by monomeric catalysts have been published.[37,38]

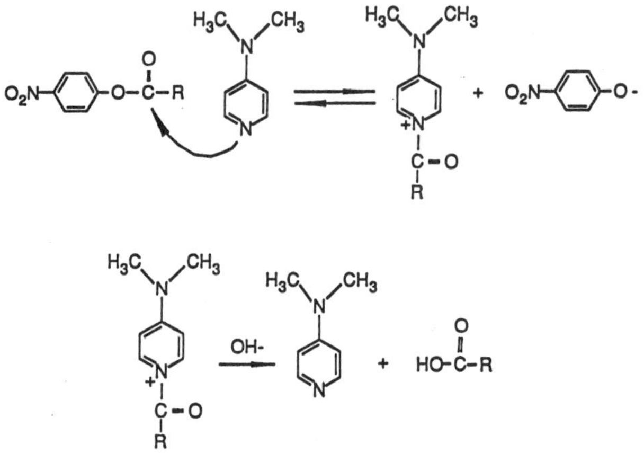

Scheme 4. Esterolysis of p-nitrophenyl esters.

3A. Initial Comparison Between Poly(DAAP) and PPY. Initial experiments were performed using p-nitrophenyl caproate (PNPC). The pKa of the polymer was expected to be around 1 to 1.5 pKa units lower than PPY, which has a pKa of approximately 10. A tris(ammonium) buffer at pH 7.8 was used as the aqueous medium. Figure 7 shows the initial results obtained for the reaction catalyzed by the polymer and PPY compared to the uncatalyzed reaction. The polymer-catalyzed reaction is significantly faster than that catalyzed by PPY and both catalysts speed the reaction beyond the uncatalyzed rate. These data represent the first experimental evidence for a supernucleophilic polymer more active than its monomeric analog in the excess substrate regime.

There are several explanations for the enhanced activity of the polymer. First, due to the already mentioned neighboring group effect, the polymer has a lower pKa than PPY. At a given concentration of the supernucleophile, this lower pKa implies that fewer catalytic sites are deactivated by protonation in the polymer than in PPY. The higher activity of the polymer could be the result of a greater concentration of active supernucleophilic centers. The second explanation arises from research on polymer bound imidazoles. In this system it has been shown that increasing apolar character of the polymer backbone resulted in higher activity towards the hydrolysis of activated ester containing long carbon chains.[39] These two explanations would only be valid if the first step of the hydrolysis reaction (the generation of the acylpyridinium

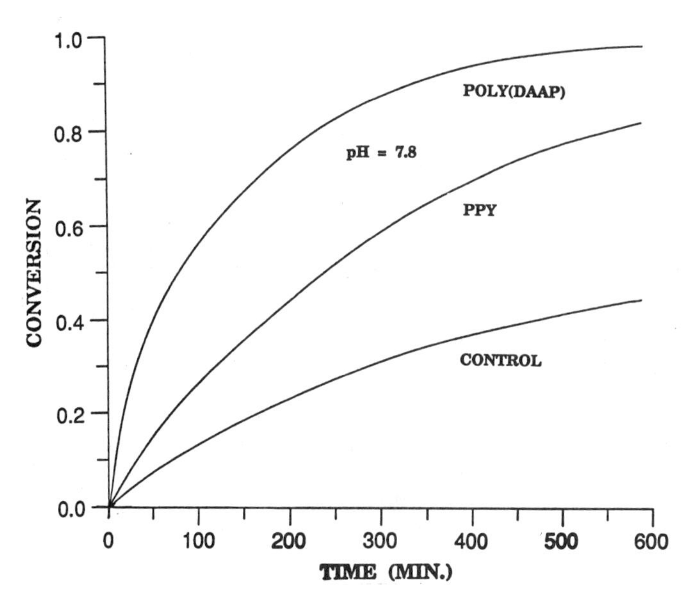

Figure 7. Rate of hydrolysis of PNPC at pH 7.8 with different catalysts.

intermediate) were rate controlling. Both factors will be examined in more detail.

3B. Hydrophobic Effects. In 1967 Overberger et al., discovered that under catalysis by poly(4(5)-vinylimidazole) the hydrolysis of long chain esters proceeded at anomalously high rates.[39] This effect was attributed to interactions between the hydrophobic backbone of the polymeric imidazole and the hydrocarbon portions of the substrate. Similar effects were later demonstrated with imidazoles grafted into poly(vinylamine) with spacers of different hydrophobicity.[36]

A series of p-nitrophenyl esters of straight chain aliphatic acids ranging from the propionate to the dodecanoate were used to study hydrophobic effects in hydrolysis catalyzed by poly(DAAP). Reactions were carried out at pH 7.8. Data were collected within the first 30 min of the reaction, where a linear dependence of conversion on time was observed. Figure 8 shows a plot of the rates obtained for the polymer- and PPY-catalyzed reactions as a function of the chain length of the acid portion of the substrate. The reaction rates for the PPY catalyzed reaction show a steady decrease with increasing chain length of the acid portion of the ester. This can be attributed to the fact that the longest substrates display reduced solubility and increasing self association and aggregation.[40] The rates for the polymer-catalyzed reactions increase with increasing substrate size, although the effect seems to level off at long chain length. These observations suggest that the backbone of the homopolymer of DAAP is hydrophobic and that the substrate prefers to be in this hydrophobic environment at long chain lengths. It is then apparent that increasing chain lengths enhance the relative activity of the polymer. Hydrophobic interaction between the substrate and the homopolymer of DAAP must therefore be an operative factor contributing to the enhanced activity of the polymer as compared to PPY.

3C. Effect of pH on Catalyst Effectivity. It is well known that polyamines have a lower pKa than the monomeric analogues. This effect is due

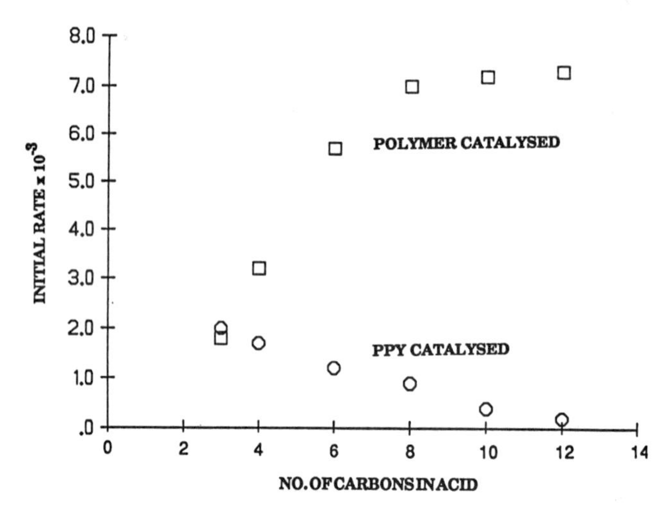

Figure 8. Effect of chain length and substrate hydrophobicity on the hydrolysis reaction.

to the fact that partial protonation of the polyamine generates a cationic environment in which further protonation is energetically less favorable. The pKa of the homopolymer was determined spectrophotometrically, as described in the experimental section. In this way we obtained a value of 7.8 for poly(DAAP) and 10.5 for PPY. The latter value is in agreement with reported values.[1,6]

As mentioned before, the lower pKa of the polymer as compared to PPY implies that at any given concentration of supernucleophile in solution, a smaller fraction is deactivated by protonation for the polymer than for PPY. In Figure 9 we observe that the ratio of concentrations of unprotonated DAAP units to that of unprotonated PPY displays a maximum at pH = 7.5. These data suggest that the polymer should show maximum rate enhancement in comparison to PPY at a pH of 7.5 if only the concentration of free nucleophile were significant in determining the rate of hydrolysis. Table 4 lists the ratio of the corrected initial rates for the polymer catalyzed reaction to those for the PPY catalyzed reaction at different pH values.

As we see, the ratio of rates shows a maximum at pH 7.8, which coincides with the maximum for the ratio of active catalyst units. Thus, the lowered pKa of the polymer contributes strongly to its enhanced activity.

Table 4. Ratio of inital reaction rates and active catalyst concentrations for polymer and PPY at different pH values.

pH of buffer	Ratio of corrected reaction rates polym/PPY	Ratio of active catalyst concn. polym/PPY
7.2	8.08	4.99
7.8	21.2	12.6
8.3	12.9	7.42
9.2	11.5	4.7

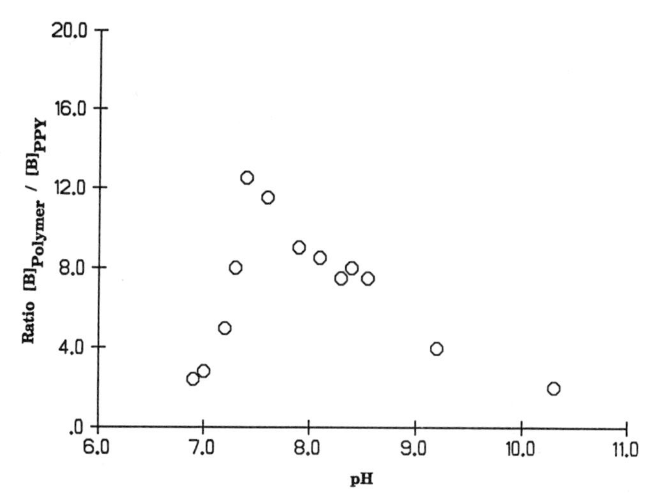

Figure 9. Comparison between active catalyst in the polymer and PPY at different pH values.

However, if the rate enhancement were due only to the increased number of non-protonated catalytic units, then the ratios of hydrolysis rates and active catalytic concentrations should be similar. We see that the former values are always higher. This supplies further evidence for the hypothesis of hydrophobic interaction between substrate and the polymeric catalyst.

3D. The Rate Determining Step. The esterolysis reaction takes place in two steps and it is necessary to determine if the generation of the acylpyridinium species or its disappearance is the rate determining step. The latter has been identified to be rate limiting in imidazole-catalyzed esterolyses of active esters.[41] If deacylation is rate determining, the acylpyridinium intermediate should build up in the reaction medium. The acylpyridinium species generated with PPY absorbs at approximately 312 nm in the ultraviolet spectrum. PPY itself and its protonated form absorb at 258 and 280 nm, respectively. Thus the detection of absorbance at 312 nm should distinguish between the two possibilities. Previous research on supernucleophiles bound to poly(ethyleneimine) employed the hydrolysis of p-nitrophenyl caproate under excess catalyst conditions as the model reaction.[6] A bathochromic shift from 258 to 315 nm confirmed the presence of the acylpyridinium species as an intermediate in the hydrolysis of PNPC under these conditions. These reaction conditions were very similar to those of our experiments except for the fact that we operated always in the excess substrate regime. Experiments to observe the peak at 315 nm failed consistently, even after substrate and catalyst concentrations were increased by a factor of 10. Even when the substrate was replaced by hexanoyl chloride, the acylpyridinium absorption could not be detected. From these data we concluded that in our system the acylpyridinium formation is the rate determining step.

3E. Copolymers of DAAP as Catalysts. Preliminary runs showed that the reaction rates depend on substrate concentration as given by Equation 4, where AB is the substrate PNPC and k_{obs} the pseudo-first order or observed rate constant.

$$\frac{d[AB]}{dt} = - k_{obs}[AB] \qquad \text{(Equation 4)}$$

The observed rate constant, k_{obs}, is given by Equation 5, where k_N is the normalized second order rate constant, [B] is the molar concentration of the free DAAP monomer units in the copolymer, and k_o is the reaction rate constant measured in the absence of added catalyst.[38]

$$k_{obs} = k_N[B] + k_o \qquad \text{(Equation 5)}$$

The rate constant of the uncatalyzed reaction, k_o, was determined for each pH and subtracted from the corresponding k_{obs}. [B] is given by the injected catalyst amount, the DAAP mol fraction, and the ratio [B]/[BH+], which is computed with the Henderson-Hasselbalch equation (Equation 3).

The catalytic activity of poly(DAAP) was compared with that of copolymers of DAAP with dimethylacrylamide (DMAM), dimethyldiallylammonium chloride (DMDAACl) and acrylic acid. Figure 10 shows that the homopolymer has in general a higher activity than the copolymers when the values of k_{obs}/(g of polymer) are compared. However, the calculation of the normalized (second order) rate constant yields quite different results (Figure 11). The copolymer with acrylic acid displays in all cases a specific activity that is similar or superior to that of copolymers with neutral or cationic monomers. The high k_N of that catalyst is partly due to the low concentration of free DAAP monomer units: as shown in the former section, at the pH values of the study most of the DAAP units are protonated, that is, not available as catalytic centers. Another factor has to be mentioned: as we said before, the rate limiting step in the esterolysis reaction is the acylpyridinium intermediate formation. Being a positive charged species, it becomes stabilized by vicinal negative carboxylate anions from the acrylic acid units (mostly free at that pH). If the hydrolysis, and not the formation of acylpyridinium cation were rate determining, then the opposite effect would be expected and positive monomer units like DMDAACl would accelerate the reaction as a whole.

To evaluate the relationship between catalytic activity and type and distribution of the comonomers it is necessary to know if there is rate

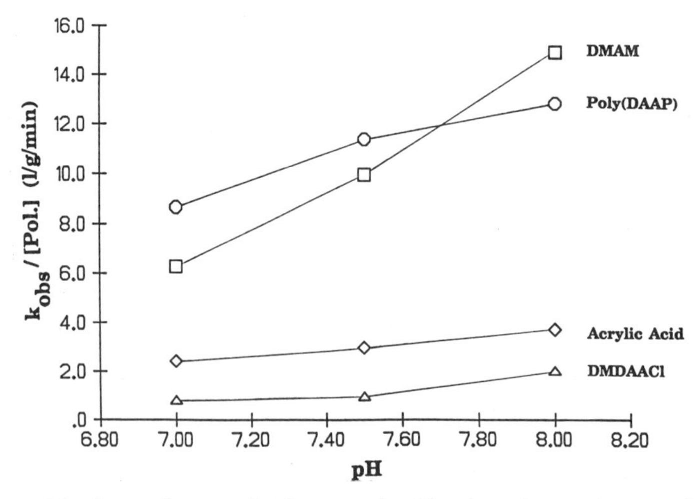

Figure 10. Dependence of the pseudo-first order rate constant of hydrolysis of p-nitrophenyl caproate on pH for poly(DAAP) and DAAP copolymers (labeled as the corresponding comonomers).

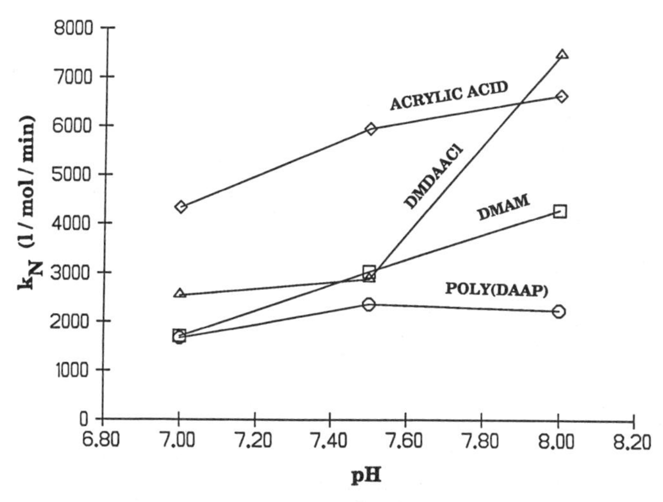

Figure 11. Dependence of the normalized rate constant of hydrolysis of p-nitrophenyl caproate on pH for poly(DAAP) and some DAAP copolymers.

enhancement due to the comonomer itself, that is, independent of DAAP monomer units. We studied the esterolysis of PNPC with poly(DMAM), poly-(DMDAACl) and poly(acrylic acid) under the same conditions as the experiments. Table 5 shows that both poly(DMAM) and poly(DMDAACl) moderately accelerate the esterolysis reaction. Poly(DMDAACl) appears to be a better catalyst than its DAAP copolymer at pH 7.5.

Table 5. The pH dependence of the esterolysis of PNPC for poly(DMAM), poly(DMDAACl) and poly(acrylic acid) at 30°C.

	(a)	(b)	(c)
pH 7.0			
k_{obs} (min^{-1})	8.02×10^{-4}	1.21×10^{-3}	3.70×10^{-4}
st. dev.	9.2×10^{-5}	6.5×10^{-5}	2.5×10^{-5}
k_{obs}/[polym.] (l/g/min)	2.90	6.24	1.28
pH 7.5			
k_{obs} (min^{-1})	1.17×10^{-3}	2.26×10^{-3}	1.20×10^{-3}
st. dev.	1.4×10^{-4}	8.7×10^{-5}	1.2×10^{-4}
k_{obs}/[polym.] (l/g/min)	4.22	11.65	4.17
pH 8.0			
k_{obs} (min^{-1})	1.47×10^{-3}	1.71×10^{-3}	1.52×10^{-3}
st. dev.	5.7×10^{-5}	3.9×10^{-4}	1.4×10^{-4}
k_{obs}/[polym.] (l/g/min)	5.31	8.81	5.28

(a) Poly(DMAM), [polym.] = 2.77×10^{-4} g/L
(b) Poly(DMDAACl), [polym.] = 1.94×10^{-4} g/L
(c) Poly(Acrylic Acid), [polym.] = 2.88×10^{-4} g/L

The high activity of poly(DMDAACl) at pH 7.5 indicates that catalysis of esterolysis may also involve a different mechanism. In spite of being a polyelectrolyte, the polymer provides a hydrophobic backbone, where binding of a non-polar substrate is possible. Furthermore, the cationic monomer units attract anions like Cl- and OH-. Therefore, although the measured pH is 7.5, [OH-] in the surroundings of the polymer is higher than $10^{-6.5}$ M. OH- groups can accelerate esterolyses by acid-base catalysis, as in saponification reactions.[42,43]

4. Effect of Copolymer Composition on Esterolytic Activity

The results of the previous part indicate that the copolymers of DAAP have a higher activity per active DAAP unit than poly(DAAP). For a more profound evaluation of the catalyst it is of great interest to determine compositions that maximize efficiency and to relate it with the sequence distribution in the copolymer. Figure 11 shows that copolymers with acrylic acid are the most promising for such a study. However, we did not study these copolymers for two reasons: (a) some of these copolymers are insoluble in water, as observed before, and (b) although a high k_N is a sign of high activity per active catalytic center, it is necessary to know how much polymer is necessary to obtain a certain acceleration of the reaction rate. This is given by kobs/[Polym.]. Copolymers with DMAM are therefore more favorable from the practical point of view. A similar study on copolymers with DMDAACl makes it possible to draw conclusions about the magnitude of both beneficial and deleterious effects of comonomer type and distribution on the catalytic activity.

<u>4A. Poly(DAAP-co-DMAM).</u> Copolymers of DAAP and DMAM of different composition were compared at pH 8 and 30°C. [C], the total concentration of DAAP monomer units (free and protonated) was approximately the same in all experiments. Table 6 summarizes the results. For comparison, k_N for poly(DAAP) at these conditions was 2.25×10^3 L/mol/min.

Table 6. Dependence of the esterolysis rate of PNPC on the composition of copolymers of DAAP and DMAM at pH 8 and 30° C.

	(a)	(b)	(c)	(d)	(e)
k_{obs} (min^{-1})	6.44×10^{-3}	4.55×10^{-3}	4.29×10^{-3}	3.50×10^{-3}	3.57×10^{-3}
st. dev.	1.4×10^{-4}	5.1×10^{-4}	3.2×10^{-4}	3.0×10^{-4}	4.2×10^{-4}
k_{obs}/[polym.] (l/g/min)	6.63	11.49	14.16	11.95	14.94
[C] (M)	1.08×10^{-6}	9.50×10^{-7}	1.08×10^{-6}	1.15×10^{-6}	1.20×10^{-6}
k_N (l/mol/min)	1.41×10^4	9.74×10^3	7.36×10^3	4.44×10^3	4.29×10^3

(a) Polymer A_2, x_{DAAP} = 0.12, pKa = 8.72, [polym.] = 9.72×10^{-4} g/L
(b) Polymer A_3, x_{DAAP} = 0.29, pKa = 8.55, [polym.] = 3.96×10^{-4} g/L
(c) Polymer A_7, x_{DAAP} = 0.48, pKa = 8.2 (x) [polym.] = 3.03×10^{-4} g/L
(d) Polymer A_8, x_{DAAP} = 0.55, pKa = 8.01 [polym.] = 2.93×10^{-4} g/L
(e) Polymer A_9, x_{DAAP} = 0.80, pKa = 7.97 [polym.] = 2.39×10^{-4} g/L
(x) Estimated value.

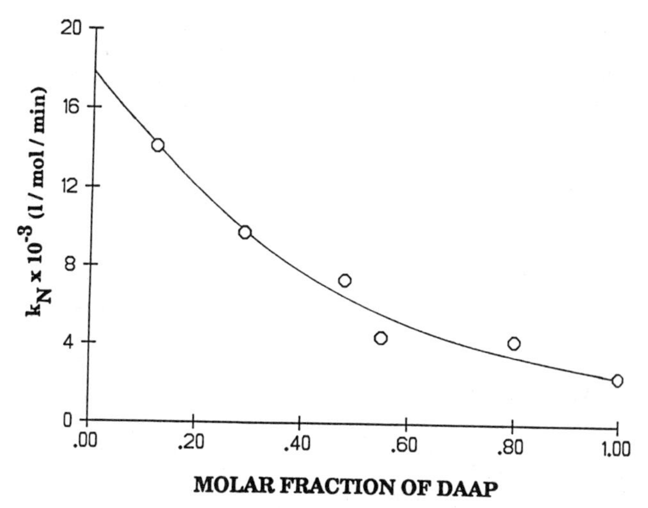

Figure 12. Esterolytic activity of poly(DAAP-co-DMAM) at pH 8 as a function of copolymer composition. k_N of poly(DAAP·co·DMAM) at pH 8.

Figure 12 shows the significant increase in catalytic activity per DAAP monomer unit with decreasing DAAP molar fraction in the copolymer. This behavior has been already observed by Klotz and coworkers[7] in polymeric catalysts based on derivatized polyethyleneimine. Moreover, polysiloxane containing pendant pyridine-1-oxide groups as catalytic sites was also more active than poly(4-diallylaminopyridine-1-oxide) in the hydrolysis of diphenyl chlorophosphate.[44] Several reasons have been forwarded for this behavior:[7]

(a) too many catalytic sites may reduce the availability of binding sites on the polymeric backbone.

(b) catalytic sites may contribute to coil shrinkage, which makes it difficult for the catalytic units to react. Poly(DAAP) has been found to be not so soluble in water as poly(DMAM). The copolymers display an intermediate behavior.

(c) cooperation between binding and catalysis may be only possible with few catalytic sites.

<u>4B. Poly(DAAP-co-DMDAACl).</u> In this case, the dependence of catalytic activity on composition was studied at pH 7. Table 7 summarizes the results. For comparison, k_{obs} for poly(DAAP) is 1.43×10^{-3} min^{-1}, k_N is 1.67×10^{3} L/mol/min and k_{obs} for poly(DMDAACl) is 1.21×10^{-3} min^{-1}.

Figure 13 shows that the dependence of catalyst effectivity on composition for these copolymers is rather different from that observed in the copolymers with DMAM. It seems reasonable that cationic comonomer units (DMDAACl) could destabilize the positively charged intermediate adduct, which explains the slightly increasing catalytic activity with DAAP molar fraction, until a k_N value of 1.67×10^{3} L/mol/min is reached for poly(DAAP). This effect should lead to minimum activity when all DAAP monomer units have DMDAACl units on both sides. No further destabilization would be then possible: additional DMDAACl units are too far away to interact electrostatically with the DAAP unit in a rigid polymer like this. A calculation of the approximate copolymer composition resulting in all DAAP units having DMDAACl monomer units on both sides is possible

Table 7. Dependence of esterolysis of PNPC on composition of copolymers of DAAP and DMDAACl at pH 7 and 30°C					
	(a)	(b)	(c)	(d)	(e)
k_{obs} (min^{-1})	$2.36x10^{-3}$	$1.49x10^{-3}$	$1.21x10^{-3}$	$1.31x10^{-3}$	$1.34x10^{-3}$
st. dev.	$1.3x10^{-4}$	$7.1x10^{-5}$	$6.0x10^{-5}$	$4.0x10^{-5}$	$6.8x10^{-5}$
k_{obs}/[pol.] (l/g/min)	1.19	1.00	3.47	4.77	5.68
[C] (M)	$8.6x10^{-7}$	$1.1x10^{-6}$	$1.0x10^{-6}$	$9.9x10^{-7}$	$1.1x10^{-6}$
k_N (l/mol/min)	$5.87x10^3$	$1.89x10^3$	$1.26x10^3$	$1.53x10^3$	$1.50x10^3$

(a) Polymer B_1, x_{DAAP} = 0.07, pKa = 8.15, [polym.] = $1.99x10^{-3}$ g/L
(b) Polymer B_2, x_{DAAP} = 0.12, pKa = 8.13, [polym.] = $1.49x10^{-3}$ g/L
(c) Polymer B_6, x_{DAAP} = 0.48, pKa = 8.13, [polym.] = $3.49x10^{-4}$ g/L
(d) Polymer B_7, x_{DAAP} = 0.60, pKa = 8.14, [polym.] = $2.79x10^{-4}$ g/L
(e) Polymer B_9, x_{DAAP} = 0.80, pKa = 7.96, [polym.] = $2.36x10^{-4}$ g/L

because, as Figure 5 shows, this system follows closely the first order Markow or terminal model of copolymerization. n1, the number-average sequence length of monomer units of 1 is calculated from Equation 6,[45] where $[M_1]$ and $[M_3]$ are concentrations of the monomers in the feed. Figure 14 gives the dependence of n1 on feed composition for the system DAAP-DMDAACl. It is then possible to calculate at which monomer fraction of DAAP copolymer is obtained where most of the DAAP monomer units have only vicinal DMDAACl units. Using these values and the copolymerization diagram of this system (Figure 5) we find that at a DAAP mol fraction of 0.3 there must be already very few vicinal DAAP units left.

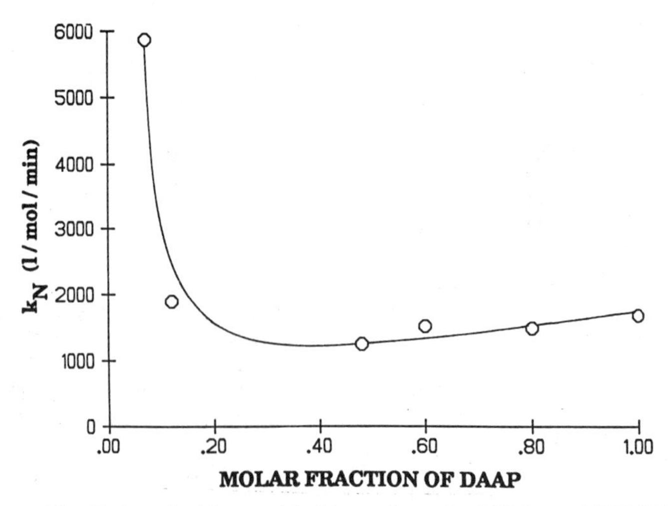

Figure 13. Esterolytic activity of poly(DAAP-co-DMDAACl) at pH 7 as a function of copolymer composition. k_N of poly(DAAP·co·DMDAACl) at pH 7.

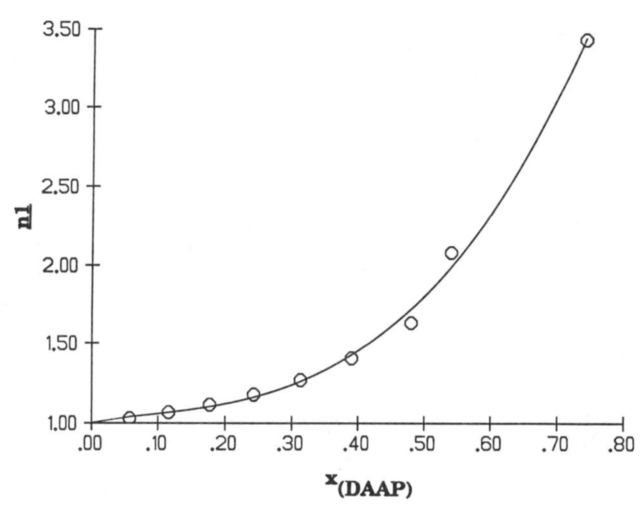

Figure 14. Number average sequence length of DAAP units (n1) in the copolymerization with DMDAACl, x(DAAP) = Molar fraction of DAAP monomer units (Ref. 8).

$$\underline{n1} = \frac{r_{13}[M_1] + [M_3]}{[M_3]} \qquad \text{(Equation 6)}$$

In the absence of another effect, no further change in specific activity of the catalytic centers would be expected. A different mechanism is then operational, probably the one involved in the increase of catalytic activity with DMAM molar fractions in DAAP-DMAM copolymers (see above). It must also be kept in mind that poly(DMDAACl) itself catalyzes the esterolysis. However, pure poly(DMDAACl) has a pseudo first order rate constant $k_{obs} = 1.21 \times 10^{-3}$ min^{-1}. This value is similar to that of the copolymers with DAAP. Probably both increased [OH-] and availability of nucleophilic centers (so spaced that they do not interfere with each other) provide optimum conditions for these copolymers.

5. Mechanism of the Catalyzed Hydrolysis

The importance of binding site availability on the catalytic activity was studied further. Our previous results suggest that the esterolysis of p-nitrophenyl esters obeys the same mechanism present in enzyme catalyzed reactions (Michaelis-Menten mechanism), where a complex between substrate and enzyme is formed prior to enzymatic attack on the substrate.[46] In the case of poly(DAAP) the increased reactivity observed for esters of larger aliphatic carboxylic acids suggests that association is due to hydrophobic interaction between the non-polar substrate moiety and the polymeric backbone, as already reported for imidazole-containing polymeric catalysts.[36,47] In the Michaelis-Menten mechanism a catalyst-substrate complex is initially formed, which then reacts and dissociates yielding products and the catalyst. In the case of enzymes, complex formation can be due also to hydrogen bond formation, electrostatic attraction and dipole-dipole interaction. The rate expression is given by Equation 7,[48] where V_o is the initial reaction rate, V_∞ is the reaction rate at infinite substrate concentration (actually when $[AB]_o \gg K_M$) and K_M is the Michaelis-Menten constant. V_∞ and K_M are obtained by measuring V_o for different substrate concentrations and plotting V_o^{-1} against $[AB]_o^{-1}$ (Lineweaver-Burk plot).

$$V_o = \frac{V_\infty}{1 + K_M/[AB]_o} \qquad \text{(Equation 7)}$$

A linear plot strongly supports the Michaelis-Menten mechanism. Due to the fact that in general the experimental points are unevenly distributed, calculation of the most probable straight line by the least square procedure assigns too much weight to points far removed from the others. These correspond to the lowest substrate concentration. It has also been shown that the error envelope (the region of the plot inside which the experimental points are supposed to lie with a given degree of probability) is not confined by two parallel lines but is funnel-shaped, with increasing uncertainty moving away from the origin and a corresponding increase in slope error.[49] Equation 7 may be rewritten in the form of Equation 8.

$$\frac{[AB]_o}{V_o} = \frac{[AB]_o}{V_\infty} + \frac{K_M}{V_\infty} \qquad \text{(Equation 8)}$$

Plotting $[AB]_o/V_o$ against $[AB]_o$ (giving the so-called half-reciprocal plot), V_∞ is obtained from the reciprocal slope and K_M from the product of V_∞ and the intercept of the least square straight line. Although the point distribution is not improved with respect to the Lineweaver-Burk procedure, the error envelope is found to be parallel to the least squares straight line.[49] Therefore this graphical method was chosen for parameter calculations. Applying this procedure to the reaction catalyzed

Table 8. Dependence of the rate of PNPC hydrolysis at 30°C on its concentration with A_7 catalyst (polyDAAP-co-DMAM of ca. 50:50 composition).

$[AB]_o$ (mol/L)	pH 7	pH 8	pH 9
$1x10^{-5}$			
V_o (mol/L/min)	$2.64x10^{-8}$	$4.60x10^{-8}$	$7.30x10^{-8}$
st. dev.	$5.3x10^{-9}$	$1.9x10^{-9}$	$1.8x10^{-8}$
$2x10^{-5}$			
V_o (mol/L/min)		$1.10x10^{-7}$	$1.98x10^{-7}$
st. dev.		$1.3x10^{-8}$	$3.3x10^{-8}$
$4x10^{-5}$			
V_o (mol/L/min)	$7.94x10^{-8}$	$1.71x10^{-7}$	$2.75x10^{-7}$
st. dev.	$8.5x10^{-9}$	$1.7x10^{-8}$	$3.9x10^{-8}$
$8x10^{-5}$			
V_o (mol/L/min)	$1.19x10^{-7}$	$3.07x10^{-7}$	
st. dev.	$4.0x10^{-9}$	$1.4x10^{-8}$	
$1x10^{-4}$			
V_o (mol/L/min)		$3.29x10^{-7}$	$2.15x10^{-7}$
st. dev.		$3.2x10^{-8}$	$2.0x10^{-9}$
$2x10^{-4}$			
V_o (mol/L/min)	$1.12x10^{-7}$	$4.79x10^{-7}$	$2.85x10^{-7}$
st. dev.	$8.0x10^{-9}$	$7.4x10^{-8}$	$8.1x10^{-8}$

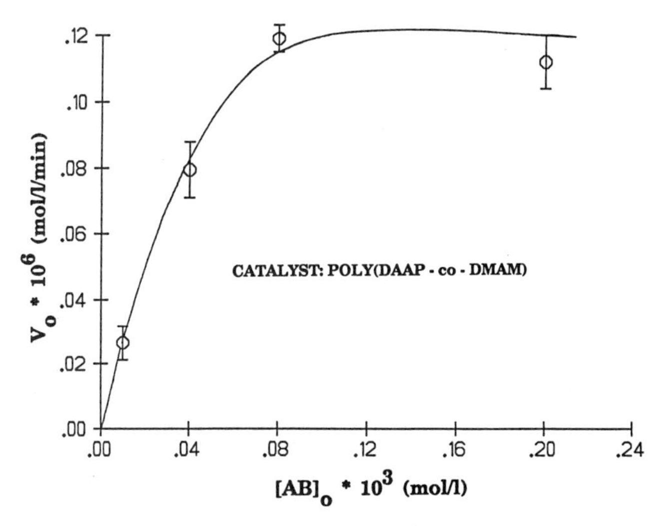

Figure 15. Dependence of the esterolysis rate on PNPC concentration. Catalyst = poly(DAAP-co-DMAM); DAAP molar fraction = 0.46 (pH = 7).

by polymer A_7, poly(DAAP-co-DMAM) with x_{DAAP} = 0.46, a change is observed in the kinetic parameters from pH 7 to pH 9 (Table 8). In all cases the catalyst concentration was 1.01 x 10^{-6} M in DAAP units and the substrate concentration was varied from 1 to 20 x 10^{-5} M (excess substrate conditions). The Michaelis-Menten plot (Figure 15) shows that saturation of the catalyst occurs in the pH range under investigation, this being shown by the flattening of the curve at higher substrate concentrations. The most probable values of K_M and V_∞ were calculated as slope and intercept of the straight lines in the plot of $[AB]_o/V_o$ against $[AB]_o$ (Figure 16).

Table 9. Michaelis-Menten constants for the esterolysis of PNPC catalyzed by poly(DAAP-co-DMAM) at 30°C.

(a)	pH 7	pH 8	pH 9
Slope	7.61x10^6	1.21x10^6	3.45x10^6
Intercept	222	181	59.8
V_∞ (mol/L/min)	1.3x10^{-7}	8.3x10^{-7}	2.9x10^{-7}
K_M (mol/L)	2.9x10^{-5}	1.5x10^{-4}	1.7x10^{-5}
Correlation coefficient	0.987	0.980	0.981
(b)	**pH 7**	**pH 8**	**pH 9**
Slope	321	202	102
Intercept	5.5x10^{-6}	6.6x10^{-5}	2.2x10^{-6}
V_∞ (mol/L/min)	1.8x10^{-7}	1.5x10^{-6}	4.5x10^{-7}
K_M (mol/L)	5.8x10^{-5}	3.1x10^{-4}	4.6x10^{-5}
Correlation coefficient	0.995	0.993	0.925

(a) From half reciprocal plot.
(b) From Lineweaver-Burk plot.

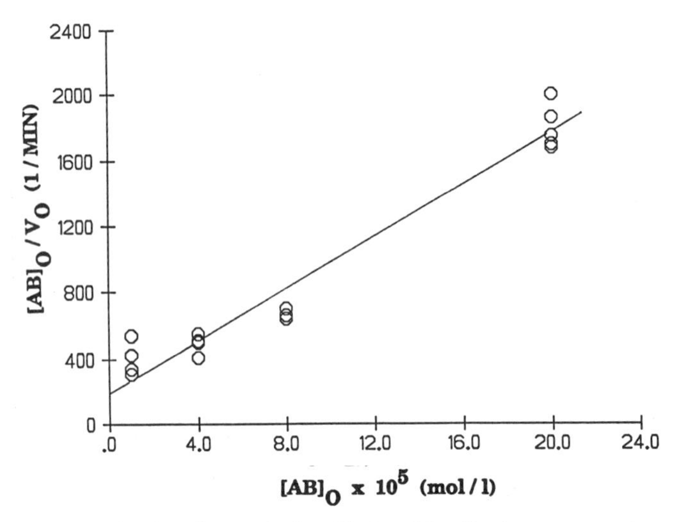

Figure 16. Calculation of the Michaelis-Menten parameters for poly(DAAP-co-DMAM). Rate of esterolysis of p-NPC at pH 7.

The linearity of this plot strongly supports the assumption that the catalyzed esterolysis actually proceeds by the Michaelis-Menten mechanism. While V_∞ is related to catalyst activity, K_M is inversely proportional to the substrate binding ability of the catalyst. Table 10 shows that the maximum of both magnitudes occurs at pH 8. Assuming a Michaelis-Menten behavior, this would indicate not only an increased catalytic activity but also a lower stability of the catalyst-substrate complex at pH values close to the pKa of the catalytic units (pKa close to 8.2). However, K_M is given by Equation 9,[20] where k_1 is the rate constant of complex formation, k_{-1} the reverse rate constant and k_2 the rate constant of product formation. Therefore, an increasing value of K_M may be due not only to an increased ratio k_{-1}/k_1 (decomposition into reagents) but also to a higher k_2 value (product formation). If this is true, it follows that the trend observed in both Michaelis-Menten parameters corresponds to a maximum of the catalytic effectivity at pH values close to pKa.

$$K_M = (k_2 + k_{-1})/k_1 \qquad \text{(Equation 9)}$$

The hypothesis of a Michaelis-Menten behavior is strongly supported by the experiments carried out at growing substrate concentration. However, more complicated mechanisms cannot be discarded. This is the case of the mechanism proposed by Bender et al.,[50] which also involves acyl-enzyme complex formation and which was useful for describing the kinetics of chymotrypsin-catalyzed esterolyses and amidolyses.

The observation of maximum values of the Michaelis-Menten parameters at pH values close to pKa parallels the behavior of biological systems. Jencks[51] observed that enzymes acting at pH close to 7 utilize groups such as imidazole (pKa close to 7) as nucleophilic centers. This yields the maximum ratio of nucleophilicity with respect to basicity of the catalytic group.

While the enzyme cannot change the pH of the environment, it selects the most effective nucleophilic group available at that pH. In our case, the pH and not the nucleophile is the variable, but the system showed the same behavior, that is, maximum effectivity at pH values close to pKa. This provides further evidence of enzyme-like behavior of these polymeric catalysts.

6. Esterification of Linalool

The reaction under study is summarized below as Scheme 5. A tertiary alcohol proved to be the best choice due to the greater rate difference between catalyzed and uncatalyzed reactions.

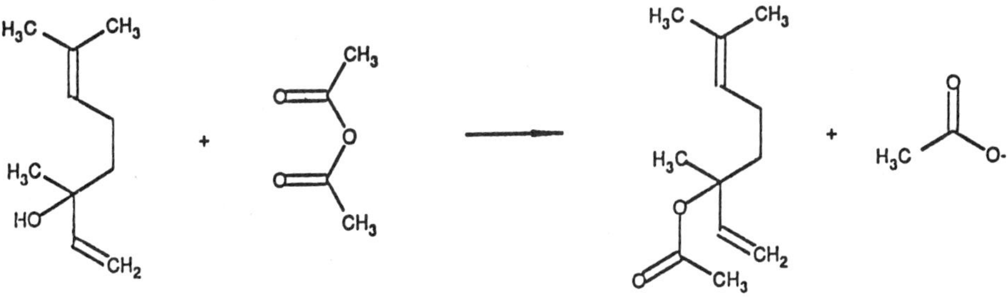

Scheme 5. Esterification of linalool at 80°C in toluene.

6A. DAAP Copolymers as Catalysts. Preliminary experiments with the homopolymer and some DAAP copolymers showed the range of the rates displayed by this system. The copolymers were added to the reactor as acetic anhydride solutions to yield the desired final catalyst concentrations. Figure 17 shows several conversion vs time curves from which the rates can be calculated as the initial slope of the kinetic curves. All polymeric catalysts were less effective than DMAP in this reaction although the opposite was seen for esterolysis of PNPC. Copolymers of DAAP with DMAM showed a higher absolute and per-unit efficiency than poly(DAAP), while copolymers with DMDAACl were also more effective than poly(DAAP) on a per DAAP monomer unit. These results are similar to those obtained in the investigation of the catalyzed esterolysis of p-nitrophenyl esters in that the activity of a given DAAP unit seems to be reduced by the reaction of neighboring units with substrate. Polymer bound DMAP has also been previously found to display higher catalytic activity (per catalytic center) with decreasing number of units in the same reaction.[59] We therefore examined a series of copolymers of DAAP with DMAM and with DMDAACl with low DAAP content, expecting to find a maximum of the normalized effectivity of the polymeric catalyst in this region.

Solutions of DAAP-DMAM copolymers in acetic anhydride were prepared to obtain a final concentration of approx. 2×10^{-4} mol DAAP monomer units in the reactor. Figure 18 summarizes the results at that concentration with rates plotted after normalization to a catalyst concentration of 2×10^{-4} mol DAAP units. The interpolation procedure used seems to indicate that activity of the catalyst at a DAAP mol fraction of approximately 0.4 is similar to that displayed by DMAP. The same investigation was performed with DAAP-DMDAACl copolymers. Their reactivity proved to be much lower than that of DAAP-DMAM copolymers.

The results obtained with 1×10^{-4} DAAP monomer units were extrapolated to 2×10^{-4} M. Figure 19 shows that there is only a very slight dependence of effectivity on copolymer composition with 2×10^{-4} mol DAAP units. Catalyst activity seems then not to be strongly dependent on con-

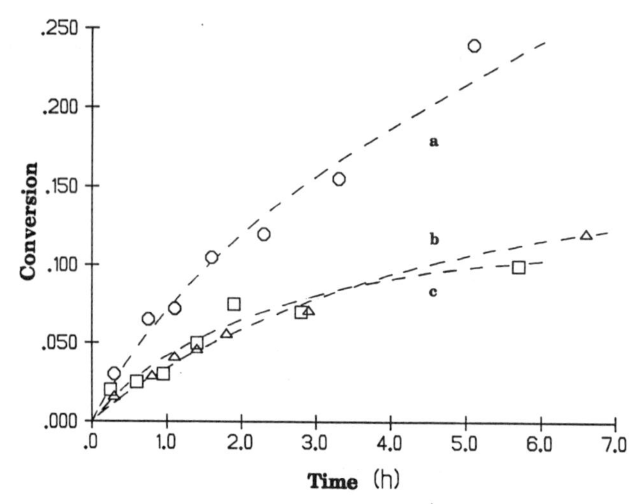

Figure 17. Kinetic curves for the esterification of linalool catalyzed by the homopolymer and copolymers of DAAP. (a) = Poly(DAAP-co-DMAM), x(DAAP) = 0.24; (b) = Poly(DAAP); (c) = Poly(DAAP-co-DMDAACl), x(DAAP) = 0.07.

centration; when the amount of catalyst is reduced to 2 x 10^{-5} mol there is no significant reduction of the rate of esterolysis. This is more clearly seen when we compare the runs with 2 x 10^{-4} mol DAAP units in copolymer B_{10} ($x_{(DAAP)}$ = 0.06) and 2 x 10^{-5} mol DAAP units in copolymer B_{15} ($x_{(DAAP)}$ = 0.07). The reason for this is probably the low solubility of these copolymers in the reaction mixture. At the higher concentration cloudiness appears as soon as the polymer-acetic anhydride solution is added to the reaction mixture. After a period of approx. 2 hrs a brown precipitate is formed which cannot be redissolved although it is swellable in methanol and water. This precludes reaction of most of the DAAP monomer units. At a concentration of 2 x 10^{-5} mol DAAP units the solubility limit is not reached, and almost all DAAP units remain active. This results in a higher reactivity per catalytic center.

CONCLUSIONS

The efficiency of the catalysts used so far can be compared in a qualitative way as shown in Table 10.

Table 10. Relative catalytic activity rankings of the materials studies, in decreasing order down the table, for esterolysis and esterification reactions.

Esterolysis	Esterification
DAAP copolymers	DMAP and Poly(DAAP-co-DMAM) (x_{DAAP} = 0.4)
Poly(DAAP) PPY	Other copolymers of DAAP and Poly(DAAP)

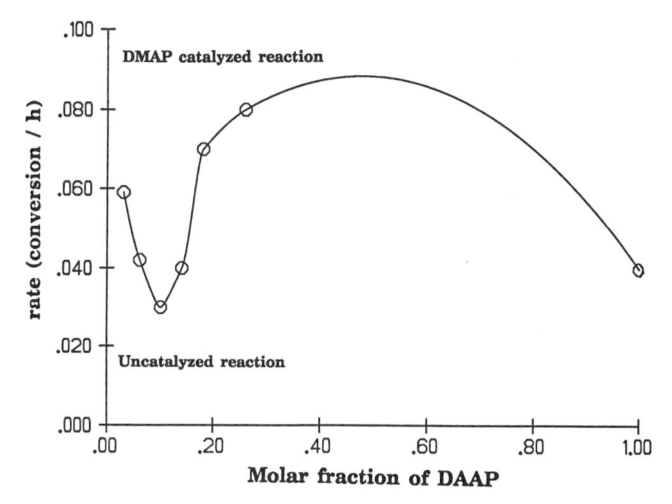

Figure 18. Dependence of the catalytic activity of copolymers of DAAP and DMAM on DAAP content in the esterification of linalool at 80°C with acetic anhydride. Catalyst is poly(DAAP·co·DMAM).

The success in esterolysis reactions is mainly due to the choice of comonomers. They are all hydrophylic and allow substrate interaction with the catalytic center. In one case a copolymer of DAAP and acrylic acid proved to be the most active, giving the highest value for the specific esterolysis rate constant k_N. The high specific activity of these copolymers can be explained assuming stabilization of the acyl-DAAP unit intermediate (which is positively charged) by carboxylate anions.

The increase of efficiency with isolation of the catalytic center in copolymers with both DMAM and DMDAACl is remarkable. Several reasons have been given for this behavior, which leads in both copolymers to substantial increase with respect to the homopolymer.

The k_N was introduced as a measure of the specific activity, that is, per active catalyst unit. Factors like copolymer composition and pKa of the solution have to be considered for its calculation. Copolymers with low DAAP content and specially those displaying a low pKa show higher k_N values than the homopolymer because the activity is concentrated in few active centers. Grading the catalyst according only to its k_N value in a certain reaction is, however, misleading. This is because k_N does not involve the mass of catalyst and materials that appear to be very reactive have to be added at high concentrations, which is not economical and sometimes not possible due to solubility problems. This is the case of the copolymers with acrylic acid. In opposition to this, k_{obs}/[Polym.] gives the activity of the catalyst per unit of mass, that is, comonomer and protonated catalyst units are also considered. In view of the fact that the homopolymers of DMAM and DMDAACl also accelerate the esterolysis, k_{obs}/[Polym.] seems to be more important from a practical point of view. In other words, it tells us how much of the copolymer is necessary for a certain reaction rate.

Copolymers with DMDAACl showed a complex dependence of the efficiency as esterolytic catalysts on composition: a minimum at x_{DAAP} = 0.4 indicates that two opposite effects are present. As mentioned before the positive charge of the DMDAACl unit destabilize the acyl-pyridinium intermediate. This leads to a slight increase of k_N with DAAP content in

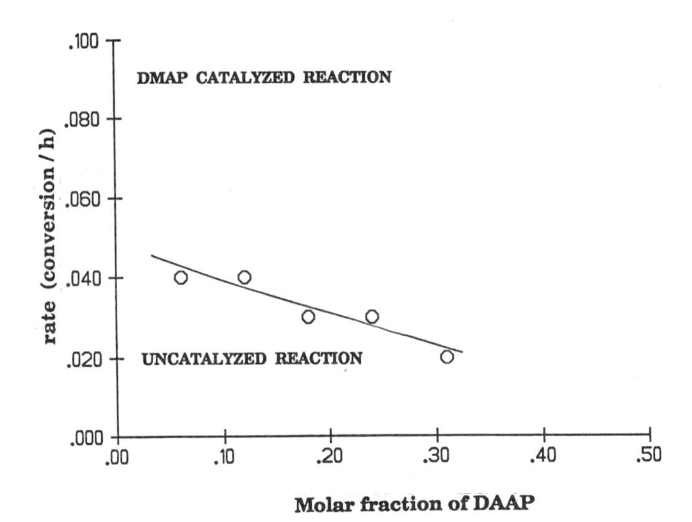

Figure 19. Dependence of the catalytic activity of copolymers of DAAP and DMDAACl on DAAP content in the esterification of linalool at 80°C with acetic anhydride. Catalyst is poly(DAAP·co·DMDAACl).

DAAP-rich compositions. The steep increase of efficiency at very low DAAP containing copolymer compositions is most probably related to an interaction of the intrinsic catalytic properties of poly(DMDAACl) (see Table 5) and the higher availability of the catalytic centers in an extended coil (most monomer units are cationic and repel each other). These provide also a higher counterion concentration, and it is possible that the [OH-] is higher around the polymer than that observed macroscopically. A higher [OH-] also causes hydrolysis: observe for instance the trend of k_N and k_{obs} with pH (Table 8) obtained after correction for uncatalyzed reaction and for protonation of part of the catalytic units.

Copolymers with DMAM most probably follow the Michaelis-Menten mechanism, similar to enzymes. Our results at different pH indicate a maximum of the Michaelis-Menten parameters at pH values close to pKa. This correspond to the behavior of enzymes that catalyze the same type of reactions.

DMAP [1] proved to be better than all DAAP copolymers investigated so far in the esterification of linalool with acetic anhydride. In this case the kinetic results are more complex (Figure 18). It seems that aggregation is the main problem. Both DMAM and DMDAACl yield hydrophilic and strongly polar polymers, with reduced solubility in the reaction medium (toluene). This leads to a tight conformation of the polymer coils and, therefore, to limited availability of the catalytic centers for the substrate. The situation with copolymers with DMDAACl is even worse because precipitation occurs. Therefore most of the reactive sites are occluded. Although it is clear than other comonomers should be tested for this reaction, some of the copolymers with DMAM showed a higher activity. The complex shape of Figure 18 shows that opposite effects are present, resulting, at certain compositions, in an activity similar to that of DMAP.

REFERENCES

1. G. Hofle, W. Steglich, & H. Vorbruggen, Angew. Chem., Int. Ed. Eng., **17**, 569 (1978).

2. A. Hassner, L. Krepski & V. Alexanian, Tetrahedron, **34**, 2069 (1978).
3. E. F. V. Scriven, Chem. Soc. Rev., **12**, 129 (1983).
4. K. K. Mathur, C. K. Narang & R. E. Williams, in: "*Polymers as Aids in Organic Chemistry*," Academic Press, New York, 1980, page 198.
5. L. J. Mathias, & C. E. Carraher, Jr., Eds., "*Crown Ethers and Phase Transfer Catalysis in Polymer Science*," Plenum, New York, 1984.
6. M. A. Hierl, E. P. Gamson & I. M. Klotz, J. Amer. Chem. Soc., **101**, 6020 (1979).
7. E. J. Delaney, L. E. Wood & I. M. Klotz, J. Amer. Chem. Soc., **104**, 799 (1982).
8. S. Shinkai, H. Tsuji, Y. Hara, & O. Manabe, Bull. Chem. Soc. Japan, **54**, 631 (1981).
9. F. Guendouz, R. Jacquier, & J. Verducci, Tetrahedron Lett., **25**, 4521 (1984).
10. F. M. Menger, & D. J. McCann, J. Org. Chem., **50**, 3928 (1985).
11. M. Tomoi, M. Goto & H. Kakiuchi, Makromol. Chem., Rapid Commun., **6**, 397 (1985).
12. M. Tomoi, Y. Akada, & H. Kakiuchi, Makromol. Chem., Rapid Commun., **3**, 537 (1982).
13. W. Storck & G. Manecke, J. Mol. Catal., **30**, 145 (1985).
14. S. C. Narang & R. J. Ramharack, J. Polym. Sci., Polym. Lett. Ed., **23**, 147 (1985).
15. Ref. 4, pp. 14-18
16. L. J. Mathias, J. B. Canterberry & M. South, J. Polym. Sci., Polym. Lett. Ed., **20**, 473 (1982).
17. L. J. Mathias & J. B. Canterberry, "*Cyclopolymerization and Polymers with Chain-Ring Structures*," G. B. Butler & J. E. Kresta, Eds., American Chemical Society: Washington DC, 1982, pp. 139-148
18. G. B. Butler, A. Crawshaw, & W. L. Miller, J. Amer. Chem. Soc., **80**, 3615 (1958).
19. G. B. Butler, in: "*Polymeric Amines and Ammonium Salts*," E. J. Goethals, Ed., Pergamon Press, New York, 1979, pp. 125-142.
20. M. L. Bender, "*Mechanisms of Homogeneous Catalysis from Protons to Proteins*", Wiley Interscience, New York, 1971, p. 505.
21. C. G. Overberger & J. C. Salamone, Acc. Chem. Res., **2**, 217 (1969).
22. P. W. Tidwell & G. A. Mortimer, J. Polym. Sci., Part A, **3**, 369 (1965).
23. D. Braun, G. Brendlein, & G. Mott, Eur. Polym. J., **9**, 1007 (1973).
24. T. Kelen & F. Tudos, J. Macromol. Sci., Chem., **A9**, 1 (1975).
25. F. Tudos, T. Kelen, T. Foldes-Berezsnich, & B. Turcsanyi, J. Macromol. Sci., Chem., **A10**, 1513 (1976).
26. A. Katchalsky & I. Miller, J. Polym. Sci., **13**, 57 (1954).
27. R. A. Vaidya & L. J. Mathias, J. Amer. Chem. Soc., **108**, 5514 (1986).
28. CRC Handbook of Chemistry and Physics, CRC Press, Inc., Boca Raton, FL, 1980, p. D-147.
29. F. M. Menger & U. V. Venkataram, J. Amer. Chem. Soc., **108**, 2980 (1986).
30. J. P. Guthrie, J. Chem. Soc., Chem. Commun., **1972**, 897.
31. A. Katchalsky & P. Spitnik, J. Polym. Sci., **2**, 432 (1947).
32. C. Tanford, "*Physical Chemistry of Macromolecules*," Wiley, New York, 1961, p. 526.
33. C. G. Overberger & Y. Okamoto, Macromolecules, **5**, 363 (1971).
34. W. Z. Kern, Phys. Chem., **181**, 249 (1938).
35. Sigma Chemical Company, Milwaukee, Wisconsin.
36. R. Tomko & C. G. Overberger, J. Polym. Sci., Polym. Chem. Ed., **23**, 279 (1985).
37. W. P. Jencks & J. Carriuolo, J. Amer. Chem. Soc., **82**, 1778 (1960).
38. T. C. Bruice & R. Lapinski, J. Amer. Chem. Soc., **80**, 2265 (1958).
39. C. G. Overberger, J. C. Salamone & S. Yaroslavsky, Pure Appl. Chem., **15**, 453 (1967).

40. J. P. Guthrie, Can. J. Chem., **51**, 3494 (1973).
41. W. P. Jencks, in: "*Catalysis in Chemistry and Enzymology*," McGraw Hill, New York, 1969, p. 67.
42. P. G. Ashmore, in: "*Catalysis and Inhibition of Chemical Reactions*," Butterworths, London, 1963, p. 45.
43. M. Polanyi & A. L. Szabo, Trans. Faraday Soc., **30**, 508 (1934).
44. M. Zeldin, W. K. Fife, C. Tian & J. Xu, Organometallics, 1987 (submitted).
45. G. Odian, "*Principles of Polymerization*," Wiley Interscience, New York, 1981, p.429.
46. M. L. Bender, in: "*Mechanisms of Homogeneous Catalysis from Protons to Proteins*," Wiley Interscience, New York, 1971, p. 397.
47. T. Kunitake, F. Shimada & C. Aso, J. Amer. Chem. Soc., **91**, 2716 (1969).
48. S. W. Benson, "*The Foundations of Chemical Kinetics*," McGraw Hill, New York, 1960, page 652.
49. C. W. Wharton & R. Eisenthal, in: "*Molecular Enzymology*," Halsted Press, New York, 1981, page 275.
50. B. Zerner, & M. L. Bender, J. Amer. Chem. Soc., **86**, 3670 (1964).
51. W. P. Jencks, in: "*Catalysis in Chemistry and Enzymology*," McGraw Hill, New York, 1969, p. 84.
52. R. B. Merrifield, J. Amer. Chem. Soc., **85**, 2149 (1963).
53. R. L. Letsinger & M. J. Kornet, J. Amer. Chem. Soc., **85**, 3045 (1963).
54. J. M. J. Frechet & M. Vivas de Meftahi, Brit. Polymer J., **16**, 193 (1984).
55. J. M. J. Frechet, P. Darling & M. J. Farrall, J. Org. Chem., **46**, 1728 (1981).
56. P. Hodge & D. C. Sherrington, Eds., "*Polymer-supported Reactions in Organic Synthesis*," Wiley, London, 1980.
57. E. H. Urruti & T. Kilp, Macromolecules, **17**, 50 (1984).
58. A. Deratani, G. D. Darling, D. Horak, J. M. J. Frechet, Macromolecules, **20**, 767 (1987).
59. M. Tomoi, M. Goto & H. Kakiuchi, J. Polym. Sci., Polym. Chem. Ed., **25**, 77 (1987).

SYNTHESIS OF PLATINUM AND TITANIUM POLYAMINO ACIDS

Charles E. Carraher, Jr., Louis G. Tissinger, Isabel Lopez and Melanie Williams

Florida Atlantic University
Department of Chemistry
Boca Raton, FL 33431
and
Wright State University
Department of Chemistry
Dayton, OH 45435

The synthesis of titanium and platinum II polyamino acids is described. The titanium products are derived through a condensation process with the products being high polymers for all but monopeptides. The platinum products are synthesized through a coordination process with chain lengths varying from low to high. The basic polymer structures are shown below.

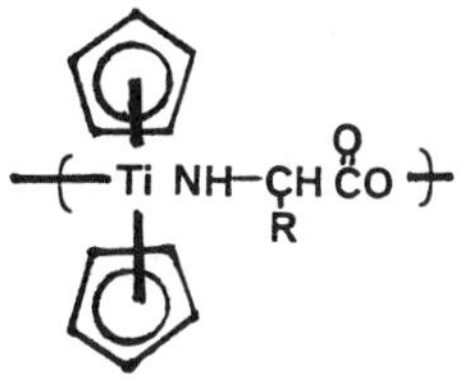

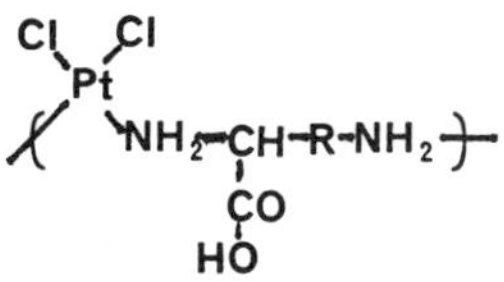

$$-\!\!\left(\mathrm{Ti(Cp)_2}\;\mathrm{NH{-}\underset{R}{CH}\overset{O}{C}O}\right)\!\!-$$

$$\left(\mathrm{Cl_2Pt}\;\mathrm{NH_2{-}\underset{\underset{HO}{CO}}{CH}{-}R{-}NH_2}\right)$$

INTRODUCTION

We have been active in the synthesis and physical and biological characterization of biomimetic polymers, that is, to develop polymeric materials having biological activities. Many definitions have been given to describe the term enzyme-mimetic polymers. One of the simplest definitions is polymeric materials having enzyme-like activities. Even so, the term "enzyme-like activities" is broadly defined including (a) end results (i.e. does the "synthetic enzyme" accomplish the same (or

Biomimetic Polymers
Edited by C. G. Gebelein
Plenum Press, New York, 1990

similar) chemical and/or biological effect) and (b) mechanism of action. Enzyme reactions are typically (a) highly stereospecific, (b) site specific and (c) rapid. They typically obey Michaelis-Menten kinetics. Here we will focus on the synthesis of polymers that may contain enzymes or that are enzyme-like in structure and/or biological activity. Recent reviews of enzyme-mimetic polymers are available[1-4].

Not all polypeptides must be large to exhibit enzyme behavior. For instance, many of the enkephalines and endorphines are polypeptides of short chain lengths. Leu-enkephalin has five amino acid units as does Met-enkephalin. The active site of many enzymes can be small. For serine enzymes the number of amino acids needed to describe chymotrypsin is five (Gly-Asp-Ser-Gly-Gly), for trypsin it is also five (Gly Asp Ser Gly Pro), for elastase it is four (Gly-Asp-Ser-Gly) and for thrombin it is three (Asp-Ser-Gly). Thus enzymatic behavior occurs with selected oligomeric polypeptides which can be placed within polymers to function as part of the secondary polymer chain and/or released to function on its own. In general, connection of such moieties through nonpolar bonds will discourage fragmentation and subsequent release of the polypeptide portion. Further, the released portion should exhibit the same activity as a "natural" polypeptide since hydrolysis acts to produce the free polypeptide. The reactivity of bound polypeptides may be modified due to the presence of the connecting portion and its mobility limited due to the overall size of the secondary polymer.

EXPERIMENTAL

Instrumentation

Infrared spectra were obtained employing the following spectrophotometers: Perkins Elmer 451, Alpha Centauri - FTIR and Nicolet 50X FTIR. Mass spectrometry was performed utilizing a direct insertion probe into a DuPont 491 or Kratos MS-50 mass spectrometer operating in the EI mode. The Kratos mass spectrometry was performed at the Midwest Center for Mass Spectrometry, Lincoln, Nebraska. Elemental analyses were performed with a Perkin Elmer 240 elemental analyses for C, H and N. Titanium analyses were obtained employing thermal analysis along with perchloric acid digestion. Platinum analyses were carried out employing thermal analysis.

Polymerization Procedures

For the Group IVB metallocenes, reactions were carried out utilizing a one quart Kimex emulsifying jar placed on a Waring Blendor (Model 1120) with a no-load speed of about 18,000 rpm (120 volts). The aqueous solution containing dissolved amino acid and sodium hydroxide was first added to the Kimex jar. The lid was screwed on and blending begun. The organic phase containing the metallocene was rapidly added employing a powder funnel placed through a hole in the jar lid. Recovery was effected employing suction filtration. The precipitated product was washed, placed in a petri dish and allowed to dry. Reactions employing potassium tetrachloroplatinate II are run by mixing two aqueous solutions together (with stirring) into a flask. Each aqueous solution contains either the Lewis base reactant (diamine) or the tetrachloroplatinate. Again, the product was obtained as a precipitate, washed, placed onto a petri dish and allowed to dry.

RESULTS AND DISCUSSION

The purposes of our studies include the following: (1) Site specific delivery of drugs (typically metal-containing organometallics) employing enzymes, polypeptides and amino acids as "magic arrows." (2) Controlled release of either or both the drug and amino acid-derived portion. (3) Studies aimed at evaluating structure-biological activity relationships. Here we will describe two general approaches illustrative of ongoing activities.

The two amino acid-containing families were chosen to illustrate the breadth and dependencies of reaction with peptides and peptide-like reactants. Dicyclopentadienyltitanium dichloride, hereafter referred to as simply the metallocene, is reactive with strong Lewis bases under mild solution conditions. An amino acid containing one acid and one amine group will have a functionality of two when reacting with the metallocene. Amino acids with two amines and one acid or two acids and one amine will exhibit a functionality of three leading to crosslinked products. Further, the metallocene is not reactive under mild conditions with tertiary nitrogens.

The reactivity of nitrogen-containing groups is much greater than carboxyl (containing RCOOH or $RCOO^-$) groups towards tetrachloroplatinate. Further, the well known trans-effect is responsible for the cis stereochemistry of the products. Thus reaction with peptides containing one amino and one acid group yields monomeric products and not polymeric products.

```
Cl   Cl                                      Cl   Cl
  \ /                                          \ /
  Pt2-     +  H2N-R-COO-  ---->                 Pt
  / \                                          / \
Cl   Cl                              -OOC-R-NH2   NH2-R-COO-

                                                 [1]
```

Thus, reactants containing two or more reactive nitrogens must be present to give polymeric products. The fact that the reaction presumably results in the substitution of only two of the four chlorides surrounding the platinum II atom is probably also a result of the trans-effect where the two chlorides are trans to two "deactivating" nitrogen-containing moieties.

Titanium-Containing Polyamino Acids

The synthesis of titanium-containing polyesters [2] and polyamines [3] has been accomplished.[5-9] The reaction with amino acids is considered an extension of this. The extension is not as straight forward as it appears. First, for aqueous interfacial systems reaction with salts of dicarboxylic acids occurs within the aqueous phase requiring the acid chloride to have some water solubility and/or stability. Thus, the analogous reaction employing organic acid chloride does not give the corresponding polyanhydrides. Further, reaction with amines occurs within the organic layer. Thus for successful polymerization to occur, the acid chloride may be required to "flutter" back-and-forth from the aqueous to organic, ect. layer. Second, reaction with an alpha-amino acid may yield cyclic [5]) rather than linear products [4].

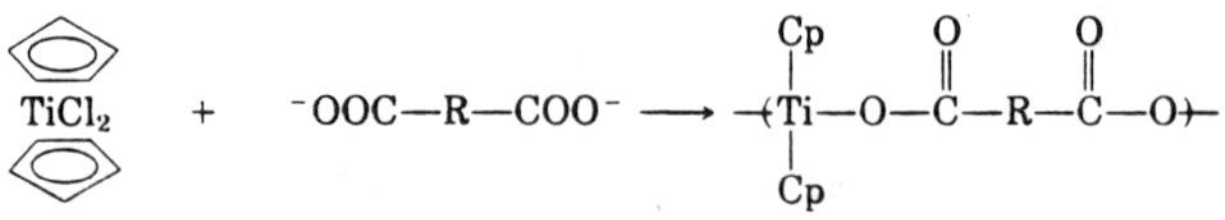

[2]

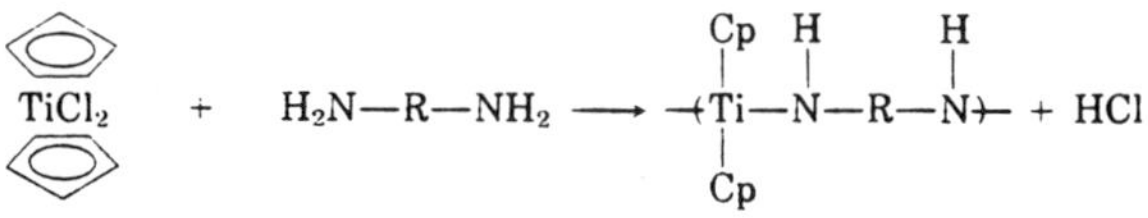

[3]

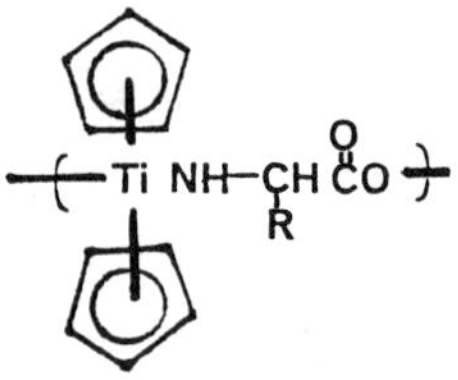

[4]

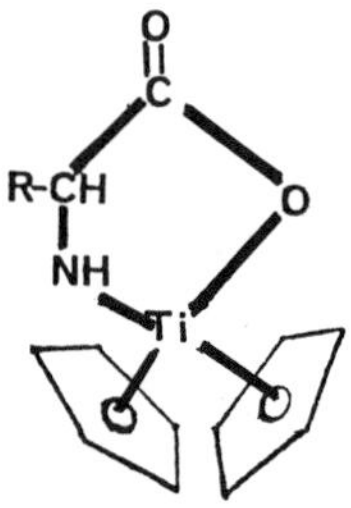

[5]

Table 1 contains results for one peptide, three dipeptides and two tripeptides. Glycine gave the lowest yield and also the lowest molecular weight possibly due to some cyclic formation such as depicted in [5] The

Table 1. Results as a function of employed amino acid.

Amino Acid	% Yield $CHCl_3/H_2O$	$\bar{M}_w$
Glycyl-Glycine	43	$1.2x10^5$
L-Alanyl-Glycyl-Glycine	40	$1.6x10^5$
Glycyl-D-Phenylalanine	50	$4.2x10^5$
L-Isolencyl-L-Alanine	65	$3.2x10^5$
Glycyl Glycyl-L-Leucine	48	$3.5x10^5$
Glycine	39	$7.8x10^2$

degree of polymerization is about three for glycine. The other products are all polymeric with degrees of polymerization about 500.

Biscylopentadienyltitanium dichloride dissolves in water forming a number of products that act in condensation reactions to insert the Cp_2Ti moiety so it can be considered to be water soluble in this respect.

The infrared spectrum of the products are consistent with a product containing a repeat unit such as [**4**]. Thus, a sharp band appears at 3015 (all bands given in cm^{-1}) characteristic of the C-H stretch for the Cp moiety. Additional bands characteristic of the Cp moiety appear at 1391, 1030 and 805. The band at 1670 is characteristic of carbonyl stretching and a new band at 1060 is assigned as being derived from the Ti-O stretch. Mass spectral data is also consistent with a repeat unit described in [**4**]. For the product derived from glycyl-glycyl-L-leucine, ion fragments assigned to the tripeptide are found at 241 (all fragments given in amu or Daltons), the dipeptide gly-gly at 114, leucine (minus CO_2) at 85 and glycine at 57. The ion fragmentation is also consistent with that expected for the Cp moiety and is given in Table 2.

Platinum-Containing Amino Acids

We have been synthesizing and evaluating the biological characteristic of polymeric derivatives of cis-dichlorodiamineplatinum II, cis-DDP [**6**].

```
Cl    Cl
  \  /
   Pt
  /  \
H3N   NH3
```

[**6**] Cis-DDP

C-DDP is a well established anticancer drug but it is also quite toxic exhibiting typical heavy metal toxicity. We have effectively lowered the toxicity of polymeric derivatives of c-DDP while retaining good anticancer activity.[11-13] Further, polymeric derivatives have also been shown to be good antiviral agents and to effectively control a virus that results in juvenile diabetes in live animal tests.[11,13] The reactivity to chelation of tetrachloroplatinate by Lewis bases varies greatly and is $H_2N \gg CO_2^-$. Thus, amino acids containing one additional amine group were utilized yielding products containing at least some free carboxylic acid groups that could be further exploited for future reactions or to effect aqueous solubility. Suitable reactants should contain one amino acid moiety that contains an additional amine group. Such amino acids include proline, tryptophan, asparagine, glutamine and lysine. The presence of

Table 2. Ion fragmentation pattern for fragments assigned to cyclopentadiene for the condensation product of Cp_2TiCl_2 and glycyl-glycyl-L-leucine.

m/z	66	65	39	40	63	67	62
John Wiley	100	49	32	27	8	6	6
Found	100	43	14	8	12	9	5

more than one of these amino acid moieties in the reactant or the presence of amino acids that contain more than two nitrogen functional groups (such as arginine and histidine) may yield crosslinked products.

It is important to again note that reactions employing reactants such as Group IVB and Group VA organometallic dihalides occur through condensation sequences. Reactions with the tetrachloroplatinate occur through a coordinate process the same as simple chelation reactions occurring between metal ions and Lewis base. For the product from tetrachloroplatinate and glycyl-L-histidine, the following IR bands are assigned (all bands given in cm^{-1}): N-H stretching, 3100 to 3000; NH_2 stretching, 3094 to 2983; N-H ring stretching, 3450 to 3400; C-O stretching (acid), 1387 and 1321; O-H bending, 1194; OH stretching (acid), 3300 to 2500; CH (aliphatic) stretching, 2931 to 2568; C=O (acid) stretching, 1773; C=O (amide) stretching, 1729; NH (amide) bending, 1610; Pt-Cl, 324 with shoulders at 329 and 317.

Mass spectral data is consistent with the proposed structure containing a moiety derived from tetrachloroplatinate II ($PtCl_2$ and/or PtCl) and

Table 3. Ion fragments derived from the product of dipotassium tetrachloroplatinate II and D-biotin heated to 340°C and then to 380°.

m/b	%-Relative Intensity 340°C	380°C	Parent and/or Ion Fragment Assignment
35	2		Cl (35)
36	7		HCl (35)
38	3		HCl (37)
39	8	19	C_2H_6
41	28	51	C_2H_6
43	3	5	CONH
44	27	34	CO_2
45		7	SCH
55	4	10	
56	6	11	C_4H_8, HNCONH, $C_2H_4CO_2H$
57	100	100	
58	5	4	
67		11	Ring-SHNHCO
69		5	
84		5	$C_3H_6CO_2$
85		6	
97		16	
98		5	Ring-SH_2
119	3	14	Ring-C_2H_4
130		10	Ring-CH_2
143		9	Biotin-$C_4H_8CO_2$
151	8	4	
153	3	7	
155	11	9	Biotin Ring
157	4		
169		27	Biotin-$C_2H_4CO_2$
181		18	Biotin-CH_2CO_2
212		6	Biotin-S
219		13	Biotin-C_2H_4
231		10	Biotin-CH_2
243		4	Biotin

the amino acid. The ion fragments derived from the product of tetrachloroplatinate II and D-biotin are given in Table 3. Ion fragments derived from the platinum-containing reactant are found at 35, 36 and 38. Ion fragments from the biotin are numerous and include biotin itself, the ring portion of biotin and fragments derived from the break-up of the aliphatic biotin chain.

While the structure for simple platinum II polyamines can be depicted as [7], the structure of polymers derived from reactants containing more than two reactive nitrogen sites probably contains several units. Thus gly-his has four active nitrogen sites with nucleophilicity of the order 1>4>3>2. The most probable average structure is form [8].

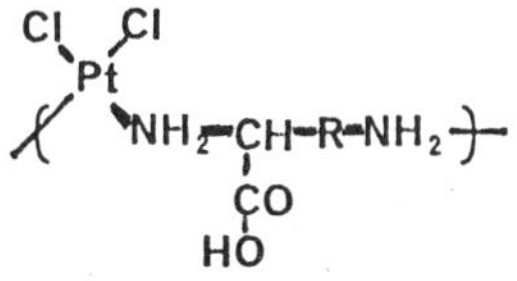

[7]

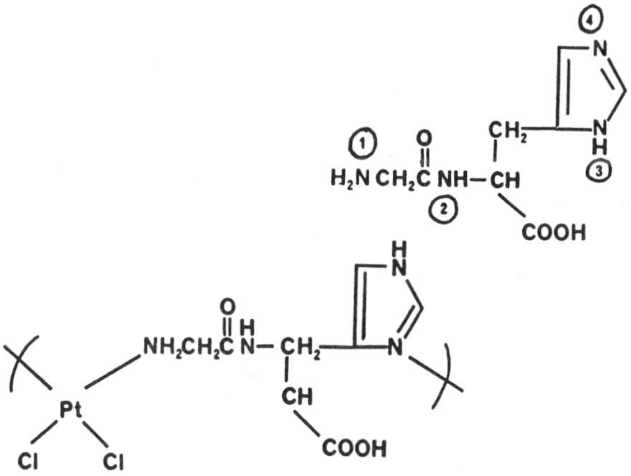

[8]

A number of studies have involved the reaction between platinum II-containing reactants and amino acids under reaction conditions favoring the monomeric product.[14-16] The reactivity of the amino group is much greater than the carboxylate (protonated or ionized form) group. The reason cited for the favored coordination with nitrogen-donors is that the nitrogen-donors are at the strong field end of the spectrochemical series. Thus, they cause crystal-field stabilization in square-planer platinum II complexes , i.e., the enthalpies of bonding are enhanced by the crystal field stabilization energy of the d^8 configuration.

Though the reactivity of the nitrogen-containing moiety is far greater compared with the acid group and there is resistance through deactivation of the chlorides by attached nitrogen ligands, amino acids containing "flexible" units that permit the carboxyl group to approach the platinum atom may allow attachment by the carboxyl group and release of a chloride ion. Reaction with glycyl-L-histidine, methotrexate, folic acid and D-biotin give Cl/Pt ratios of 2:1 since the acid group is either

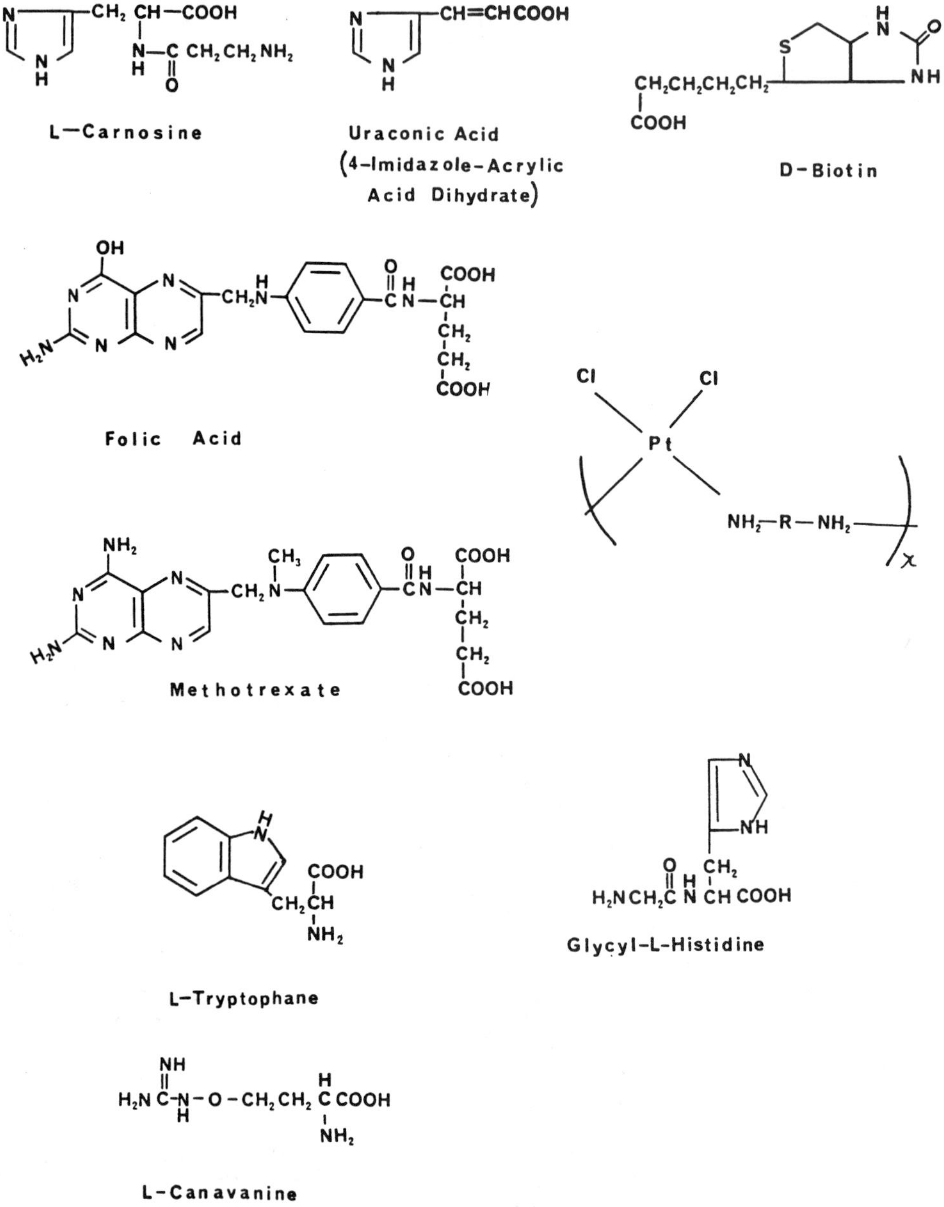

L–Carnosine

Uraconic Acid
(4–Imidazole–Acrylic Acid Dihydrate)

D–Biotin

Folic Acid

Methotrexate

L–Tryptophane

Glycyl–L–Histidine

L–Canavanine

Figure 1. Structures of amino acids used.

Table 4. Molecular Weight of Selected Platinum II Polyamino Acids.		
Lewis base	M_w	DP_w
Biotin	$1.3x10^7$	$2.6x10^4$
Methotrexate	$2.1x10^5$	$2.9x10^2$
Folic Acid	$5.6x10^4$	$8.0x10^1$
Glycyl-L-Histidine	$6.0x10^5$	$1.3x10^3$
Tryptophane	$3.7x10^7$	$8.0x10^4$
Carnosine	$8.0x10^3$	$1.6x10^1$

substantially removed from the nitrogen reactive groups or sterically hindered. By comparison, products from lysine (Cl/Pt ratio = 0.7:1), DL-histidine (1:1) and L-arginine (1.1:1) have Cl/Pt ratios near 1:1 consistent with the presence of repeat units of form [**9**] shown below.

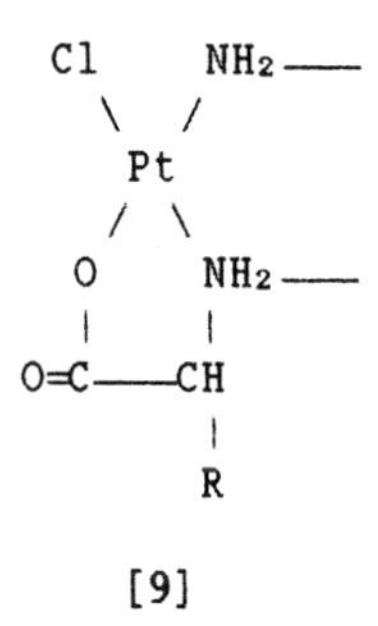

[**9**]

Structures of some of the amino acids employed in the present study are given in Figure 1. Molecular weights and degrees of polymerization are given in Table 4.

ACKNOWLEDGMENTS

We are pleased to acknowledge partial support from the American Chemical Society Petroleum Research Fund Grant #19222-B7-C.

REFERENCES

1. Y. Imanishi in: "*Bioactive Polymeric Systems - An Overview,*" C. G. Gebelein and C. E. Carraher, Jr., Eds., Plenum, NY, 1985, Chapter 17; J. Polymer Sci., Macromol. Revs., **14**, 1 (1979).
2. F. Westheimer, Chem. Tech., 748 (1980).
3. I. M. Klotz, Adv. Chem. Phys., **39**, 109 (1978).
4. G. Manecke and W. Storck, Angen. Chemie, **90**, 691 (1978).
5. C. Carraher, Makromol. Chemie, **166**, 31 (1973).
6. C. Carraher, J. Polymer Sci., **A-1**, 9 (1971).
7. C. Carraher and J. L. Lee, J. Macromol. Chem., **A9**(2), 191 (1973).
8. C. Carraher, European Polymer J., **8**, 1339 (1972).
9. C. Carraher, J. Chem. Ed., **58**, 11 (1981).
10. C. Carraher, W. Scott and D. Giron in: "*Bioactive Polymeric Systems -*

An Overview," C. G. Gebelein and C. E. Carraher, Jr., Eds., Plenum, NY, 1985, Chapter 20.
11. C. Carraher, I. Lopez and D. Giron, Polymeric Materials, **53**, 644 (1985).
12. D. Siegmann, C. Carraher and A. Friend, Polymeric Materials, **56**, 79 (1987).
13. C. Carraher, I. Lopez and D. Giron, "*Advances in Biomedical Polymers,*" C. Gebelein, Ed., Plenum, NY, 1987, pp. 311-324.
14. H. C. Freeman, "*Inorganic Biochemistry,*" Vol. 1, G. Eichhorn, Ed., Elsevier, NY, 1973, Chapter 4.
15. L. Volshtein and L. Dickanskaya, Russian J. Inorg. Chem., **13**(9), 1304 (1968).
16. J. Altman, M. Wilchek and A. Warshawsky, Inorganica Chim. Acta, **107**, 165 (1985).

ZEOLITES AS INORGANIC ANALOGS OF BIOPOLYMERS

Norman Herron

E. I. duPont de Nemours and Company
Central Research and Development Department
P. O. Box 80328
Wilmington, DE 19880-0328

While it is a non-traditional viewpoint, zeolites can be considered as rigid replacements for biopolymer proteins and can result in a novel series of biomimetic catalysis materials. The only requirements are that the synthetic chemists view the zeolite as a sterically demanding support for included materials and that zeolite chemists view zeolites as something other than acid catalysts or ion-exchangers. On these bases, there can be considerable opportunities for zeolites in all aspects of inclusion chemistry. This paper and similar articles by the author attempt to communicate this conviction.

INTRODUCTION

The "effortless" ease with which nature performs many of the difficult transformations essential to the life process has long been admired by synthetic chemists. By a judicious choice of catalysts (metal ion complexes) and supports (protein "biopolymers"), nature has built an impressive array of enzymes which perform such complex tasks as reversible oxygen binding (hemoglobin, myoglobin), selective partial oxidation of unactivated hydrocarbons (cytochromes P-450) and electron transport (ferridoxins). It has become increasingly clear that many of these natural systems can be viewed as a simple metal ion complex which can perform the basic non-selective chemistry of the desired type, which then has its reactivity modified by steric effects imposed by the polymer (protein) surrounding the active site.

When the term "polymer is invoked it most usually conjures up the idea of an organic plastic, such as polyethylene or PVC. However, this does a grave injustice to the vast arena of inorganic polymers of which much of the inanimate natural world is constructed. Oxides such as silica and associated aluminosilcates, as a generic class, represent some of the most common inorganic polymers and one sub-class of these materials are the open framework materials known as zeolites.

Our concept of an enzyme as a non-selective metal ion active site

Biomimetic Polymers
Edited by C. G. Gebelein
Plenum Press, New York, 1990

embedded in a very selective protein tertiary structure led us to utilize this model. Instead of using the conventional carbon based polymer frameworks as our protein mimic, we decided to investigate the inorganic polymers based on zeolites. The widespread use of their ion-exchange properties in the water-softening/detergent industry and their strong acid catalysis properties in petroleum-refining have made zeolites the workhorse materials in both applications. While this has attracted many researchers to the zeolite field, it has also had the effect of typecasting these remarkable materials into a limited number of chemical roles. The work reported in this chapter is a brief review of our own work here at Du Pont which attempts to dispel this stereotype. Using a perspective where zeolites are viewed as a highly ordered host "supermolecule" capable of sieving and directing substrate molecules, we have developed a series of novel guest-host and catalyst species. There are many remarkable similarities between zeolite structures and those of the protein portions of natural enzymes.[1,2] By recognizing and taking advantage of these similarities one can develop exciting new catalysts which combine the attractive features of the robust, chemically inert zeolite with the tremendous selectivity and activity of enzymes.

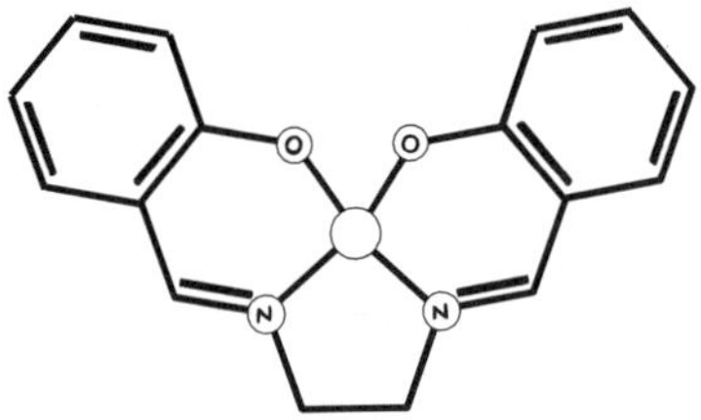

Structure of a "supermolecule".

What is a zeolite?[3] In generic terms, a zeolite is a crystalline, yet porous, solid where the sizes of the pores are all identical and of a size comparable to small molecules (e.g., water, benzene, etc.). When dictated by relative polarities, zeolite materials can act as sorptive hosts for guest molecules and form tightly interacting host-guest complexes. This property has been used extensively for applications in water removal for drying solvents and gases, for example. In the present discussion, only the silicoaluminate derived family of zeolites will be described and these are constructed from an open framework of SiO_4 and AlO_4 tetrahedra linked through oxygen bridges. The open framework has pores and cavities of molecular dimensions 3-13 A, making them the ultimate extrapolation of the quest for even smaller reaction vessels. These "nano-bottles" are such that chemistry carried out inside them is itself affected by the confines in which it is being performed. One negative charge per aluminum is present on the framework and is compensated by loosely attached, and hence ion-exchangeable, cations. These cation-exchange sites allow the straightforward introduction of active metal sites for catalysis (for a comprehensive discussion of ion-exchange and metal active sites in zeolites please see reference 3). A few representative zeolite structures are depicted in Figure 1.

The following sections review systems which we have developed to explore the generality of these ideas of inclusion in zeolites as a route to novel enzyme-mimic species. A reversible oxygen carrier, based on a ship-in-a-bottle inclusion complex, cytochrome P-450 mimics using an included active site and using zeolites to direct the host-guest interaction and a system utilizing the restrictive yet highly organized void spaces of the zeolite to control growth of size-confined semi-conductors,

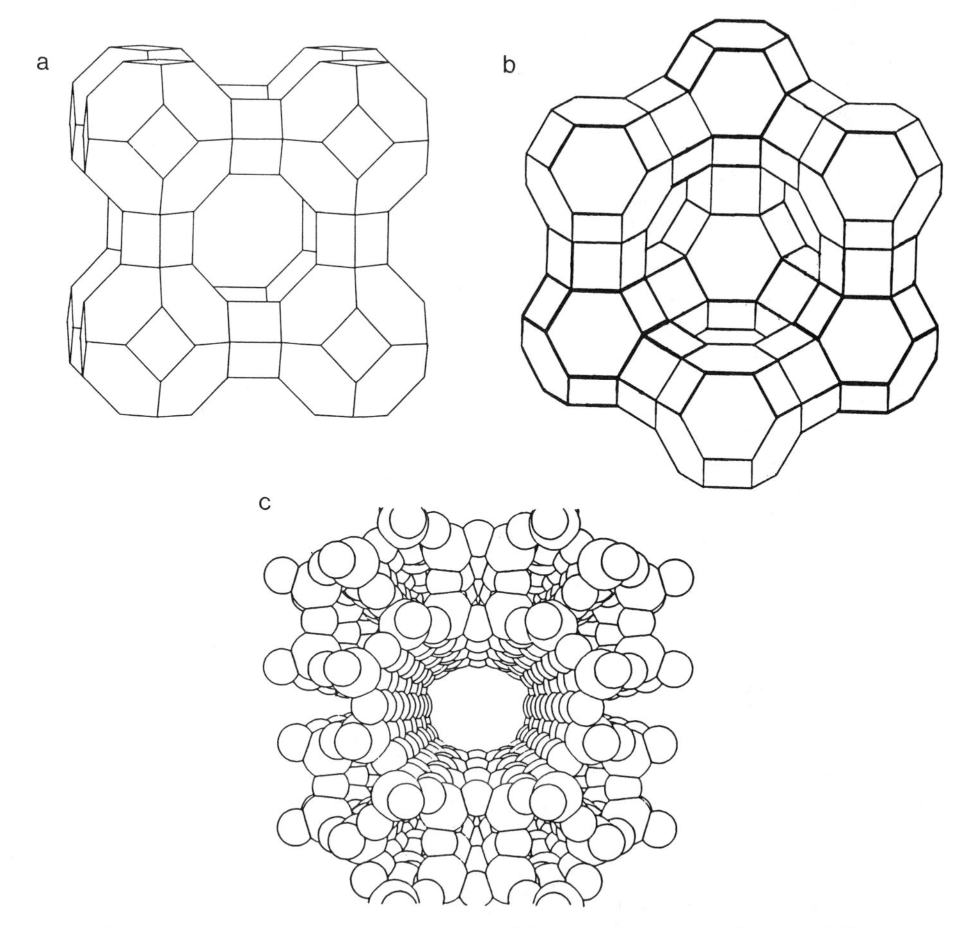

Figure 1. Some representative zeolite structures: (a) zeolite A supercage, (b) zeolite Y supercage and (c) zeolite mordenite viewed down the c-axis. (a) and (b) are stick representations where lines connect the Si and/or Al atoms and the intervening O atoms are omitted. (c) is a space filling view with perspective down a linear channel and where the O atoms are now included.

are all described. While the use of zeolites as host-guest materials is an under-studied area it is certainly not new.[1] The intent here is to emphasize the general concepts of zeolite inclusion phenomena related to biomimicry, and to do so by highlighting the key results we have obtained. Interested readers are referred to the detailed publications reference below for full experimental procedures and characterization details.

A SHIP-IN-A-BOTTLE

Binding and transport of molecular oxygen is the first step in the respiratory chain, and its function is performed in mammals by the iron proteins hemoglobin and myoglobin. Many synthetic models and mimics of this chemistry have been prepared using coordination complexes of iron, cobalt, manganese and other metal ions, but all suffer the same problem as the natural systems - autoxidation of the active sites with concurrent loss of oxygen binding behavior. Many autoxidation mechanisms identify oxo or peroxo dimer formation as the deleterious reaction. In addition to retarding this dimerization severely, the protein polymer of hemoglobin also engenders a fascinating property known as cooperatively. Hemoglobin is a 4-subunit enzyme where each subunit has an iron oxygen binding site and these 4 sites interact with each other such as binding of oxygen at the first subunit facilitates the subsequent binding of oxygen to the remaining 3-subunits of the tetramer.

One of the first challenges we set for ourselves in our zeolite inclusion chemistry research was to prepare a so-called "ship-in-a-bottle" material.[4] This, as the name implies, is a material where a guest molecule is included within the zeolite void spaces and is in fact entrapped there only on the basis of its physical size. To do this a molecule must be prepared *in situ* within the pores from its smaller components and the assembled molecule must be large and rigid enough to be unable to escape from the bottle. As a prototypical system we chose the reversible oxygen carrier molecule cobalt-salen-I, largely because of the useful handle for characterization provided by EPR.[5,6] In addition, the catalytic action of this molecule in reversibly binding molecule oxygen provided reaction chemistry with which to assess the role of the zeolite host in affecting the thermodynamic and kinetic behavior of the included guest.

Ion-exchange of cobalt(11)ions into the crystallite voids of zeolite Y at a concentration of ~1Co per 2 supercages, followed by sublimation of the flexible free ligand Salen into the same voids, leads to assembly of the rigid complex. This complex involves square-planar coordination of the tetradentate salen ligand such that the complex has dimensions greater than the window size of the zeolite. Once constructed into the pores, the complex is physically trapped on the basis of its size: a true ship-in-a-bottle (Figure 2). This complex, as its pyridine adduct, forms an oxygen adduct as judged by the characteristic EPR signal (Figure 2) generated upon exposure to dioxygen. There are two remarkable features of this zeolite entrapped adduct.

Firstly, compared to the same complex in free solution or as a crystalline solid, the zeolite encapsulated material displays quite remarkable stability towards autoxidation and peroxo-dimer formation.[14] For example the half-life for the oxygen adduct EPR signal at room temperature in air is ~4 weeks when entrapped in the zeolite compared to several minutes in free solution.[7-9] In solid, crystalline Co-salen, peroxo-dimers are formed exclusively.[7-9] This is a manifestation of the extremely effective site-isolation achieved by entrapping the complexes inside the pores of the zeolite, leading to elimination of the normal autoxidation mechanism (peroxo and oxo dimer formation).

Secondly, an examination of the thermodynamics of oxygen binding as plotted in a Hill plot (Figure 3) at 4 different temperatures shows evidence of a negative cooperatively between cobalt binding sites (slope < 1).[10] This result means that binding of oxygen to the first few cobalt sites makes it progressively more difficult for subsequent oxygen to bind to other sites. This is the reverse of the behavior displayed by hemoglobin.[11-13] The explanation for this behavior is still unclear, but may well result from progressively more difficult oxygen binding as one progresses from the exterior to the interior of each zeolite crystallite. Binding of oxygen at the cobalt sites in the outermost cages of the crystallite is easiest, but binding becomes more difficult as the oxygen has to diffuse toward the interior sites. A van't Hoff plot of the data reveals an enthalpy of binding of -11.4 Kcal/mol and an entropy of -51 e.u. which compare with values of -12.4 Kcal/mol and -47 e.u. for the same complex in pyridine solution.[14] The lower oxygen binding constants inside

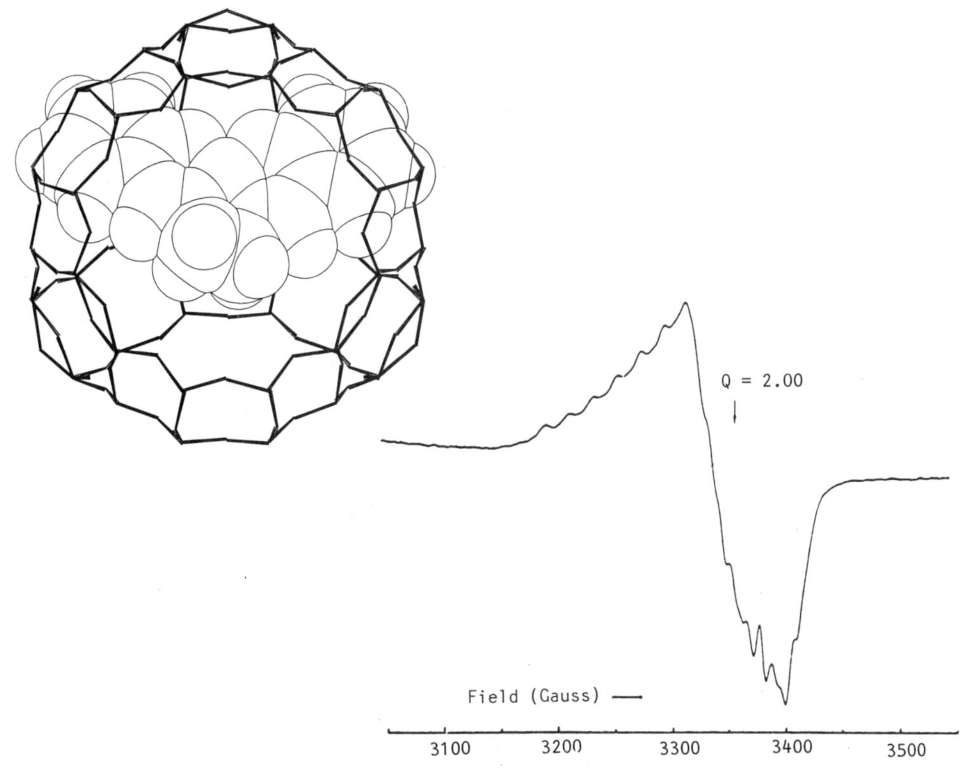

Figure 2. Representation of Co-salen inside zeolite super-cage. Zeolite framework has been reduced to stick bonds for clarity, Co atom is shaded. Inset shows EPR spectrum of oxygen adduct at 298°K and 760 torr oxygen.

the zeolite, therefore, appear to result from a reduced exothermicity consistent with the oxygen being bound at a restrictive site where the steric interaction of bound oxygen with the zeolite cavity walls is important.

This example demonstrates the ability of zeolites to act as effective host media for inclusion of reactive guest species and to affect the chemistry occurring at the guest. This type of control is typical of enzyme proteins which dictate both the interactions between active sites and the thermodynamics of reactions at these active sites. The next example demonstrates the zeolite's ability to control product selectivity kinetically in an oxidation reaction typical of the monoxygenase enzymes.

PARTIAL OXIDATION CATALYSTS FROM ZEOLITE INCLUSION MATERIALS

The monoxygenase enzymes, cytochromes P-450, perform the function of selectively oxidizing organic materials to usable or excretable hydrophilic compounds. Their unique ability to convert, for example, unactivated alkanes to alcohols with unusual selectivities using only molecular oxygen and a reducing cofactor (while in aqueous solution and at room temperature) has made these materials the envy of synthetic organic chemists. There have been numerous attempts to model the heme-iron active site of the enzymes so as to reproduce some of their selectivity by in-

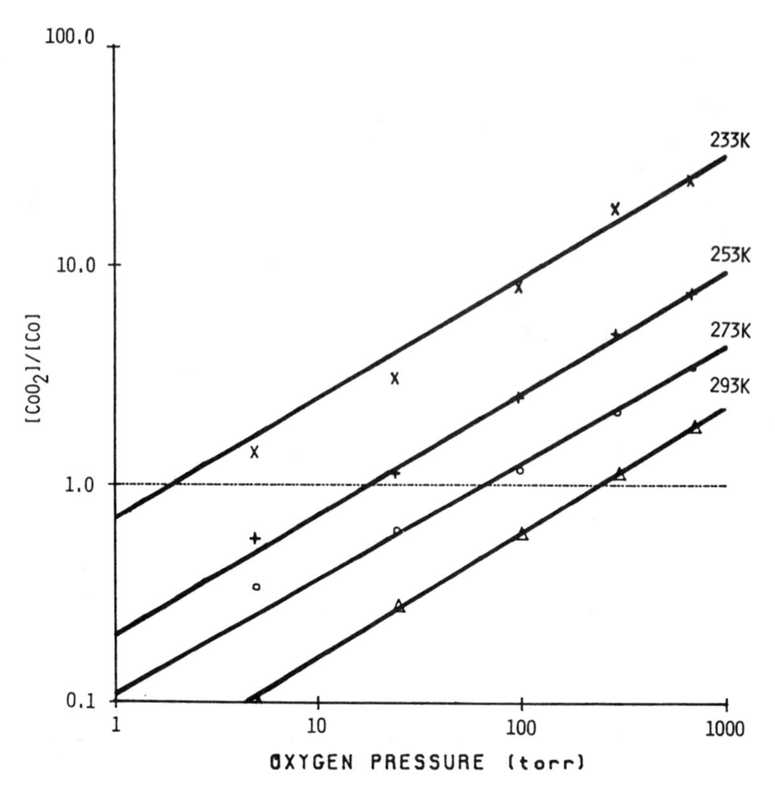

Figure 3. Hill plot of oxygen binding to Co-salen py in zeolite Y.

cluding multiple bulky peripheral substituents upon the basic porphyrin nucleus.[15] While these models have undoubtedly contributed tremendously to an understanding of the mechanistic features of the monoxygenases, they have yet to demonstrate the phenomenal selectivities.

The enzyme cycle is represented in Figure 4, emphasizing only the redox and coordination at the iron center. If one accepts the tenet that the FeO^{3+} species is the active potent oxidant, then the selectivity of the enzyme is dictated by how the protein sieves and directs substrates toward this indiscriminate oxidant.

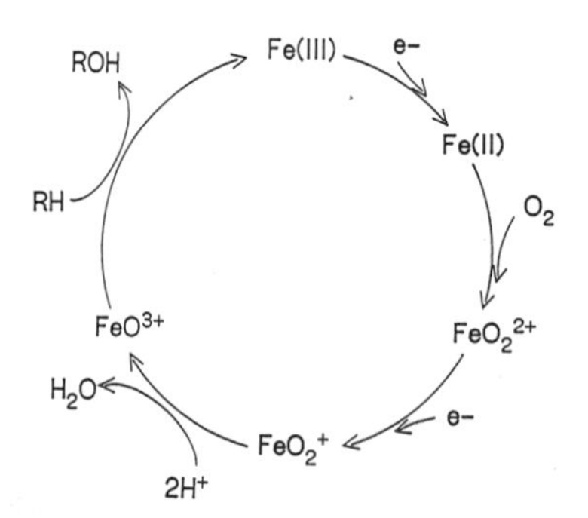

Figure 4. Enzyme cycle of Cytochrome P-450.

How does nature product this oxidant? She takes molecular oxygen, two electrons and two protons and eliminates water! We have developed two approaches to mimicing this chemistry based on inclusion materials.

1. A Phthalocyanine "Ship-in-a-Bottle"

The first mimic follows closely the example above.[2,16] A coordination complex of the active metal ion (in this case iron II) is constructed within the void space of the zeolite host so as to be entrapped there on the basis of its size. The complex is chosen to mimic the heme prosthetic group of Cy-P-450 and thus is a 16-member tetradentate macrocycle having 4 imine type nitrogen donors in an aromatic ring just like the porphyrin ligand of heme. The phthalocyanine (Pc) complex is constructed by template condensation of 1,2-dicyanobenzene around the iron ions of Fe(II) ion exchanged zeolite Y at 200°C under nitrogen.[16] The resultant ship-in-bottle material is depicted in Figure 5.

Using this material as a catalyst in the now well established oxidation system where iodosobenzene is the oxygen atom transfer reagent results in oxidation of alkanes.[16] While the activity is low (~6 turnovers based on Fe), it still exceeds that of the the native FePc itself (~1 turnover). However, the rate of oxidation is extremely slow, taking ~24 hours to reach completion in the zeolite material compared with ~30 minutes for FePc itself. Much of these problems of slow rate and low activity are a reflection of a common problem with zeolite based inclusion compounds namely pore blockage. While the zeolite provides an excellent method of isolating FePc units from one another, and thereby prevents bimolecular destruction of the active sites, it also is a very restrictive environment which can be easily blocked by substrate, product, oxidant and catalyst molecules themselves. This blocking means that the active sites are essentially limited in the rates of the chemistry they perform because of diffusion of reagents and are limited in turn-

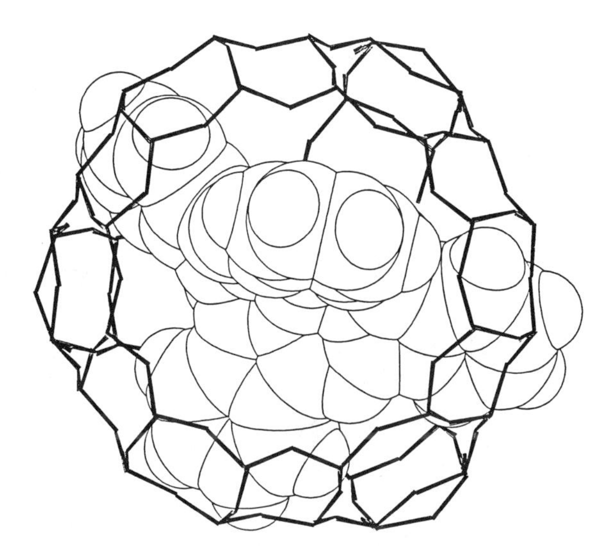

Figure 5. Representation of saddle distorted Fe-phthalocyanine in the supercage of zeolite X or Y. The zeolite framework is reduced to stick bonds for clarity and the Fe atom is shaded.

overs by eventual isolation of the active site from access by further substrate and oxidant.

Despite these problems, the selectivity of oxidations performed by the encapsulated FePcs shows the desired effect. Examples of substrate selectivity in the competitive oxidation of cyclohexane and cyclododecane are shown in Figure 6. It is apparent that the selectivity of the oxidant for the smaller substrate of the two is enhanced when the zeolite pore can be tailored to select which ion can more easily access the active sites.[16] In addition to this selection process the selected substrate is also oriented as it proceeds through the pore system so that the end of the long molecular axis is the first part of the substrate to encounter the active site. As a consequence we find that the zeolite encapsulated FePc system tends to favor regioselective oxidation towards the ends of the long molecular axis of the substrate. The orientation phenomenon can also manifest itself in stereoselective oxidation of one of a pair of diasterotopic hydrogens in a substrate such as methylcyclohexane or norbornane.[16]

While the above results tend to justify the concept of using zeolite inclusion materials as highly selective catalyst species, they also point to the problems of such materials - low rates and turnovers due to diffusional and blockage problems within the pore system. This is exacerbated when the included active catalyst molecule is so large as to contribute significantly to the pore blocking. With this in mind we went on to develop the following system in attempts to minimize pore blocking problems and to maximize selectivities using the inclusion approach.

2. A Completely Inorganic Cytochrome P-450 Mimic

Our second mimic is designed to reproduce the enzyme cycle of Figure

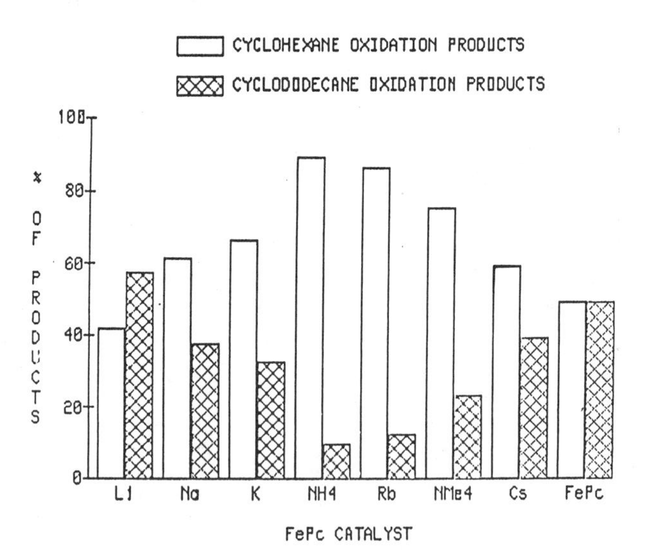

Figure 6. Bar-graph representation of substrate selectivities in a competitive cyclohexane/cyclododecane oxidation over various cation/exchanged FePc zeolite catalysts. Products are a combination of alcohols and ketones.

4 even more closely by taking the two electrons and two protons together as molecular hydrogen and combining it with oxygen over Pd(0) to generate hydrogen peroxide.[17,18] This hydrogen peroxide can then be reacted at an iron site to eliminate water and give the desired FeO^{3+} unit. If all of this is done inside the stereochemically demanding pores of a zeolite, then any substrate which is simultaneously in those pores should suffer selective oxidation by the FeO^{3+} as directed by the zeolite framework.

This mimic is prepared by sequential ion-exchanges with iron(II) and Pd(II) tetramine cations followed by calcinations and reduction of the Pd(II) to Pd(0), as previously described.[17,18] A material with ~2 wt% Fe(II) and 1 wt% Pd(0) is used by immersing the dry zeolite solid in neat substrate alkane and then pressurizing the reaction vessel with a 3:1 mixture of oxygen:hydrogen. After shaking this mixture at room temperature for 4 hours, the products are analyzed by capillary GC. As a control to assess the intrinsic selectivity of such a Pd/Fe system in the absence of steric effects of the zeolite, catalysts prepared with amorphous alumino-silicate supports were run for comparison. In these cases, all reactions must take place at the particle surface since there is no interior pore structure available. In addition, comparison of reaction selectivities of this catalysts with our zeolite materials allows us to ascertain that the Fe active sites must be actually inside (and not on the exterior surface) of the zeolite crystallites.

The first host explored was zeolite A which is a very selective absorbent of linear alkanes to the exclusion of branched or cyclic hydrocarbons.[19] It was therefore expected that extreme examples of substrate selectivity could be achieved in competitive oxidation of linear alkanes vs. cyclic alkanes.

Figure 7a confirms this expectation. While the selectivity for oxidation of n-octane in the presence of cyclohexane is slight (55:45) over the control silicoaluminate support, the selectivity over the A zeolite is tremendous (>200:1) in favor of the linear alkane (plots are of the total of all oxidation products from each substrate). This indicates that the zeolite is exerting its sieving effect and, therefore, the desired chemistry is occurring inside the zeolite pores. Indeed, not only is the chemistry occurring there but the vast majority of the octanol products are remaining there at the end of the reaction and can only be released for analysis by complete destruction of the framework using conc. sulfuric acid! This trapping of products is more than simple absorption since they are not released simply by displacement with a more polar molecule (e.g., water). The production of secondary alcohols as part of the product mix means that the pores of the zeolite rapidly become filled with molecules which are too large to escape from the interior and remain trapped. These molecules then act as plugs for escape of even the linear alcohol products so that the entire zeolite interior becomes saturated with products and the catalytic activity is shut down.

The incredible substrate selectivity of this system pales in comparison to the regioselectivity displayed by the octane oxidation products. In Figure 7b the dramatic increase of products derived from terminal methyl group oxidation in the zeolite system is apparent (plots are normalized to a "per hydrogen basis and represent the total of products [alcohols and ketones] at each position in the chain). Over the control catalyst, the primary/secondary oxidation ratio is 0.05 while in the zeolite A this ratio is 0.6. This selectivity for oxidation towards the end of the alkane chain probably arises from the very close fit between the alkane and the A pore size, which essentially constrains it to have an extended "linear" conformation. It is therefore the methyl groups which are the first to encounter and so be oxidized by the iron active

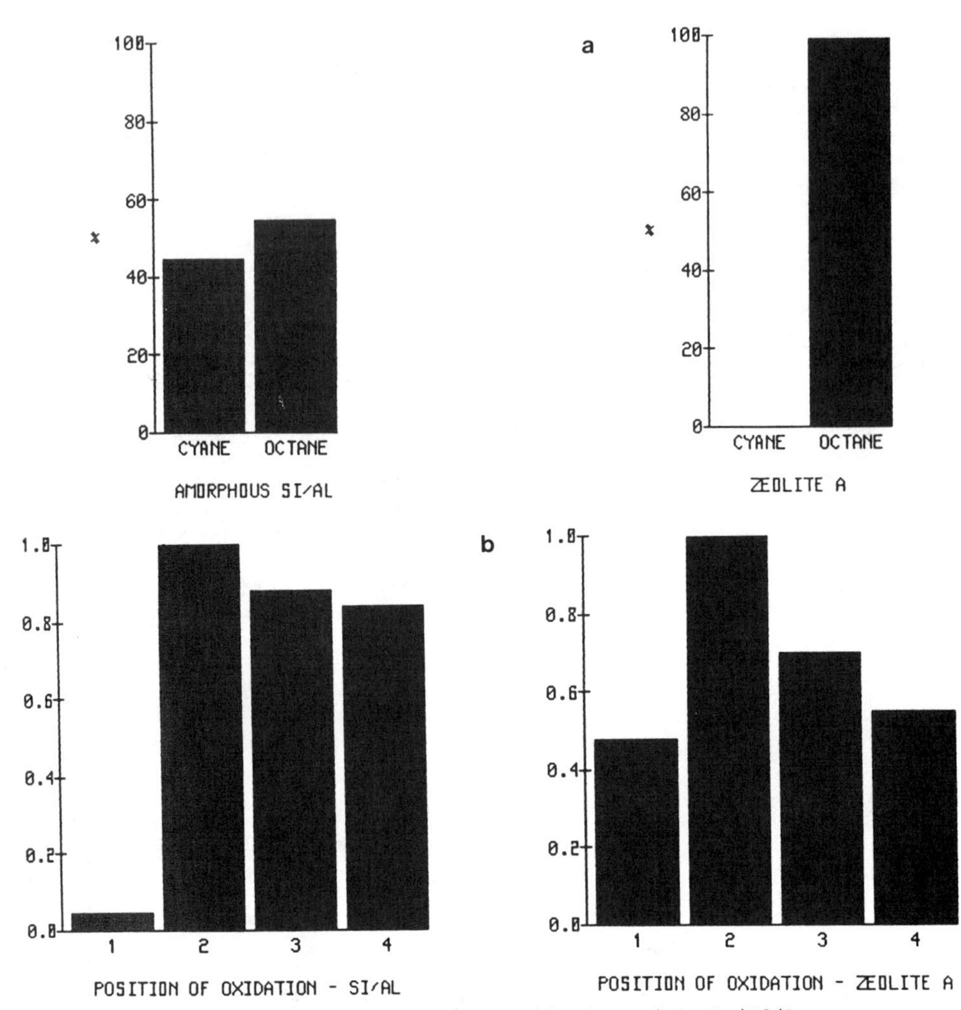

Figure 7. Selectivity of alkane oxidation with Fe/Pd/A zeolite and H_2/O_2. (a) Substrate selectivity between n-octane and cyclohexane and (b) regioselectivity of n-octane oxidation.

sites which tend to be located in the sixth ring faces of the zeolite cage (illustrated schematically in Figure 8).

While the selectivities of this system are dramatic, the activity is low (~1 turnover on Fe in a 4 hr. batch run), and the necessity of running all reactions in an explosive rated environment (because of the H_2/O_2 mixtures) combined with the need to dissolve away the zeolite to reclaim products makes this a very impractical synthetic method for alkane oxidation. These problems can be addressed as follows. The low turnover and product trapping are really the same problem since we believe that it is the plugging of the pore system by products which leads to the shutdown of the reaction since at that point no further substrate can get to the active sites. This is a familiar problem with zeolite catalysts and should be solved by two changes.[16] (1) A larger pore system will permit egress of the secondary alcohol products (at the obvious expense of lowered substrate selectivity); (2) a higher Si/Al ratio zeolite is more hydrophobic and at a ratio above ~20 begins to

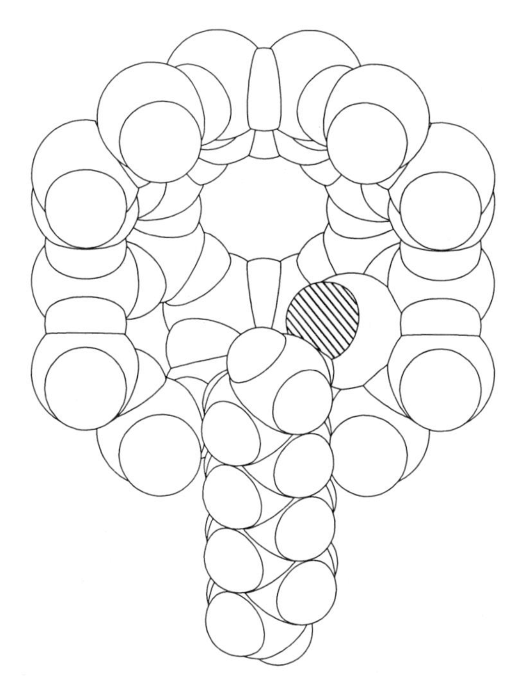

Figure 8. Cutaway representation of n-octane proceeding through the 8-ring window of zeolite A towards the iron active site for oxidation. Crosshatched atom is the active oxygen on iron.

favor absorption of the reactant alkane over the product alcohols (at the expense of ion exchange sites). These criteria can be met by the ZSM-5 zeolite with its 10-ring channels and high Si/Al ratio. The inconvenience of using hydrogen/oxygen mixtures can be circumvented by using preformed hydrogen peroxide rather than making it *in situ* by combination over Pd(0). The ideal system is therefore a simple Fe ion-exchange high silica ZSM-5 zeolite fed hydrogen peroxide (either aqueous or in organic solvents at low concentration). Oxidation of n-octane with this system does indeed lead to all products being recovered from the supernatant solution without zeolite dissolution, and the system becoming truly catalytic with ~4 turnovers on Fe in 4 hours. Remarkably, the regioselectivity of the ZSM-5 system is even more pronounced for oxidation towards the terminus of the alkane chain with a primary/secondary oxidation ratio of 3.3. This material is now in the ballpark of a viable omega-hydroxylase mimic. This enzyme is capable of regioselective oxidation of terminal methyl groups of alkanes or linear carboxylic acids with a primary/secondary oxidation ratio of ~10.

ELECTRON TRANSFER PROTEIN ANALOG

One of the essential cofactor enzymes in the cytochrome P-450 network provides electrons to reduce iron and or oxygen to generate the active oxidant (see Figure 4). These natural electron-transfer materials are typified by the iron-sulfur proteins.[20] Work by Holm and others has modeled the basic iron-sulfur cores of these proteins and a variety of structure types have been reproduced. One of these types, an iron-sulfur cubane like cluster, consists of interpenetrating tetrahedra of iron and sulfur ions. In all cases these proteins consist of this electron transferring inorganic core which is surrounded by the insulating biopolymer which holds the core together and facilitates the electron transfer step by orienting the method of approach between the protein and its electron transfer partner.

Our fascination with using zeolites as very small reaction vessels for the production of novel species led us to explore the possibility of synthesizing extremely small semiconductor materials within the pores. Semiconductors are well known to undergo similar kinds of electron transfers to biological systems when irradiated with light above their band gap energy. In addition, small pieces of a bulk semiconductor lattice cannot fully develop the normal semiconductor band structure and so reside in the so-called size quantized or quantum confined regime. This is when the electron-hole pair of a photo-excited semiconductor particle has a radius larger than the actual particle radius.[21] The electron then behaves like a particle in a box and novel optical properties result.

Cadmium ion-exchange of zeolite Y, followed by calcination in flowing oxygen, leads to materials having from 0-90% of the original sodium ions replaced by Cd. Treatment of the dry, calcined zeolite with hydrogen sulfide gas at 100°C and 760 torr generates the CdS clusters inside the zeolite pores.[22,23] Quantum confinement effects are manifest in that the material is typically pale cream or white rather than the yellow orange of bulk CdS. The actual structure of the CdS clusters is revealed by detailed x-ray power diffraction and EXAFS analysis. Figure 9 depicts the Cd_4S_4 units and their location within the sodalite cavities of the zeolite. This unit is very reminiscent of the Fe_4S_4 core of ferredoxin proteins discussed above.

Since all of the CdS clusters reside in the sodalite cages of the Y framework, the larger supercages of the structure are still available for absorption of substrate molecules - in this case we studied olefins for photo-oxidation via electron transfer. Colloidal CdS in free solution has been used for such oxidations previously,[24] and in a competitive oxidation of styrene and 1,1-diphenylethylene we find that unconfined CdS will

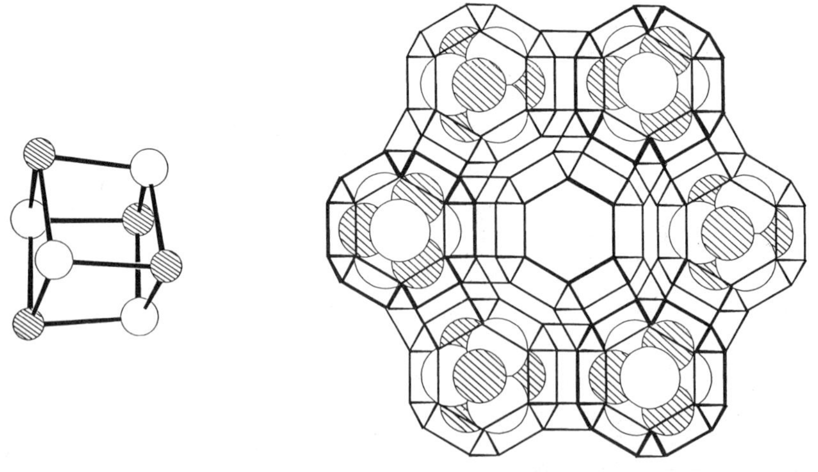

Figure 9. CdS clusters in the sodalite cages of zeolite Y. The zeolite has been reduced to sticks connecting only the tetrahedral atoms for clarity. Inset shows the structure of an individual CdS cubane-like cluster. The Cd atoms are cross hatched.

effect oxidation in a ratio of 1:2 for these two olefins (irradiation at 365 nm). In the zeolite confined case however, we find a ratio of 1:1; i.e., a slight shift in favor of the smaller substrate, as may be expected on the basis of size/diffusion effects. We have here a system which mimics the core structure of iron-sulfur proteins while also undergoing electron transfer (photo-stimulated) with substrate molecules in analogy with the same proteins.

CONCLUSION

The above examples have demonstrated that less traditional perspectives of zeolites - and particularly viewing them as rigid replacements for biopolymer proteins - can result in a novel series of biomimetic materials for catalysis. It is only necessary that synthetic chemists view zeolites as sterically demanding supports for included materials and that zeolite chemists view zeolites as something other than acid catalysts or ion-exchangers. If that can be communicated, by this and similar articles, there may be considerable opportunities for zeolites in all aspects of inclusion chemistry.

ACKNOWLEDGMENTS

The technical assistance of J. B. Jensen, S. Harvey, J. D. Nicholson, and J. E. Macdougall was invaluable while contributions of Drs. C. A. Tolman, Y. Wang, D. R. Corbin, R. D. Farlee, G. D. Stucky, M. M. Eddy, W. E. Farneth, T. Bein, and K. Moller to various aspects of this work were likewise dramatic.

REFERENCES

1. A. Dyer, G. G. Hayes, G/ O. Phillips, R. P. Townsend, ACS Symp. Ser. 1973, No. 121, 299.
2. N. Herron, C. A. Tolman, G. D. Stucky, J. C. S. Chem. Commun., 1521 (1986).
3. D. W. Breck, "*Zeolite Molecular Sieves,*" Wiley, New York, 1974.
4. This term was first coined in N. Herron, G. D. Stucky, C. A. Tolman, Inorg. Chim. Acta., **100**, 135 (1985).
5. N. Herron, Inorg. Chem., **25**, 4717 (1986).
6. W. K. Wilmarth, S. Aranoff, M. Calvin, J. Am. Chem. Soc., **68**, 2263 (1946).
7. C. Floriani, F. Calderazzo, J. Chem. Soc. A., **946** (1069).
8. E. I. Ochiai, J. Inorg. Nucl. Chem., **35**, 1727 (1973).
9. C. H. Barkelew, M. Calvin, J. Am. Chem. Soc., **68**, 2257 (1946).
10. A. V. Hill, J. Physiol. (london), **40**, IV-VII (1910).
11. See R. D. Jones, D. A. Summerville, F. Basolo, Chem. Revs., **79**, 139 (1979).
12. J. O. Alben, W. H. Fuchsmau, C. A. Beaudreau, W. S. Caughey, Biochemistry, **7**, 624 (1968).
13. J. M. Baldwin, Brit. Med. Bull., **32**, 213 (1976).
14. G. Lauzher, G. Amiconi, E. Antonini, M. Brunori, G. Costa, Nature (London) New Biol., **241**, 222 (1973).
15. See for example, B. R. Cook, T. J. Reinert, K. S. Suslick, J. Am. Chem. Soc., **108**, 7281 (1986).
16. N. Herron, J. Coord. Chem., **19**, 25 (1988).
17. N. Herron, C. A. Tolman, J. Am. Chem. Soc., **109**, 2837 (1987).

18. N. Herron, New Journal of Chemistry, **13**, 761 (1989).
19. R. M. Barrier, in "*Zeolite and Clay Minerals as Sorbents and Molecular Sieves,*" Academic Press, New York, 1978.
20. J. C. M. Tsibris, R. D. Woody, Coord. Chem. Revs., **5**, 417 (1970).
21. L. E. Brus, J. Phys. Chem., **90**, 2555 (1986).
22. N. Herron, Y. Wang, G. D. Stucky, M. M. Eddy, D. E. Cox, T. Bein, K. Moller, J. Am. Chem. Soc., **111**, 530 (1989).
23. Y. Wang, N. Herron, J. Phys. Chem., **91**, 257 (1987).
24. M. A. Fox, Acc. Chem. Res., **16**, 314 (1983).

CATALYTIC ANTIBODIES

Donald Hilvert

Department of Molecular Biology
Research Institute of Scripps Clinic
10666 North Torrey Pines Road
La Jolla, CA 92037

Attention has focused on the mammalian immune system as a source of highly specific, tailored catalysts. The construction of enzyme-like immunoglobulins involves synthesizing compounds that mimic the transition state structure of a particular reaction, eliciting an immune response against such substances, and screening the resulting antibodies for the desired activity and specificity. The recent and rapid progress that has been made in generating catalytic antibodies for diverse chemical transformations, from hydrolytic to concerted pericyclic processes, is reviewed.

INTRODUCTION

Chemists and biologists have long been fascinated by enzymes, nature's catalysts. These proteins make complex life possible. With very few exceptions, each of the tens of thousands of chemical reactions that sustain living systems takes place quickly and smoothly through the action of a specific enzyme. Clearly, if scientists could make enzymes, many of the difficult, inefficient, and expensive chemical reactions that are still required today in the production of pharmaceuticals, industrial chemicals, and synthetic materials of all kinds would become straightforward. Engineered biocatalysts could also find extensive application in medicine as therapeutic and diagnostic agents.

Enzymes work because each has a unique three-dimensional shape that provides a distinctive surface upon which other molecules can bind and undergo chemical alteration. Binding events are, in fact, at the heart of biological catalysis.[1] By exploiting binding energy, enzymes organize chemical reactants and stabilize rate limiting transition states. The first step in designing a protein biocatalyst must therefore address the challenge of constructing a suitable binding pocket.

Unfortunately, it is not yet practical to design and synthesize protein active sites from first principles. Scientists do not know the rules that govern protein folding and cannot predict amino acid sequences that will yield stable, well-defined binding pockets with high affinity for

Biomimetic Polymers
Edited by C. G. Gebelein
Plenum Press, New York, 1990

small molecules. Lacking this knowledge, the only way to obtain useful protein catalysts has been to search nature for enzymes with desirable properties.[2] This was true, in fact, until recent advances in molecular biology and immunology provided tools for creating new active sites from existing structures.

Nature provides us with a wealth of proteins, each with a precise function and specificity. When these molecules have been well characterized, their binding pockets can serve as convenient starting points for the construction of new active sites. For example, site-directed mutagenesis using recombinant DNA techniques can be used to replace, add or delete amino acids in a given enzyme and thereby alter its chemical activity or specificity.[3,4] Alternately, selective chemical reactions can be used to modify proteins posttranslationally and thereby introduce non-natural amino acids or catalytic prosthetic groups, including metals, vitamins and other cofactors, into their binding sites,[6,7] If properly designed, the resulting semisynthetic hybrids couple the unique chemistry of the prosthetic group with the binding specificity of the protein template.

Although modification of existing proteins is a powerful tool for creating new enzyme functions, the relatively small number of structurally well-characterized protein templates limits its general application, as does the need for highly selective modification protocols. Ideally, one would like to be able to construct unique receptor sites with appropriate topologies for catalyzing any chemical transformation with any designated substrate molecule. Although this is not yet feasible in the laboratory, one of the body's most vital functions, the immune system, can create high affinity receptors with new and unique shapes for a wide range of ligands.[8-10] Recently, scientists have shown that this remarkable protein-generating system can be exploited to produce antibody proteins with enzymic properties. This article reviews the rapid progress that has been made in this exiting field and outlines its potential for yielding completely new catalytic activities for use in research, industry and medicine.

ANTIBODIES AND THE IMMUNE SYSTEM

The mammalian immune system is certainly the most prolific source of specific receptor molecules known.[8-10] To protect the body against disease-causing microorganisms and toxins, the immune system can generate a virtually limitless number of unique and highly specific antibody molecules that bind foreign substances (antigens) with high affinity. It has been estimated that the immune repertoire contains up to 10^{12} unique specificities.[8] Such tremendous diversity is achieved in the cell by a process of genetic recombination and somatic mutation at the level of DNA.[11-12] Moreover, natural selection drives the immune system to produce antibodies with increasingly high affinity for the target antigen. Dissociation constants for typical antibody-antigen complexes fall in the range 10^{-4} to 10^{-12} M.[8]

Scientists have learned to exploit the immune system to raise antibodies, on demand, against a wide range of materials, including natural and man-made compounds, charged and neutral species, small organic molecules, inorganic complexes, and large polymers.[8-10] These induced receptors are enormously important as diagnostic and imaging agents, in drug delivery, and for affinity purification of diverse materials.[13] Given the great diversity inherent in the immune system, it seemed reasonable to expect that antibodies could also be found, or engineered, that possess intrinsic catalytic activity.

The similarity between antibodies and enzymes was recognized by Linus Pauling forty years ago.[14] Antibodies are structurally well-characterized proteins (M_r 160,000) that bind ligands tightly and specifically.[9,10,15] They are bivalent molecules made up of two sets of two polypeptides (Figure 1), the heavy and light chains, with relative molecular weights of roughly 55,000 and 25,000, respectively. The antigen binding site, or complementarity determining region, is formed by the N-terminal subunit region (Fv) of both chains. The amino acid sequence of these segments is unique to each different immunoglobulin and determines the specificity of the antibody molecule. The combining site is roughly the same size as that of a typical enzyme: roughly 5 or 6 amino acids or a similar number of glucose units can be accommodated.[10] Moreover, ligand binding apparently involves the same factors that are important for substrate binding by enzymes:[8] (1) hydrophobic interactions, (2) hydrogen-bonding, (3) ion-pairing, and (4) dispersion forces. Nevertheless, antibodies do not normally catalyze reactions; enzymes do.

Chemical reactions proceed through the formation of high energy, short-lived transition states that are intermediate in structure between substrate and product (Figure 2). To achieve catalysis, an enzyme's active site must be complementary to this high energy species and stabilize it relative to the bound ground state (Figure 2).[14] Antibodies normally leave the molecules they bind chemically unchanged because their combining site matches a ground state species not a transition state structure. Catalysis cannot occur if the ground state is stabilized relative to the transition state. In order to produce antibody proteins with catalytic activity, it is therefore necessary to use as immunogens stable molecules which mimic the unique geometry and electrostatics of the ephemeral transition state for the targeted chemical reaction. Although this notion was suggested by Jencks in 1969,[16] it was reduced to practice only recently.

EARLY ATTEMPTS TO RAISE CATALYTIC ANTIBODIES

Over the years a number of reports have appeared in the literature describing stoichiometric reactions between antibodies and labile esters. An anti-dihydrotestosterone IgG molecule, for example, augmented the cleavage of an umbelliferone ester of testosterone,[17] and stoichiometric hydrolysis of 7-[*N*-(2,4-dinitrophenyl)-6-amino-hexanoyl]-coumarin was accelerated by a monoclonal IgG antibody directed against the 2,4-dinitrophenyl group.[18] Evidence was presented suggesting that a nucleophilic group in the antibody combining site was responsible for substrate cleavage, and, in each case, the reaction was specific with respect to the nature of the substrate. However, catalytic turnover was not observed in these systems.

Raso and Stollar were among the first researchers to design a hapten specifically for the purpose of eliciting antibody sites with catalytic activity.[19,20] They prepared *N*-(5-phosphopyridoxyl)-3'-amino-L-tyrosine as a mimic of the Schiff's base intermediate that is formed during the pyridoxal catalyzed transamination of tyrosine (Figure 3) and showed that it was a site-directed inhibitor of the pyridoxal-dependent enzymes tyrosine transaminase and tyrosine decarboxylase. Partially purified polyclonal antibodies, elicited against γ-globulin conjugates of this Schiff's base analog, recognized both the cofactor and tyrosine portions of the hapten. Although the polyclonals did not enhance the rate of Schiff's base formation between pyridoxal and tyrosine, they accelerated

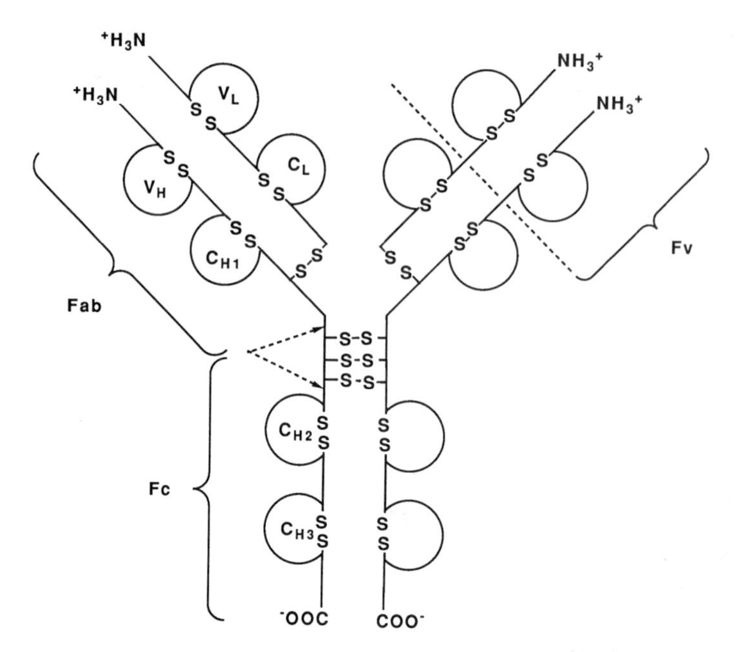

Figure 1. Structure of a typical mouse IgG antibody.

the rate of tyrosine transamination by about 5-fold.

These examples demonstrated that antibodies were capable of doing chemistry, but the reactions described are stoichiometric or only weakly catalytic. As described below, identification of truly efficient immunoglobulin-based catalysts required the development of better transition state mimics,[21-23] the use of monoclonal rather than polyclonal antibodies,[24,25] and more effective screening of the entire immune response.

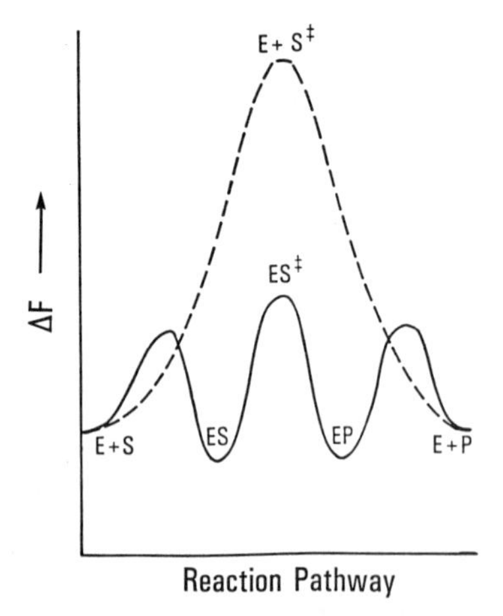

Figure 2. An energy coordinate diagram comparing the free energy change for a catalyzed ($E \cdot S \rightarrow E \cdot S^\ddagger$) chemical reaction.

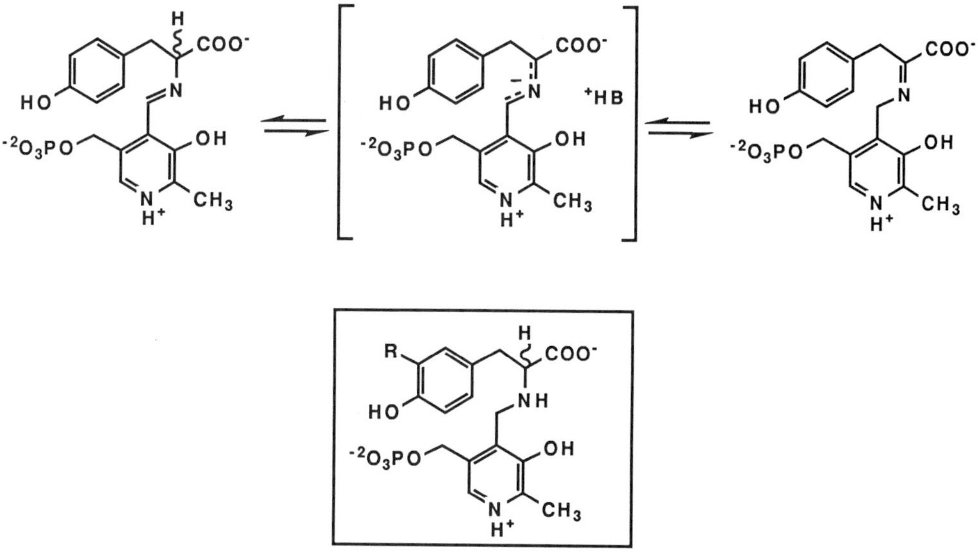

Figure 3. Pyridoxal-mediated transamination of tyrosine requires conversion of an aldimine into a ketimine. The reduced Schiff's base formed by pyridoxal and tyrosine (box) is a stable analog of the key intermediate.

HYDROLYTIC ANTIBODIES

Aryl and alkyl phosphonates and phosphonamidates have been employed as high affinity inhibitors of naturally occurring proteases for a number of years.[25,26] The tetrahedral phosphorous is apparently an excellent mimic of the tetrahedral intermediates and transition states that occur along the reaction coordinate for ester and amide hydrolysis (Figure 4). Consequently, such molecules were excellent candidates for use as haptens to elicit antibodies with hydrolytic activity. Because these molecules differ substantially from the planar starting materials and products of ester and amide hydrolysis, catalysis should result mainly through differential binding of the transition state and ground state by the immunoglobulin.[1,16] Furthermore, the significant difference in structure of the reactants/products and transition state should minimize complications due to product inhibition.

1. Ester-Cleaving Antibodies

Tramontano, Lerner and their colleagues demonstrated that aryl phosphonate esters could in fact be used to generate truly catalytic esterolytic antibodies.[27] Schultz and coworkers subsequently showed that antibodies against such compounds could also promote carbonate hydrolysis.[28] The catalyzed reactions underwent multiple turnovers and exhibited saturation kinetics. The latter observation is consistent with a kinetic scheme in which the antibody and its substrate form a Michaelis-type complex prior to reaction. In aqueous buffer the observed rate accelerations over the uncatalyzed hydrolyses were typically *ca.* 10^3-fold. In one particularly favorable case,[29] a k_{cat} value of 20 s^{-1} for cleavage of an aryl ester was measured, corresponding to a rate enhancement greater than 6 million fold.

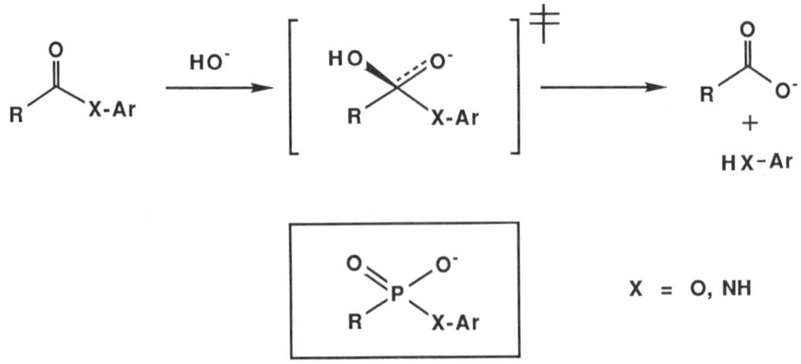

Figure 4. Hydrolysis of an ester or amide occurs through a tetrahedral transition state. Phosphonate esters or phosphonamidates represent stable analogs of this high energy species.

A number of controls verified that the results obtained with the antibodies were not artifactual. Heat denaturation of the immunoglobulins abolished catalytic activity, and chemical modification studies of several systems suggested the presence of specific amino acid side chains in or near the catalytic site. The finely tuned substrate selectivity exhibited by the antibody catalysts provided further compelling evidence that the observed activity was due to the induced binding pockets. For instance, esterolytic antibodies were able to discriminate between substrates bearing acetyl and trifluoroacetyl groups[27] and between *o*-, *m*-and *p*-substituted phenyl esters.[28,30] Furthermore, unrelated antibodies with no affinity for the target substrate did not promote substrate cleavage.

Detailed structural and mechanistic studies have not yet been carried out on the esterase antibodies. However, pH-rate data[27,30] suggest that the least efficient catalysts simply bind their substrate and polarize its scissile carbonyl, thereby activating the molecule for attack by hydroxide. The combining site of these proteins may thus resemble the "oxyanion hole" of serine proteases. In fact, a genetically engineered variant of the bacterial protease subtilisin, which lacks the catalytic triad, still hydrolyzes *p*-nitroanilide peptide substrates 10^3-fold faster than the spontaneous background reaction.[31] The rate enhancements seen in these cases may reflect the intrinsic contribution to catalysis to be expected from transition state stabilization alone. Consistent with this interpretation, the transition state analog used to immunize was generally found to be a competitive inhibitor of hydrolysis. Values of K_i were typically 10^2-10^4 fold smaller than the K_m values for the structurally analogous substrate molecule, and the ratio of K_i to K_m could often be roughly correlated with the observed rate accelerations.[27,28] With the most efficient esterolytic antibody, on the other hand, general acid/base catalysis may further augment the observed rate. A catalytic group with a pK_a of ~9 was detected in the pH-rate profile, and preliminary chemical modification studies[29] have implicated a catalytically essential active site tyrosine residue.

Although the first esterolytic antibodies to be reported operated only on activated aryl ester substrates, recent studies[32] demonstrate that immunoglobulins can also be prepared that hydrolyze unactivated alkyl esters. The properties of these antibodies resemble those of naturally occurring lipases and consequently may have considerable practical and commercial value.

2. Lactonizations

Enzymes frequently employ catalytic groups (i.e. general acids, general bases and nucleophiles) to enhance the rate of catalysis. While evidence suggests that specific amino acid side chains can be elicited in an antibody combining site complementary to charged or hydrophobic residues in the hapten, the requirements for tight binding and catalysis are not necessarily the same. Especially for energetically difficult reactions for which several catalytic groups may be necessary, the probability of generating the necessary constellation of side chains during immunization may be low. The early attempts[19,20] to create a transaminase antibody discussed above, for instance, may well have failed because no provision was made for the general base that is required to effect the protonations and deprotonations essential for efficient transamination.

The requirement that the antibody provide a specific and properly oriented catalytic group can be circumvented if the substrate carries the necessary functionality into the combining site every time it binds ("substrate assisted catalysis").[33] An example of substrate-assisted catalysis in antibodies was provided by Napper et al.[34] A cyclic phosphonate ester was designed as a transition state analog for lactonization of phenyl 6-acetamido-5-hydroxyhexanoate (Figure 5). One of 24 antibodies that bind the hapten enhanced the rate of the targeted intramolecular cyclization by a factor of 170. The observed rate acceleration probably reflects the ability of the antibody combining site to restrict the degrees of freedom available to the bound substrate and thereby reduce rotational entropy. Since the entropy of activation will increase as the size of the product lactone increases, even larger rate enhancements might be achieved for antibody-catalyzed macrolactonizations. Given that many important antibiotics and therapeutic agents are macrocyclic lactones, immunoglobulins that promote macrocyclizations could have practical utility.

Rate accelerations are only one aspect of enzyme catalyzed reactions. In practical terms, the ability of an enzyme to distinguish between closely related substances and steer a reaction along a single regio- and stereochemical pathway is far more interesting and important. Antibodies are chiral molecules and might therefore be expected to exert considerable stereochemical control over the reactions they promote. Use of a racemic hapten, the cyclic phosphonate ester in Figure 5, allowed this notion to be examined. In reactions with the induced catalytic antibody only 50% of racemic δ-hydroxyacid was converted to product.[35] ^{1}H-NMR spectroscopy of the δ-lactone in the presence of a chiral lanthanide *shift reagent confirmed that it was a single enantiomer. The lactone could be isolated with an enantiomeric excess of about 94%, demonstrating the potential for using antibodies to promote reactions that require stereochemical control.

3. Amide-Cleaving Antibodies

As amide hydrolysis is energetically more demanding than ester hydrolysis, its catalysis by antibodies represents a formidable challenge. Phosphonamidates, like phosphonates, have been used as transition state analog inhibitors of hydrolytic enzymes and are presumed to mimic the tetrahedral intermediates, or transition states, that occur along the reaction coordinate for amide bond cleavage (Figure 4). In recent experiments, one of 44 antibodies raised against an arylphosphonamidate was shown to promote the hydrolysis of the structurally analogous *p*-nitroanilide.[36] The catalyzed reaction occurred with high specificity and a 250,000-fold rate acceleration relative to the uncatalyzed process.

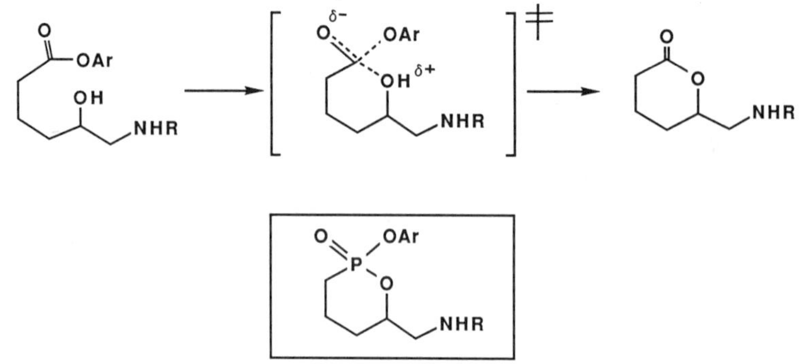

Figure 5. Conversion of a hydroxy ester into a δ-lactone proceeds through a product-like transition state. The cyclic phosphonate ester is a stable mimic of this species.

The phosphonamidate used to immunize was shown to be a tight binding inhibitor of the reaction with a K_i value of roughly 10 μM.[35] Since the K_m value for the substrate is 370 μM, less than two orders of magnitude of the 10^5-fold rate acceleration can be directly attributed to transition state stabilization. Chemical catalysis presumably accounts for the remainder. The pH rate profile for the reaction of the nitroanilide with antibody has a sigmoidal shape with a plateau above pH 9, indicating the presence of an ionizable residue in the active site. This group, which could participate in acid-base catalysis, might be a tyrosine as found in the active site of the most efficient esterolytic antibody.[28]

Because of the relatively low pK_a of its leaving group, a *p*-nitroanilide is substantially easier to cleave than the amide bond found in peptides. Anti-phosphonamidate antibodies have not yet been prepared that cleave unactivated amides, and, because of the difficulty of the reaction, the probability of isolating efficient antibody amidases using such haptens may be low. An alternate strategy takes advantage of extensive modeling studies which show that the rate of amide cleavage is substantially enhanced by metal ion catalysis.[36-38] Recently, an antigen containing a peptide-like fragment complexed to a substitutionally inert metal complex was prepared and used to generate antibodies that cleave unactivated amide linkages.[39]

Iverson and Lerner induced a number of monoclonal antibodies with a Co(III)(trien)-peptide hapten.[39] When reconstituted with a trien metal complex, these antibodies promoted the hydrolysis of the Gly-Phe bond of peptide substrates at neutral pH (Figure 6). A trien complex of Zn(II) was the optimal metal cofactor, giving a turnover number of 6×10^{-4} s^{-1}. Interestingly, trien complexes of numerous other metals, including Ga(II), Fe(II), In(III), Cu(II), Ni(II), Lu(II) and Mn(II), were also effective. This strategy is currently being extended to develop proteolytic antibodies that can catalyze sequence specific polypeptide hydrolyses. Such molecules could have considerable importance in research and medicine as "restriction enzymes" for proteins or as agents to dissolve blood clots or attack disease causing microorganisms.

4. Acyl Transfer to Amines

A true catalyst does not change the equilibrium position of a chemi-

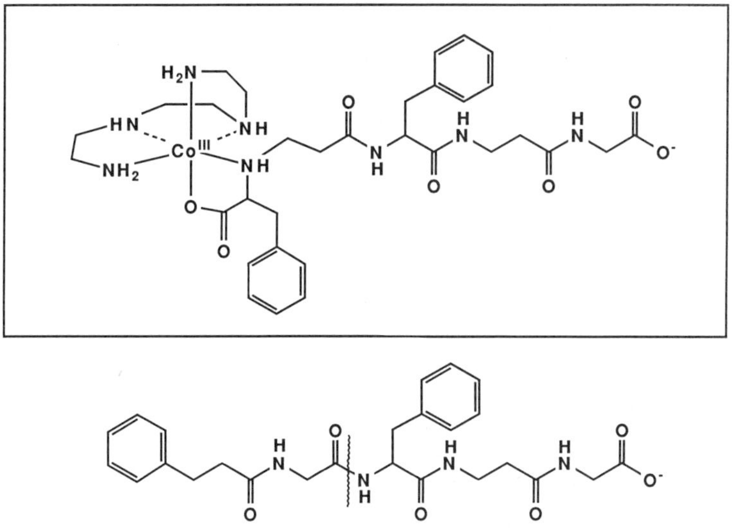

Figure 6. In the presence of trien metal complexes antibodies raised against the boxed Co(III)-peptide complex cleave the related peptide substrate at the indicated Gly-Phe bond.

cal reaction; it must accelerate both the forward and reverse reactions to an equal extent. Therefore, antibodies designed to hydrolyze amides and esters could be employed, in principle, to synthesize amides and esters. Efficient catalysts of acyl transfer would be useful, for example, as peptide ligases[40] for the total synthesis of polypeptide hormones and large proteins. Two reports describing antibody catalyzed acyl transfer to amines have appeared.

The compound designed as a mimic for lactonization of phenyl 6-acetamido-5-hydroxyhexanoate[34] also resembles the anticipated transition state for esterolysis or aminolysis of the product δ-lactone (Figure 7). In fact, the antibody that catalyzed the cyclization step was also shown to promote the stereospecific reaction between lactone and 1,4-phenylenediamine.[41] The kinetic mechanism of the bimolecular process involved random equilibrium binding of lactone and amine, and the observed turnover rate could be approximated from the measured difference between the binding of reactants and the transition state analog. The latter fact suggests that entropic factors are largely responsible for the observed rate acceleration, with minimal contribution from specific catalytic groups in the active site.

Catalysis of an acyl transfer reaction between an aryl ester and an amine has also been achieved using anti-phosphonamidate antibodies.[42] The immunoglobulins in question were induced with a quinoline phosphonamidate. Although the antibodies did not hydrolyze structurally analogous aryl amides, they facilitated both hydrolysis and aminolysis of a related aryl ester. Saturation kinetics were observed in the case of amide synthesis with both ester and amine substrate, but detailed kinetic studies of this system were not undertaken. An effective molarity of about 10.5 M was calculated from the value of k_{cat} at pH 8.0. Improved design of the transition state analog, possibly incorporating both leaving group and attacking nucleophile, may lead to even more impressive rate enhancements.

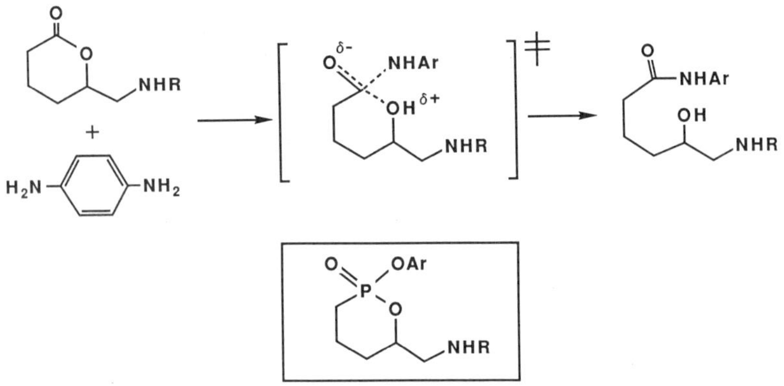

Figure 7. Aminolysis of a δ-lactone proceeds through a transition state similar in structure to that for lactone hydrolysis, mimicked successfully by the cyclic phosphonate ester in Figure 5.

ANTIBODIES FOR CONCERTED CHEMICAL REACTIONS

Shape-selective reactions that do not require the participation of specific binding site catalysts should be especially good targets for acceleration by antibodies. Natural selection drives the immune system to provide high affinity binding sites, but there is no special selection for activity. Induction of properly oriented catalytic groups within the active site is not impossible, as discussed above, but it remains an unpredictable and uncontrollable event, which presumably becomes less probable as the number of groups needed to act in concert increases. On the other hand, generation of a binding site with the approximate shape, or topology, for a chemical reaction will be a relatively easy task. Binding energy can be utilized directly to do chemical work: strain or distortion effects, proximity effects, and desolvation are all used be enzymes to lower reaction barriers and are made possible by substrate complexation.[1,16] These factors are exactly those that an antibody is likely to impart most effectively.

Concerted pericyclic reactions formally involve simple reorganization of electrons in the transition state. They do not require chemical catalysis but should be sensitive to strain, proximity and desolvation effects. They are, moreover, of enormous theoretical and practical interest to chemists and biologists alike. For example, in vitamin D metabolism, ergosterol undergoes electrocyclic ring opening to give precalciferol. Precalciferol, in turn, undergoes a 1,7-sigmatropic hydrogen atom shift to give vitamin D_2. On the other hand, sigmatropic rearrangements and Diels-Alder cycloadditions are among the most useful processes in the armamentarium of synthetic organic chemists for the preparation of complex natural products and therapeutic agents. We have targeted this class of transformations for catalysis by antibodies because of their practical utility and to develop unique tools to probe the role of strain, proximity and desolvation in protein catalyzed reactions.

1. Chorismate Mutase Antibodies

An example of a biologically relevant 3,3-sigmatropic process is the Claisen rearrangement of chorismate into prephenate (Figure 8). This transformation is the committed step in the biosynthesis of the aromatic amino acids phenylalanine and tyrosine in plants and lower organisms.[43] The enzyme chorismate mutase speeds it more than 2 million times over the background rate.[44] Although it is not understood how the enzyme achieves this large rate enhancement, chemical catalysis probably plays a minor role. Extensive inhibitor[45,46] and stereochemical[47] experiments suggested that the reaction proceeds through a transition state with pseudo diaxial geometry. Consistent with this notion, an oxabicyclic inhibitor molecule (Figure 8) designed by Bartlett and Johnson[48] to mimic this structure is currently the best inhibitor of the enzyme, binding roughly two orders of magnitude more tightly than chorismate itself. Significantly, this inhibitor also proved effective in generating antibodies that promote the stereospecific conversion of chorismate into prephenate.

We prepared 45 monoclonal antibodies against Bartlett's compound; two were shown to speed the rearrangement of (-)-chorismate substantially, relative to the spontaneous uncatalyzed process.[49] The observed rate enhancement (*ca.* 2 x 10^2 fold) resulted entirely from a reduction of the enthalpy of activation, suggesting the mechanistic importance of induced strain or distortion in catalysis. Independently, Bartlett, Schultz and their colleagues used the same inhibitor molecule to elicit an antibody with a roughly 10^4 fold rate acceleration for chorismate rearrangement, only two orders of magnitude less than that achieved by naturally occurring chorismate mutase.[50] This result is significant because it demonstrates that an imperfect transition state analog can be used to produce an antibody with substantial rate enhancements. It also underscores the importance of screening the immune system widely to identify catalysts with the desired specific activity.

We probed the stereospecificity of the chorismate mutase antibodies in parallel studies with racemic chorismate and the optically pure (-)-isomer.[51] Under conditions in which all of (-)-chorismate is converted to prephenate, half of the racemate rearranges rapidly and the remainder at the background rate. These results, together with steady state kinetic measurements, indicated that (+)-chorismate was not a substrate for the antibody. Studies with optically pure (+)-chorismate, obtained by kinetic resolution of the racemate in the presence of immunoglobulin, established

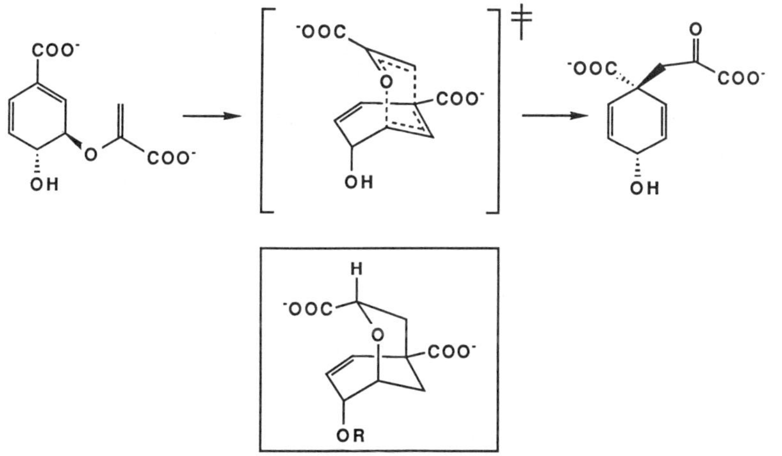

COO⁻
O
COO⁻
OH
⁻OOC
O
COO⁻
OH
‡
⁻OOC
O
COO⁻
OH
H
⁻OOC
O
COO⁻
OR

Figure 8. Chorismate rearranges into prephenate via a transition state with diaxial chair-like geometry. A stable oxabicyclic molecule (box) effectively mimics this species.

that the enantioselectivity of the catalytic antibody was greater than 90:1 at low substrate concentrations. Given that the oxabicyclic hapten was racemic, antibodies specific for rearrangement of the (+)-isomer might also have been induced in the immune response. None of the antibodies in the particular fusion that we tested, however, were able to utilize this material as a substrate. Future fusions may provide such a catalyst. Again, efficient and broad screening of the immune response will be generally necessary for identifying antibodies with the desired properties - high specific activity, substrate specificity and stereoselectivity.

The preparation of chorismate mutase antibodies demonstrates the feasibility of using rationally designed immunogens to produce antibodies that catalyze concerted carbon-carbon bond forming reactions. It now seems all the more likely that this strategy can be extended to catalyze other processes in this important class of concerted chemical reactions. For example, additional sigmatropic processes, including Cope rearrangements, and other sigmatropic shifts should be amenable to catalysis by this approach.

2. Photochemical Reactions Promoted by Antibodies

A photochemical [2+2] reaction is another example of a pericyclic process: two olefins combine in the excited state to give a cyclobutane adduct. This is illustrated by the photodimerization of methyl *p*-nitro-*trans*-cinnamate to give four isomeric cyclobutane derivatives (Figure 9). Balan et al., used one of the products of this reaction to elicit polyclonal antibodies.[52] Although this material is not strictly a transition state analog, the hope was that a combining site would be induced that could serve as a template for bringing the two reactants together. They found that the rate of *p*-nitrocinnamate disappearance was enhanced in the presence of antibody, and the specific geometric isomer used as the immunizing hapten was preferentially formed. However, the effects were modest, perhaps reflecting the fact that a polyclonal mixture of antibodies was employed in the experiment. In addition, the reactions were carried out with approximately stoichiometric amounts of immunoglobulin. Efficient turnover does not occur in this case since the product of the reaction was the immunizing hapten and binds tightly to the antibody once it is formed.

The reverse of a [2+2] cycloaddition reaction has also been facilitated by antibodies. Schultz and co-workers reported the photosensitized cleavage of a thymine cyclobutane dimer with immunoglobulins raised

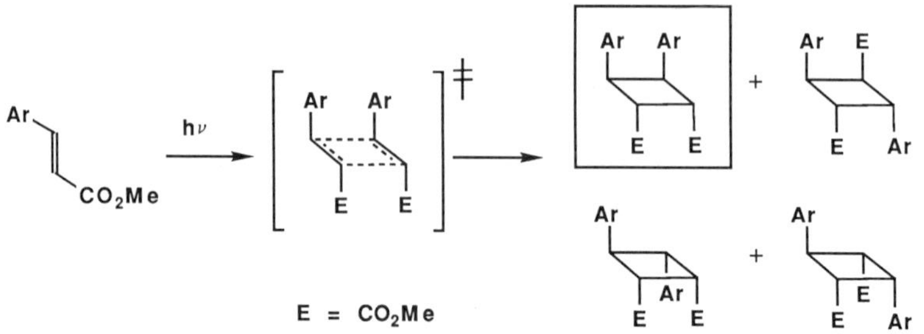

E = CO_2Me

Figure 9. Photodimerization of methyl *p*-nitro-*trans*-cinnamate gives four isomeric cyclobutanes. The transition state can be mimicked by one of the products.

against the thymine dimer isomer that is the major photolesion in DNA (Figure 10).[53] Five of six antibodies tested promoted photocleavage at 300 nm. In contrast to the previous example, product inhibition does not obviate turnover, since the substrate (hapten) is cleaved during reaction into two pieces which presumably have lower affinity for the combing site. The Michaelis constants k_{cat} and K_m for one of the induced antibodies were determined to be 1.2 min^{-1} and 6.5 μM, respectively. For comparison, the turnover number for *E. coli* DNA photolyase, an important naturally occurring enzyme involved in DNA repair, is 3.4 min^{-1}. Evidence was presented that a tryptophan residue in the antibody photosensitizes the reaction of the bound substrate.

3. Diels-Alderase Antibodies

The Diels-Alder reaction involves concerted [4 + 2] cycloaddition of a conjugated diene to an olefin to give a cyclohexene derivative.[69] Although unknown in nature, it is one of the most useful carbon-carbon bond forming reactions available to organic chemists, exploited extensively in the preparation of complex natural products, therapeutic agents and synthetic materials of all kinds. The bimolecular version of the Diels-Alder has a substantial entropic barrier (typical in the range of -30 to -40 cal K^{-1} mol^{-1}).[69] By binding the two substrates together in the active site of an antibody, we hoped to "pay for" this large loss in translational and rotational entropy and thereby accelerate the reaction greatly.

We recently succeeded in catalyzing the bimolecular Diels-Alder reaction depicted in Figure 11.[70] Tetrachlorothiophene dioxide and N-ethylmaleimide undergo [4 + 2] cycloaddition to give a bicyclic adduct that rapidly eliminates sulfur dioxide yielding dihydro-(N-ethyl)-tetrachlorophthalimide as the final product. The latter is further oxidized under the reaction conditions to tetrachlorophthalimide. The Diels-Alder reaction has a highly ordered transition state structure that resembles products more than substrates, and we chose a hexachlororbornene derivative as our hapten for eliciting antibodies (Figure 11). This material mimics the initially formed product of the cycloaddition, the unstable bicyclic intermediate, with a CCl_2 moiety replacing the SO_2 bridge. We expected that antibodies generated against this compound would have the correct shape for binding the two substrate molecules, together in the proper orientation for reaction. In addition, because the norbornene has a very different shape than the final products of the reaction (phthalimide and SO_2), complications due to product inhibition would be minimized.

A number of antibodies raised against the hexachloronorbornene hapten substantially accelerated the reaction shown in Figure 11 (after methylation of the surface lysine residues to prevent undesired side reactions). The best of these, 1E9, was investigated in some detail.[70] Catalysis was severely inhibited by the transition state analog, confirming that the reaction takes place in the induced binding pocket. The catalyst was shown to undergo multiple turnovers without significant product inhibition, vindicating the hapten design. The effects of minor structural variation in the substrates on reaction rate were also in accord with expectations based on the hapten structure. Thus, the alkyl side chain on the maleimide is vital, with the catalyzed rates increasing in the order: H$\ll$ methyl$<$ ethyl$<$ propyl$\approx$ butyl. Finally, kinetic analysis revealed that 1E9 was quite efficient, with an effective molarity of at least 110 M per binding site. The true effective molarity must be substantially higher than this value, but low solubility of the tetrachlorothiophene dioxide

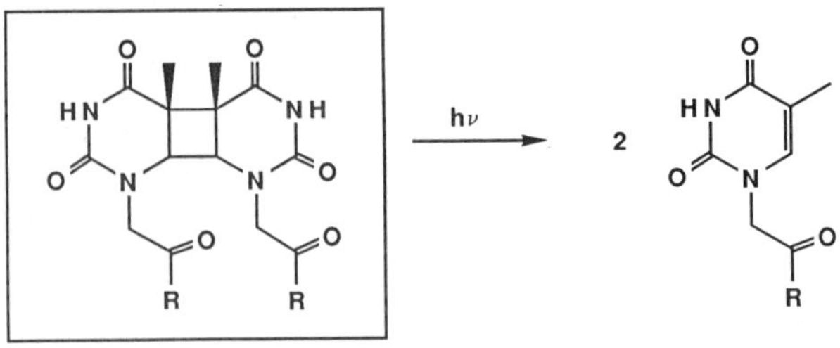

Figure 10. Photocleavage of a thymine dimer is the reverse of the transaction illustrated in Figure 9. Again, the transition state resembles the dimer more closely than the monomer.

in aqueous buffer prevented its determination.

The antibody accelerated Diels-Alder reaction provides an excellent demonstration of the feasibility of exploiting the immune system to catalyze important non-physiological processes. Future studies on Diels-Alder antibodies will certainly enhance our understanding of catalysis by approximation. Moreover, given the ubiquity of Diels-Alder reactions in organic synthesis, extension of the design strategy to other [4 + 2] cycloadditions, and the development of second-generation transition state analogs, is now necessary. Success in this endeavor will have significant practical ramifications. In particular, use of antibodies to influence the regioselectivity of Diels-Alder reactions, or to induce asymmetry into products obtained, has great promise.

COFACTOR DEPENDENT CATALYSIS BY ANTIBODIES

Vitamins and cofactors are used by enzymes to catalyze a wider variety of reactions than is possible with the protein side chains alone. Important prosthetic groups include nicotinamides, flavins, pterins, porphyrins, cobalamine, pyridoxal and thiamin. Many of these same agents could be employed in conjunction with antibody combining sites to effect selective redox and other reactions. In fact, a number of antibody-cofactor complexes have been known for some time. Vitamin K_1OH, for example, binds to the combining region of Fab New, and the crystal structure of the resulting complex has been resolved.[15] Anti-riboflavin antibodies were isolated from a patient suffering from multiple myeloma.[55] However, the catalytic properties of these immunoglobulin-cofactor complexes have never been investigated.

The early work of Raso and Stollar[19,20] has already been mentioned as an intentional attempt to create a cofactor-dependent antibody catalyst using a pyridoxal derivative as the hapten. More recently, researchers have turned to some of the redox cofactors. For example, an immune response has been elicited with a synthetic flavin analog.[55] The resulting monoclonal antibodies modulated the redox potential of the flavin by preferentially binding the oxidized over the reduced dihydro form, making redox processes possible that are not thermodynamically accessible to free flavin. In another study,[56] monoclonal antibodies were prepared against Co(III) and Fe(III) complexes of *meso*-tetra-(4-carboxyphenyl)-porphine. While the chemical reactivity of the antibody-porphyrin complexes has not yet been elucidated, interesting metal binding specificities were noted.

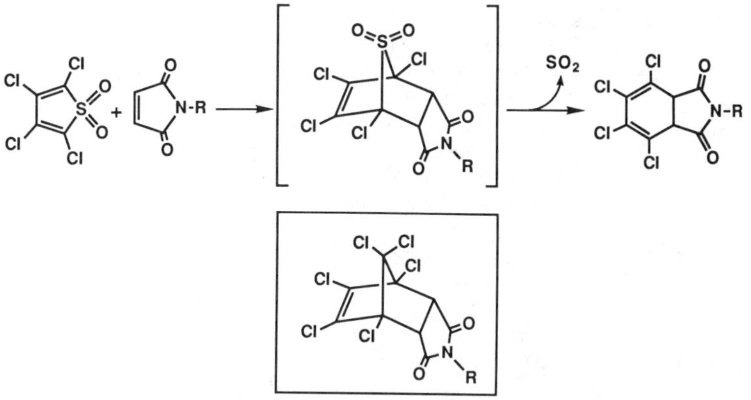

Figure 11. The Diels-Alder cycloaddition of tetrachlorothiophene dioxide and N-alkylmaleimides yields an unstable bicyclic intermediate that subsequently eliminates sulfur dioxide to give a dihydrophthalimide. The hexachloronorbornene (box) mimics the bicyclic adduct which in turn resembles the transition state structure for the Diels-Alder reaction.

The utilization of vitamins, cofactors and metals to increase the number of reactions accessible to antibody combining sites has a great deal of potential. In many cases, steps will have to be taken to insure that the induced binding pocket can complex the cofactor <u>and</u> the targeted substrate, and additional catalytic groups may be necessary to facilitate essential proton transfers, etc. These demands will require the design of sophisticated transition state analogs. However, the practical importance of many cofactor-dependent reactions, and the fact that convenient chemical routes to the same products are often unavailable, will make such an exercise worthwhile.

PERSPECTIVES AND PROSPECTS

The work reviewed in the foregoing sections amply demonstrates the feasibility of preparing tailored antibody catalysts. It seems probable that many more chemical transformations will prove amenable to catalysis by immunoglobulins. Site-specific cleavage of peptides and proteins, peptide ligation, hydrolysis and synthesis of polysaccharides and phosphate esters, preparation of synthetic polymers, aldol condensations, stereospecific cation and radical cyclizations, asymmmetric Diels-Alder cycloadditions and other pericyclic processes, as well as many cofactor mediated reactions represent only some of the transformations that are likely to be developed in the next few years.

Catalytic antibody technology thus represents a powerful and potentially general strategy for designing new biocatalysts. As such, it complements other current approaches to this problem, including (1) total synthesis of proteins from their constituent amino acids;[57,58] (2) chemical modification of existing protein binding pockets;[6,7] (3) site-directed mutagenesis of proteins using recombinant DNA techniques;[3,4] (4) random mutagenesis on microorganisms and selection for "improved" enzymes;[2,59] and (5) development of low molecular-weight artificial enzymes based on cyclodextrins[60] or purely synthetic binding cavities.[61] The development of catalytic antibodies is more versatile and, in many respects, easier than many of these alternate schemes. As we have seen, it is relatively straightforward to obtain a first generation binding pocket with modest catalytic activity if a suitable transition state analog can be designed and synthesized. Hence, the catalytic antibody approach is not limited by the availability of a pre-existing active site with the desired substrate selectivity or by our ability to construct such sites from scratch.

While the field of catalytic antibodies has generated considerable excitement in the scientific community, the general implementation of the transition state analog approach to the production of practical catalysts is still restricted by a number of factors. Large rate accelerations have been achieved with some catalytic antibodies, but even the best cases fall several orders of magnitude short of the enhancements obtained with analogous enzymes. Since all conceivable transition state analogs are inherently imperfect mimics, antibodies complementary to them are unlikely to provide optimal stabilization of true transition states. Consequently, the skill of the chemist in devising suitable transition state analogs ultimately limits the kinds of catalytic antibodies that can be made.

Other problems with the current methodology for making catalytic antibodies have already been touched upon. Severe inhibition can result if the immunogen resembles the substrate or product of the targeted reaction too closely, preventing effective turnover of the catalyst. The difficulty of generating an optimal constellation of catalytic groups (general acids, general bases, nucleophiles) within the antibody binding site during the immunization event is another significant problem. Catalytic residues will be especially important in energetically more demanding reactions than those examined to date. While effective screening of large numbers of hybridomas could overcome some of these limitations, the numbers of molecules that can be practically assayed by conventional techniques is small compared to the enormous diversity of the immune system. Identification of immunoglobulins with more than modest activity for many of the more interesting, and difficult, chemical transformations therefore remains a major challenge. The improvement of first generation antibody catalysts by genetic engineering represents another.

Future efforts will certainly be directed toward making existing antibody catalysts better. A detailed understanding of the structure and function of the first generation catalysts will be required to guide these attempts. In depth characterization of a number of the antibodies described above is currently underway. As with existing enzymes, chemical modification,[62] DNA-directed mutagenesis,[63,64] and genetic selection[58] can all be employed to improve the chemical efficiency of catalytic antibodies. Recent successes with the expression of antibodies in microorganisms will facilitate such studies.[65-68] A truly multidisciplinary approach to catalyst design - involving chemistry, immunology, molecular biology and genetics - promises to be intellectually exciting and extremely powerful. Most importantly, it will provide the means to create more sophisticated catalysts than is currently possible for use in research, industry and medicine.

Although the field of catalytic antibodies is still in its infancy, numerous practical applications of these molecules can be envisaged. Catalytic antibodies have enormous potential as research tools for enhancing our understanding of the basic mechanisms by which proteins bind ligands and catalyze chemical reactions. Using carefully selected model systems, for example, it may be possible to assess the relative importance of strain, proximity and desolvation in catalysis. Detailed biochemical and structural characterization of specific antibodies may ultimately lead to general rules that relate protein structure and function. Comparison of catalytic antibodies resulting from different fusions may shed light on the multitude of individual "solutions" to the topological problem posed by a particular chemical transformation and yield insights into the evolution of catalytic efficiency.

Possible medical applications of catalytic antibodies include their use as therapeutic or diagnostic agents. For example, antibodies might be fashioned that dissolve blood clots or that attack and destroy viruses or other disease causing organisms. Because of their high specificity, catalytic antibodies could be targeted against cancer cells or used to repair enzyme deficiencies in cells. Immunoglobulin catalysts could also be fashioned to destroy environmentally harmful pesticides or other toxins.

The industrial production of fine chemicals and synthetic materials of all kinds requires catalysts with exacting substrate specificity and stereoselectivity. These are exactly the features that distinguish the best catalytic antibodies. Highly selective immunoglobulin catalysts might therefore be effective in the commercial preparation of drugs, hormones and other molecules. In organic synthesis, for instance, they could be used to prepare chiral synthons by kinetic resolutions of racemic mixtures. Catalytic antibodies could also be employed to carry out especially difficult or expensive steps in a commercial total synthesis.

The development of catalytic antibodies represents a new frontier in enzyme engineering, and much progress can be expected in the next few years. Ultimately, this technology may indeed bring us a little closer to our goal of being able to tailor practical protein catalysts for virtually any chemical transformation.

ACKNOWLEDGMENT

This work was supported in part by the American Cancer Society (JFRA-45195) and the National Institutes of Health (GM38273).

REFERENCES

1. W. P. Jencks, Adv. Enzymol., **43**, 219 (1975).
2. H. Yamada & S. Shimizu, Angew. Chem. Int. Ed. Engl., **27**, 622 (1988).
3. J. A. Gerlt, Chem. Rev., **87**, 1079 (1987).
4. J. R. Knowles, Science, **236**, 1252 (1987).
5. W. V. Shaw, Biochem. J., **246**, 1, (1987).
6. D. Hilvert & E. T. Kaiser, Biotech. & Gen. Eng. Rev., **5**, 297 (1987).
7. E. T. Kaiser, D. S. Lawrence & S. E. Rokita, Ann. Rev. Biochem., **54**, 565 (1985).
8. D. Pressman & A. Grossberg, "*The Structural Basis of Antibody Specificity*," Benjamin, New York, 1968.
9. A. Nisonoff, J. Hopper & S. Spring, "*The Antibody Molecule*," Academic Press, New York, 1975.
10. E. A. Kabat, "*Structural Concepts in Immunology and Immunochemistry*," Holt, Reinhart and Winston, New York, 1976, p 227.
11. F. W. Alt, T. K. Blackwell & G. D. Yancopoulos, Science, **238**, 1079 (1987).
12. K. Rajewsky, I. Förster & A. Cumang, Science, **238**, 1088 (1987).
13. F. R. Seiler, P. Gronski, R. Kurrle, G. Lueben & H. P. Harthus, Angew. Chem. Int. Ed. Engl., **24**, 139 (1985).

14. L. Pauling, Amer. Sci., **36**, 51 (1948).
15. L. M. Amzel & R. J. Poljak, Ann. Rev. Biochem., **48**, 961 (1979).
16. W. P. Jencks, "*Catalysis in Chemistry and Enzymology,*" McGraw Hill, New York, 1969, p 288.
17. F. Kohen, J.-B. Kim, G. Barnard & H. R. Lindner, Biochim. Biophys. Acta, **629**, 328 (1980).
18. F. Kohen, J. B. Kim, H. R. Lindner, Z. Eshar & B. Green, FEBS Lett., **111**, 427 (1980).
19. V. Raso & B. D. Stollar, J. Am. Chem. Soc., **95**, 1621 (1973).
20. V. Raso & B. D. Stollar, Biochemistry, **14**, 584 (1975).
21. R. Wolfenden, Accts. Chem. Res., **5**, 10 (1972).
22. G. E. Lienhard, Science, **180**, 149 (1973).
23. R. Wolfenden, Ann. Rev. Biophys. Bioeng., **5**, 271 (1976).
24. G. Köhler & C. Milstein, Nature (London), **256**, 505 (1975).
25. P. A. Bartlett & C. Marlowe, Biochemistry, **22**, 4618 (1983).
26. S. A. Bernhard & L. Orgel, Science, **130**, 625 (1959).
27. A. Tramontano, K. D. Janda & R. A. Lerner, Science, **234**, 1566 (1986).
28. J. Jacobs, P. G. Schultz, R. Sugasawara & M. Powell, J. Am. Chem. Soc., **109**, 2174 (1987).
29. A. Tramontano, A. A. Ammann & R. A. Lerner, J. Am. Chem. Soc., **110**, 2282 (1988).
30. S. J. Pollack, J. W. Jacobs & P. G. Schultz, Science, **234**, 1570 (1986).
31. P. Carter & J. A. Wells, Nature, **332**, 564 (1988).
32. K. D. Janda, S. J. Benkovic & R. A. Lerner, Science, **244**, 437 (1989).
33. P. Carter & J. A. Wells, Science, **237**, 394 (1987).
34. A. D. Napper, S. J. Benkovic, A. Tramontano & R. A. Lerner, Science, **237**, 1041 (1987).
35. K. D. Janda, D. Schloeder, S. J. Benkovic & R. A. Lerner, Science, **241**, 1188 (1988).
36. P. A. Sutton, D. A. Buckingham, Accts. Chem. Res., **20**, 357 (1987).
37. L. M. Sayre, J. Am. Chem. Soc., **108**, 1632 (1986).
38. J. T. Groves & R. R. Chambers, Jr., J. Am. Chem. Soc., **106**, 630 (1984).
39. B. L. Iverson & R. A. Lerner, Science, **243**, 1184 (1989).
40. C. F. Barbas, J. R. Matos, J. B. West & C.-H. Wong, J. Am. Chem. Soc., **110**, 5162 (1988).
41. S. J. Benkovic, A. D. Napper & R. A. Lerner, Proc. Natl. Acad. Sci. USA, **85**, 5355 (1988).
42. K. D. Janda, R. A. Lerner & A. Tramontano, J. Am. Chem. Soc., **110**, 4835 (1988).
43. U. Weiss & J. M. Edwards, "*The Biosynthesis of Aromatic Amino Compounds,*" Wiley, New York, 1980, pp. 134-184.
44. P. R. Andrews, G. D. Smith & I. G. Young, Biochemistry, **18**, 3492 (1973).
45. P. R. Andrews, G. D. Smith & I. G. Young, Biochemistry, **16**, 4848 (1977).
46. H. S.-I. Chao & G. A. Berchtold, Biochemistry, **21**, 2778 (1982).
47. S. G. Sogo, T. S. Widlanski, J. H. Hoare, C. E. Grimshaw, G. A. Berchthold & J. R. Knowles, J. Am. Chem. Soc., **106**, 2701 (1984).
48. P. A. Bartlett & C. R. Johnson, J. Am. Chem. Soc., **107**, 7792 (1985).
49. D. Hilvert, S. H. Carpenter, K. D. Nared & M.-T. M. Auditor, Proc. Natl. Acad. Sci. USA, **85**, 4953 (1988).
50. D. Y. Jackson, J. W. Jacobs, R. Sugasawara, S. H. Reich, P. A. Bartlett & P.G. Schultz, J. Am. Chem. Soc., **110**, 4841 (1988).
51. D. Hilvert & K. D. Nared, J. Am. Chem. Soc., **110**, 5593 (1988).
52. A. Balan, B. P. Doctor, B. S. Green, M. Torten & H. Ziffer, J. Chem. Soc., Chem. Commun., 106 (1988).
53. A. G. Cochran, R. Sugasawara & P. G. Schultz, J. Am. Chem. Soc., **110**, 7888 (1988).
54. M. Farhangi & E. F. Osserman, N. Engl. J. Med., **294**, 177 (1976).

55. K. M. Shokat, C. J. Leumann, R. Sugasawara & P. G. Schultz, Angew. Chem. Int. Ed. Eng., **27**, 1172 (1988).
56. A. W. Schwabacher, M. I. Weinhouse, M.-T. M. Auditor & R. A. Lerner, J. Am. Chem. Soc., **111**, 2344 (1989).
57. W. F. DeGrado, Z. R. Wasserman & J. D. Lear, Science, **243**, 622 (1989).
58. E. T. Kaiser, Accts. Chem. Res., **22**, 47 (1989).
59. A. L. Demain, Science, **214**, 987 (1981).
60. R. Breslow, Science, **218**, 532 (1982).
61. H. Dugas & C. Penny, "*Bioorganic Chemistry,*" Springer-Verlag, New York, 1981.
62. S. J. Pollack, G. R. Nakayama & P. G. Schultz, Science, **242**, 1038 (1988).
63. M. Verhoeyen, C. Milstein & G. Winter, Science, **239**, 1534 (1988)
64. S. Roberts, J. C. Cheetham & A. R. Rees, Nature (London), **328**, 731 (1987).
65. A. Skerra & A. Plückthun, Science, **240**, 1038 (1988).
66. M. Better, C. P. Chang, R. R. Robinson & A. H. Horwitz, Science, **240**, 1041 (1988).
67. A. H. Horwitz, C. P. Chang, M. Better, K. E. Hellstrom & R. R. Robinson, Proc. Natl. Acad. Sci. USA, **85**, 8678 (1988).
68. J. R. Carlson, Mol. Cell. Biol., **8**, 2638 (1988).
69. J. Sauer, Angrw. Chem. Int. Ed. Engl., **5**, 271 (1966).
70. D. Hilvert, K. W. Hill, K. D. Nared & M.-T. M. Auditor, J. Am. Chem. Soc., **111**, 9261 (1989).

GENERATION OF CATALYTIC ACTIVITY BY PROTEIN MODIFICATION

Melvin H. Keyes and David E. Albert

Anatrace, Inc.
1280 Dussel Drive
Maumee, OH 43537

A unique method to generate catalytic activity is described. The method consists of perturbation of a host protein, addition of a modifier, and crosslinking of the conformationally modified protein. The preparation and properties of three types of catalytic conformationally modified proteins (CCMP) are described. CCMP with amino acid esterase, glucose isomerase, and fluorohydrolase activities have been prepared from several host proteins, crosslinking reagents and modifiers. Although a fundamental understanding of this process is not known, the ease of preparation and unique properties of CCMP suggest that they will find numerous applications in biotechnology.

INTRODUCTION

Enzymes have a long history of use in processing biochemicals and as reagents in the clinical laboratory. The uses of enzymes have been extended by the numerous techniques now available to derivatize the enzyme molecule. In the clinical arena, enzyme conjugates allow the development of enzyme immunoassays which are being utilized in a wide variety of tests. In addition enzymes are immobilized for use in several analytical tests and incorporated into biosensors. Some applications of immobilized enzymes exist in the processing of biochemicals, such as, the conversion of glucose to fructose in the production of high fructose corn syrup.

It is apparent to those knowledgeable in the field that techniques to extract, purify, utilize and derivatize enzymes are readily available. These techniques do not limit the use of enzymes as much as the narrow scope of reactions catalyzed by the naturally occurring, readily available enzymes. Of the approximately three thousand enzymes, only a handful are available commercially in large quantities.

Several methods are being developed to produce enzyme-like catalysts or to modify the properties of known enzymes. Protein engineering offers the advantage of being able to synthesize almost any amino acid sequence.[1-3] Of course considerable time and expense is required and the properties of the new "protein/enzyme" can seldom be predicted.

Biomimetic Polymers
Edited by C. G. Gebelein
Plenum Press, New York, 1990

Another intriguing method to generate catalytic activity is the modification of antibodies.[4-6] By using a model for the transition state, some antibodies have been prepared which possess catalytic activity. As in protein engineering, considerable effort and expense is required to produce these catalytic antibodies.

In this article the authors will discuss a conformational modification method for generating catalytic activity in proteins. The research efforts are based on a three step process. First the protein tertiary structure is perturbed followed by the addition of a modifier. With the protein in this conformation the structure is "frozen in" with bifunctional or multifunctional reagents.

During the last ten years, this relatively simple procedure has resulted in numerous examples of new catalytic activity in host proteins.[7-13] Three types of catalysts which have been studied in some detail will be discussed in this paper. Detailed methods of preparation are included along with the properties of these catalytic conformationally modified proteins (CCMP).

EXPERIMENTAL PROCEDURES

1. Materials

Bovine serum albumin (BSA), casein, glucose oxidase, hexokinase and sorbitol dehydrogenase were obtained from Sigma Chemical Co., St. Louis, MO. Concanavalin-A and glucose dehydrogenase were obtained from U.S. Biochemical Corp., Cleveland, OH. Bovine ribonuclease (RNase) was obtained from either Biozyme Laboratories International, San Diego, CA or Sigma Chemical Corp. Indole, indole-3-propionic acid (IPA), 5-hydroxyindole-3-acetic acid (HIAC), 5-hydroxyindole-3-acetamide (HIAA), 5-hydroxyindole-3-carboxylic acid (HICA), L-tryptophan agarose gel, alpha-N-benzoyl-L-arginine ethyl ester (BAEE) and L-tryptophan ethyl ester (L-TrEE) were purchased from the Sigma Chemical Company or U.S. Biochemical Corp. Hexamethyl phosphoramide was purchased from Fisher Scientific Company, Fairlawn, NJ. Dimethyl suberimidate dihydrochloride was obtained from Anatrace, Inc., Maumee, OH and 8% glutaraldehyde from Polysciences, Inc., Warrington, PA. Sephadex G-15 and Sepharose 4 B gels were purchased from Pharmacia Fine Chemicals, Piscataway, NJ. Bio-Gel A 1.5 m was obtained from Bio-Rad Laboratories, Richmond, CA. Glucose diagnostic kit No. 510, β-nicotinamide adenine dinucleotide, reduced form (NADH) and β-nicotinamide adenine dinucleotide phosphate (NADP) were also purchased from Sigma Chemical Corp., St. Louis, MO. All other chemicals were reagent grade.

2. Preparation of CCMP Procedures

2A. Amino Acid Esterases

2Aa. Bovine Serum Albumin (BSA). Semisynthetic amino acid esterases are generated by lowering the pH of a BSA solution (10 mg/mL) to 3. While gently stirring at room temperature 1 mL of indole (1 mg/mL) solution in methanol is added to 10 mL of the BSA solution. After 2 hours, the pH is slowly raised to 7.0 with dilute NaOH, and 30 µL of 8% glutaraldehyde is added. After reacting for 30 minutes at room temperature, the solution is cooled to 5°C and allowed to react for 17 hours.

One mL of this crosslinked modified BSA is applied to a column (2.5 x 30.5 cm) of G-15 Sephadex gel which has been equilibrated with 1mM Tris-HCl buffer, pH 7.0. Elution is accomplished with the same buffer at a flow rate of 1.0 mL per minute and the protein is detected by monitoring the absorbance at 280 nm. The modified protein is assayed for esterase activity using alpha-N-benzoyl-L-arginine ethyl ester (BAEE) as the substrate. The typical activity obtained is 1-10 U/g.

Another approach involves the modification of BSA bound to L-tryptophan-agarose gel. A small disposable glass column (1.5 x 7.5 cm) is packed with L-tryptophan-agarose gel. The column is equilibrated to pH 4.4 with 0.01M sodium acetate buffer and a 2 mL sample of 1% BSA is applied to the column at a flow rate of 1 mL/min. Approximately 3.4 mg of BSA is bound to the column.

The bound protein is crosslinked by circulating a (6.4×10^{-4}M) glutaraldehyde solution in 0.01M sodium acetate buffer, pH 4.4 through the column for 90 minutes. Next, the column is washed with 20 mL of 0.01M sodium acetate buffer, pH 4.4 to remove any excess glutaraldehyde from the solid support. The crosslinked modified protein is eluted with 0.02M glycine HCl buffer, pH 3.0 and assayed for activity using BAEE as the substrate.

Some of the CCMP have been prepared by crosslinking with dimethyl suberimidate instead of glutaraldehyde. The activity for the material reacted with glutaraldehyde was 6.8 U/g while with dimethyl suberimidate the initial activity was 17 ± 3 U/g.

<u>2Ab. RNase.</u> The initial procedure for generating CCMP from RNase consists of an aqueous solution (0.06%) of RNase adjusted to pH 3 and incubated at room temperature for one hour. To 100 mL of this solution is added 20 mg of IPA and the solution is stirred for nearly four hours, until the later is completely dissolved. The pH is then adjusted to 8.5 with 0.1N NaOH. The solution is cooled to 5°C, and 300 µL glutaraldehyde solution (8% solution packed in sealed vials under nitrogen) is added. The resulting solution is stirred overnight, then dialyzed for 48 hours against 5mM acetate buffer, pH 5.5, with 6 changes of dialysate. After dialysis, the modified protein is assayed for esterase activity as function of pH using L-tryptophan ethyl ester as the substrate. Two pH optima are observed: one at 6 and the other at 7.5.

The initial procedure is limited to about 1mM modifier concentrations due to the low solubility of IPA. By dissolving IPA in methanol prior to adding it to the RNase solution, the modifier concentration is increased from 1 to 6mM. When the concentration of IPA is 3mM, it is necessary to have approximately 10% methanol in the RNase solution; and for an IPA concentration of 6mM, the methanol concentration is 20%.

Another approach is the modification of RNase bound to L-tryptophan-agarose gel. A glass column (1.5 x 7.5 cm) is packed with 5 mL of L-tryptophan-agarose gel. The column is washed extensively with 1mM Hepes buffer, pH 8.4. Three mg of native RNase is applied and 2.5 mg bound to the solid support. This bound protein is crosslinked with 1.6 to 8.0 mM glutaraldehyde in 0.001M Hepes buffer, pH 8.4 for 2-17 hours at 5°C by circulating through the column in a closed loop fashion. To remove any excess glutaraldehyde, the gel is washed with 0.001M Hepes buffer, pH 8.4. The eluent is changed to 0.1M potassium phosphate buffer, pH 6.9 to elute the crosslinked modified RNase.

Another approach is the modification of RNase using 5-hydroxyindole derivatives that have much greater solubilities than indole-3-propionic acid. Each modifier; 5-hydroxyindole-3-acetamide, 5-hydroxyindole-3-acetic acid (HIAC) and 5-hydroxyindole-3-carboxylic acid (HICA) is evaluated using concentrations ranging from 1 to 20 mM. The modification procedure used is the same as described above when IPA is the modifier.

In addition to using a modifier that is more soluble than IPA, we have also investigated the pH of crosslinking with glutaraldehyde. Numerous CCMP have been prepared where crosslinking was performed from pH 3.0 to 8.5. After addition of the modifier 5-hydroxyindole-3-acetamide to the perturbed RNase solution, the protein is crosslinked with 2.4mM glutaraldehyde from pH 3 to pH 8.5. Except for the pH of crosslinking, all the other steps of the basic procedure are unchanged.

2B. Glucose Isomerase

2Ba. Apoglucose Oxidase. The apoenzyme of glucose oxidase is prepared by acid-ammonium sulfate precipitation.[14,15] After dialyzing extensively in 0.002M potassium phosphate buffer, pH 8, 300 mL of apoglucose oxidase solution (1 mg/mL) is modified by adding 1.5 grams of D-cellobiose and 1.0 gram of dimethyl suberimidate dihydrochloride. The solution is maintained at pH 7.0 and 5°C for 17 hours. This crosslinked modified apoglucose oxidase is then dialyzed against 2 mM potassium phosphate buffer, pH 8.0 and measured for glucose isomerase activity using D-fructose as the substrate.

Further purification is accomplished by ammonium sulfate fractionation and gel filtration on Bio-gel A 1.5M. The column eluent is 0.1M sodium phosphate buffer, pH 8.0 and the flow rate is 1.5 mL/min.

2Bb. Hexokinase. Hexokinase is also used as the starting material for preparing CCMP having glucose isomerase activity. The perturbation step for hexokinase consists of raising the temperature of the protein solution to 37°C and adding 2-mercaptoethanol to a final concentration of 2×10^{-3}M. After 15 minutes, a small aliquot of this perturbed protein (~ 10 mg) solution is then applied to a small disposable column (1.5 cm x 7.5 cm) packed with cellobiose - Sepharose 4B gel. The column is equilibrated to pH 8.0 with 0.01M sodium phosphate buffer.

Approximately 8.5 mg of perturbed hexokinase is bound to the solid support. The bound protein is crosslinked by circulating a solution of dimethyl suberimidate in 0.01M sodium phosphate buffer, pH 8.0, through the column in a closed loop fashion for 90 minutes. The column is washed with 0.01M sodium phosphate buffer, pH 8.0 to remove any excess crosslinking reagent. The crosslinked modified hexokinase is eluted from the solid support by washing with 0.1M sodium phosphate buffer, pH 8.0.

Hexokinase is also modified without using a solid support. The hexokinase (250 mg/25 mL of 0.1M tris-HCl buffer, pH 8) is perturbed by heating to 37°C for 15 minutes and adding 30 µL of 1.0M 2-mercaptoethanol. Next, cellobiose (200 mg) is added to the protein solution. A dimethyl suberimidate dihydrochloride (50 mg in 5 mL of deionized H_2O) is added slowly over a 10 minute period, and allowed to react at 5°C for 17 hours.

A sample of this modified protein is chromatographed using a G-15 Sephadex gel column. The resulting solution is assayed for glucose isomerase activity using D-glucose and fructose as substrates.

2C. Fluorohydrolases

2Ca. RNase. The first step of the modification procedure consists of applying one or two mL of a 1% RNase solution to a G-15 Sephadex gel column with 1 mM HCl or 5 mM phosphoric acid as the eluent. Next, hexamethyl phosphoramide (HMPA) is added to a concentration of 5 mM followed by the addition of 5 mM potassium phosphate buffer, pH 7.6. This modified protein is then assayed for activity using either diisopropyl fluorophosphate (DFP) or phenylmethylsulfonyl fluoride (PMSF). In initial screening experiments for fluorohydrolase activity a crosslinking reagent to fix or maintain the new conformation is not used since the modified protein does not lose activity during the assay procedure. However, for long term maintenance of the active conformation, crosslinking is required.

2Cb. BSA. BSA (10 mg dissolved in 3 mL of deionized H_2O) is applied to a G-15 Sephadex gel column which is equilibrated with 1 mM HCl or 5 mM phosphoric acid. The final concentration of the eluted BSA is adjusted to be 0.4 mg/mL. Next, HMPA is added to give a final concentration of 5 mM, followed by potassium phosphate buffer, pH 7.6. This modified protein is then assayed for activity using either DFP or PMSF. Just as with RNase (Section 2Ca), a crosslinking reagent to fix or maintain the new conformation is not used since the modified protein does not lose activity during the assay procedure.

2Cc. Casein. A 1% bovine casein (Sigma technical grade) solution is prepared by dissolving the protein in potassium phosphate buffer, pH 7.6. Two mL of this solution is applied to a G-15 Sephadex gel column (1.5 x 100 cm) and the protein eluted in 2mM potassium phosphate buffer, pH 7.2. The pH of the eluted solution is titrated to 3.0 with 10 mM phosphoric acid. The solution is diluted to a protein concentration of 0.4 mg/mL. Next, the perturbed protein is modified by adding 5 mM hexamethyl phosphoramide and after 5 minutes the pH is raised to 5.5. The modified protein is crosslinked by addition of glutaraldehyde to a concentration of 2.4 mM. After 17 hours at 5°C, the crosslinked modified protein is chromatographed on a G-15 Sephadex gel column equilibrated with 2 mM potassium phosphate buffer, pH 7.6. Activity is determined by using DFP and/or PMSF as substrates.

3. Assay Procedures

3A Amino Acid Esterases

A multipoint assay, using L-TrEE as the substrate, is performed by adding 100 µL aliquots of control and/or esterase reaction to the ethyl alcohol assay kit. The ethanol released is measured with alcohol dehydrogenase-NAD (ethyl alcohol assay kit, Sigma, 332-UV). Typically the reaction is monitored for 20-30 minutes at 25°C, and the data analyzed by linear regression. Specific activity (U/g) is determined by subtracting the slope of the control from the slope of the reaction mixture.

Another assay procedure is a high performance liquid chromatography (HPLC) method. By injecting 20 µL aliquots of either the control or esterase reaction into a stainless steel column packed with CPG/carboxyl controlled pore glass support (Pierce Chemical Corp.) the rate of reaction is monitored. The eluent is 0.001M Tris-HCl, pH 8.0 and the flow rate is 3 mL/min. The reaction is followed by using an U.V. detector (254 nm) to monitor the increases in the peak due to tryptophan. After several injections of both solutions, the data are analyzed by linear regression. The specific activity is determined by subtracting the slope of the control from the slope of the reaction mixture.

3B Glucose Isomerases

The measurement of glucose isomerase activity has been assayed by three enzymatic methods and one HPLC method. The activity can be measured using either fructose or glucose as the substrate. When glucose is used as the substrate, the formation of fructose can be monitored by its conversion to sorbitol. A control solution is prepared by mixing 0.5 mL of 0.2M potassium phosphate buffer, pH 8.0, 0.5 mL of dialysis solution and 0.5 mL of 50% D-glucose in deionized H_2O. The reaction solution contains the same reagents except that 0.5 mL of dialyzed semisynthetic glucose isomerase is added instead of the dialysis solution. After incubating at 60°C for 30 minutes; 0.25 mL of each of these solutions is added to 1.25 mL of 0.2M sodium phosphate buffer, pH 7.5 containing 0.2 mg of NADH and 4.0 units of sorbitol dehydrogenase. The decrease in optical density at 340 nm is measured for each solution.

The second assay method using glucose as the substrate is an HPLC method. A Bio-Rad 87C Aminex carbohydrate analysis column (300 x 7.8 mm) number 125-0095 separates fructose from glucose. The column is heated to 85°C and the eluent is degassed deionized H_2O. The flow rate is 0.6 mL/min. The fructose and glucose are detected with a refractive index detector (Erma Optical Works, Model ERC-7510).

With fructose as the substrate, the activity is measured by using either glucose dehydrogenase assay method or Sigma Glucose Kit No. 510. After incubating for 30 minutes at 60°C, 0.5 mL of either the control or the reaction mixture is added to 1.0 mL of 1.0M Tris-HCl, pH 8.0 containing 0.4 mg of B-NADP and 2.0 units of glucose dehydrogenase (GDH).

Using Sigma Glucose Kit No. 510, 0.25 mL of reaction and control solution is removed after 10, 20 and 30 minutes and added to 5.0 mL of test reagent from Kit No. 510, and the absorbance at 450 mn is measured on a spectrophotometer. For each time interval, the change in absorbance is calculated by subtracting the absorbance of the control from the absorbance of the reaction solution and then analyzed by linear regression. The slope is then used to calculate specific activity as units/gram protein.

3C. Fluorohydrolases

During all activity measurements, control and assay solutions are incubated at 30°C in a circulating water bath. The hydrolysis of DFP is monitored by measuring the formation of fluoride with an Orion fluoride electrode. The hydrolysis of PMSF is measured by the fluoride electrode as well as by high performance liquid chromatography. A stainless steel column (3 mm x 25 cm) packed with an amine support (Baker #7034-0) and equilibrated to pH 4.0 with 0.04 M sodium acetate buffer containing 0.1M NaCl is used to measure the hydrolysis product phenylmethylsulfonic acid (PMSA). The PMSA is detected at 220 nm with a UV monitor.

A typical control solution contains 100 mM MOPS (3-(4-Morpholino) propanesulfonic acid), pH 7.6, 5 mM potassium phosphate buffer, pH 7.6 and 5 mM DFP or PMSF in a final volume of 8 mL. The assay solution contains the same reagents except that modified protein solution is added.

AMINO ACID ESTERASES

The conformational modification process described in the introduction and detailed in the experimental section was first applied to the preparation of CCMP possessing activity towards amino acid esters. After some initial experiments showed a modification of the specificity of trypsin, both BSA and RNase were selected as host proteins and converted to amino acid esterases. CCMP from RNase was purified as previously described in the literature.[10,11] Only a summary of these results are described here.

1. Summary of Early Results

After preparation of the amino acid esterase from RNase, the modified protein was assayed for esterase activity as a function of pH using L-tryptophan ethyl ester as the substrate. Two pH optima were observed: one at 6 and the other at 7.5. Purification of the crude reaction mixture, as shown in Table 1, demonstrated the presence of two types of amino acid esterases which account for the two pH optima. After the conformational modification process, the measured activity of RNase was reduced. Purification of the crude mixture resulted in the disappearance of the native RNase activity entirely.

The two types of amino acid esterases generated from RNase differ not only in their pH optimum, but also in their specificity to esters. Table 2 shows the relative activity to several amino acid esterases. The neutral esterase has a broader specificity than does the acid esterase. Note also that the acid esterase catalyzes the hydrolysis of the D isomer of tryptophan ethyl ester nearly as well as the L isomer.

CCMP generated from RNase was purified by the same techniques that biochemists have used for decades to extract and purify enzymes from crude biological mixtures. Even more exciting was the finding that generated amino acid esterases differ in pH optimum, specificity, and molecular weight. Yet some difficulty was encountered preparing and purifying large quantities of these catalysts. The initial activity was low and the CCMP would often precipitate during purification.

Table 1. Purification of modified RNase.

Fractions	Activity (u/g) Acid esterase	Neutral esterase	RNase
Native RNase	0	0	3000
Crude IPA-RNase	5	7.5	180
Ammonium sulfate			
40%	0	21	380
70%	2	3	1160
90%	16	4	350
Biogel P-30:			
40% peak 1	-	550	0
peak 2	-	32	330
90% peak 1	30	-	-
peak 2	-	-	300
peak 3	600	-	0

Table 2. Substrate specificity of semisynthetic esterases.

Substrate	Relative activity (%) Neutral esterase	Acid esterase
L-TrEE	100	100
D-TrEE	5	70
BAEE	150	25
L-TEE	150	25
IAEE	80	0
LyEE	20	20
GIEE	20	20

2. Modifiers

To increase the initial activity, the concentration of the modifier, indole propionic acid, was increased. The concentration of indole propionic acid typically used was 1 mM which is nearly a saturated solution. To increase the concentration further, it was necessary to use a mixed solvent of methanol and water. As the concentrations of IPA and methanol increased, the initial activity increased. Larger increases in initial activity were achieved, however, by using modifiers, such as, 5-hydroxyindole-3-acetic acid and 5-hydroxyindole-3-acetamide which are more water soluble than IPA. Results to illustrate these points are listed in Table 3.

3. Affinity Column Method

Another approach to increase initial activity is the modification of RNase while bound to a L-tryptophan-agarose gel. A solution of RNase was

Table 3. Esterase activity of modified RNase.

Modifier	Conc. (mm)	Activity[3] (u/g)	# Preps. tested
IPA	1	0.4 ± 0.2	15
IPA[1]	3	0.9 ± 0.2	3
IPA[2]	6	1.8 ± 1.4	3
HIAC	20	3.6 ± 0.9	5
HIAA	10	6.4 ± 1.2	3

IPA = Indole-3-propionic acid
HIAC = 5-Hydroxyindole-3-acetic acid
HIAA = 5-Hydroxyindole-3-acetamide

(1) Contained 10% methanol during modification process.
(2) Contained 20% methanol during modification process.
(3) Error is the standard deviation of the mean.

Table 4. Conformational modification of RNase on L-tryptophan agarose gel.

Glutaraldehyde conc (mM)	Time of crosslinking (hrs.)	Esterase activity (U/g)	% Protein eluted
8.0*	17	-	-
2.4	3	10.4 ± 1.6	41.8
2.4	2	27.4 ± 3.7	11.4
1.6	2	13.5 ± 2.7	18.4

* Protein not assayed because it did not elute from the column.

applied to the column and the bound fraction crosslinked with glutaraldehyde. When a concentration of 8.0 mM glutaraldehyde was circulated over the column for 17 hours, no protein could be eluted from the column. By reducing the time of exposure and the concentration of glutaraldehyde, a portion of the protein was eluted from the column. The results included in Table 4 indicate that the initial activity of the CCMP prepared on this affinity column is indeed higher than preparations in solution.

Since each time this experiment was performed more than half of the crosslinked protein could not be removed from the column, the protein might be covalently linked to any amines present on the support material. To neutralize amines, acetic acid activated with EDC (1-ethyl-3-(3-dimethyl aminopropyl carbodiimide) was added to L-tryptophan-agarose. When the procedure for modification of RNase was tried using this modified gel, the neutral esterase activity was only 7.4 ± 5.1 U/g and only 25% of the crosslinked protein was removed. Apparently the presence of amines on the support material was not the cause of the difficulty in removing the modified RNase. The irreversible binding of protein was more likely due to intermolecular crosslinking of protein on the gel.

Even though the specific activities of these affinity column modified RNase derivatives were good, less than half the protein could be recovered. In addition, not all the protein applied to the column bound to the column. Therefore, the total activity generated from the procedure was only about 0.02 units which is much lower than that of a typical soluble preparation in which between 0.1 to 0.5 U/g are generated. However, if the affinity column method could be optimized, it would save considerable time and effort in purification required for a soluble preparation.

4. Crosslinking Reaction Conditions

The preparations of amino acid esterases described above were all prepared using glutaraldehyde as the crosslinking agent. Although the use of glutaraldehyde has consistently resulted in active material, the derivatized RNase has poor solubility and often precipitates upon storage or during concentration. We have recently investigated methods to modify the procedure in order to eliminate precipitation during storage and purification.

The reaction time during crosslinking with glutaraldehyde is extremely important. We have demonstrated that crosslinking modified RNase for only 2 or 3 hours by the procedure described above does not produce any measurable esterase activity. Yet, if the time of crosslinking is extended to 17 to 18 hours at 5°C, esterase activity is consistently generated. Even so the initial activity does vary considerably from preparation to preparation. When the crosslinking is continued for 24 hr or longer, the solution turns cloudy from protein precipitation.

Another variable that affects the initial activity and properties of the CCMP is the reaction pH. For the experiments reported above, the pH of the crosslinking solution was adjusted to pH 8.5. The pH at the end of the 17 hr reaction period was 6.5 to 6.7. Even after only one hr of reaction the pH was 7.5. Thus most of the reaction takes place at neutral pH rather than at 8.5.

Several studies indicate that the mechanism of crosslinking proteins with glutaraldehyde changes dramatically with the pH of the reaction.[16-18] Since it has been shown that the pH drifted during the typical preparation of a CCMP, it is not surprising that initial activity varied from preparation to preparation. To determine the degree of crosslinking, a series of preparations have been analyzed by the use of a ninhydrin reagent to determine the number of primary amines.

Since ninhydrin reacts with primary amines of amino acids and polypeptides, the degree of reaction with ninhydrin gives a measure of the number of amines crosslinked on the protein surface. The protein concentration of RNase and the RNase derivatives was determined by the Lowry method. By comparing the ninhydrin-Lowry plot of native RNase to the RNase derivatives, we can get a relative measure (expressed as percent) of the effectiveness of the crosslinking reaction. Three separate derivatives were examined by this method. Derivative A was prepared as usual, while B and C were crosslinked and maintained at pH 8.5 by the slow addition of 0.01M NaOH for 90 min. Figure 1 shows a plot of ninhydrin (absorbance at 570 nm) versus protein concentration of native RNase and derivatives A, B, and C. Derivative A has the same ninhydrin value as native RNase indicating little if any crosslinking while B has 38% less available amines and C has 45% less. These results indicate that maintaining the pH at 8.5 resulted in a much greater degree of crosslinking than allowing the pH to drift lower.

Table 5. Extent of crosslinking and the effect on esterase activity.

Percent decrease in reactive amines	Esterase activity (U/g)
0	none
10-20	0.4 ± 0.7
21-30	-
31-40	0.9 ± 0.4
41-50	0.4 ± 0.3
51-60	0.3 ± 0.2
61-70	0.8 ± 0.2

Native RNase was used as the standard.

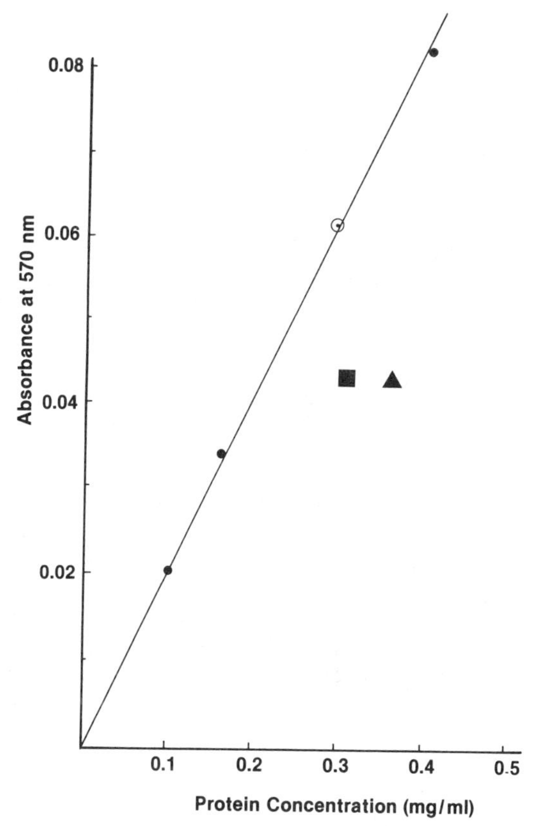

Figure 1. The degree of reaction with ninhydrin (absorbance at 570 nm) of modified RNase samples is plotted versus the protein concentration. Derivative A (○) has the same absorbance as native RNase (●) indicating no measurable crosslinking, while derivative B (■) has 38% less available amines and derivative C (▲) has 45% less.

Several RNase derivatives were prepared, assayed for esterase activity, and also assayed for reactive amines by the ninhydrin method. Each RNase derivative was placed in a group based upon the percent of reactive amines crosslinked. The ninhydrin reaction for native RNase was used as a standard (100%). Table 5 shows the results of these measurements. The one experiment in which no reduction of amines was observed also resulted in no activity. In the range of 10 to 20% decrease, some activity was observed, but with considerable spread. The individual values were 0.1, 0.1 and 1.1 U/g. Since the original procedure probably resulted in this low level of crosslinking it is not surprising that there was considerable variation observed in initial activities. No preparations fell in the range of 21 to 30% decrease in amines while the best initial activities were obtained when the crosslinking range was either 31-40 or 61-70%.

Several researchers report that glutaraldehyde reacts with amines over a wide pH range.[16-18] The nature of the crosslink is however different in the acid range than in the alkaline range. Nevertheless this reagent affords the opportunity to prepare CCMP over a broad pH environment. At a glutaraldehyde concentration used in earlier studies (2.4 mM), RNase was crosslinked at several different pH values. The modifier used was 5-

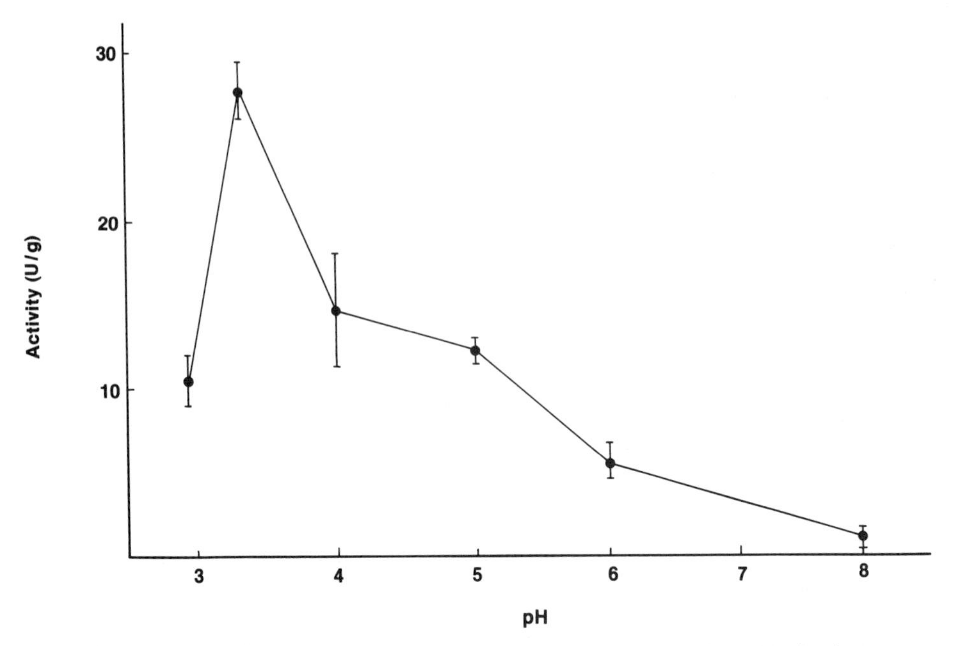

Figure 2. Esterase activity is plotted versus the pH during crosslinking with glutaraldehyde.

hydroxyindole-3-acetamide at a concentration of 10mM. Otherwise the procedure was the same as before except that just prior to adding glutaraldehyde, the pH was adjusted to a value from 3 to 8. Each preparation was assayed towards L-TrEE at pH 7.5. The activity of each esterase was measured by at least two of our four assay techniques and in most cases the activity was confirmed by three different assay measurements. A plot of the mean esterase activity versus the pH of glutaraldehyde crosslinking is shown in Figure 2. Initially, we expected that crosslinking with glutaraldehyde required an alkaline environment. However, results indicate that crosslinking in the acid region is sufficient and that the initial activity is considerably higher. This higher activity obtained at pH 3.5 is not surprising since at this pH the protein is perturbed and presumably binds the modifier in the desired conformation.

GLUCOSE ISOMERASES

1. Properties of Native Enzyme

The production of high fructose corn syrup (HFCS) is made possible by using glucose isomerase.[19] The native enzyme is not ideal however for industrial process conditions as it requires the use of transition metal ions such as Mg^{2+}. Thus in the commercial process, an ion exchange column must be employed to remove metal ions from the product.

The initial hydrolysis of starch is accomplished by the use of glucoamylase in the acid region. Unfortunately, glucose isomerase is most effective in the alkaline region; therefore the pH must be adjusted for

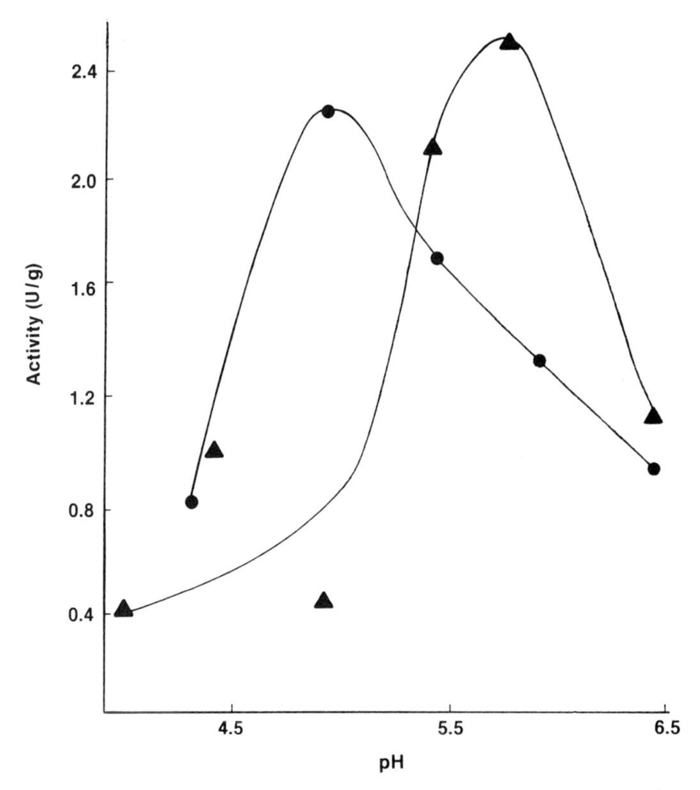

Figure 3. These data illustrate the pH profile of glucose isomerase activity of modified Concanavalin-A. Modified Concanavalin-A was reacted with either glutaraldehyde (●) or dimethyl suberimidate (▲) for 6 hours, dialyzed, and assayed for glucose isomerase activity at the pH values indicated.

the use of glucose isomerase. Producing a semisynthetic glucose isomerase that does not require metal ion cofactors and has a pH optimum in the acid region would have commercial significance.

2. Concanavalin A

Concanavalin A (Con-A) can be converted to a semisynthetic glucose isomerase by conformational modification. Both L-sorbitol and L-mannitol were effective as modifiers. Figure 3 illustrates that the pH optimum of the semisynthetic glucose isomerase can be changed by the use of a different bifunctional crosslinking reagent. Addition of metal ions such as calcium, magnesium and manganese did not increase the activity, nor did EDTA decrease the activity.

3. Hexokinase

The semisynthetic glucose isomerase prepared from hexokinase modified and crosslinked on a cellobiose solid support was assayed for activity in 25% glucose at pH 7.6 and 60°C. This CCMP had approximately 1.0 U/mg when

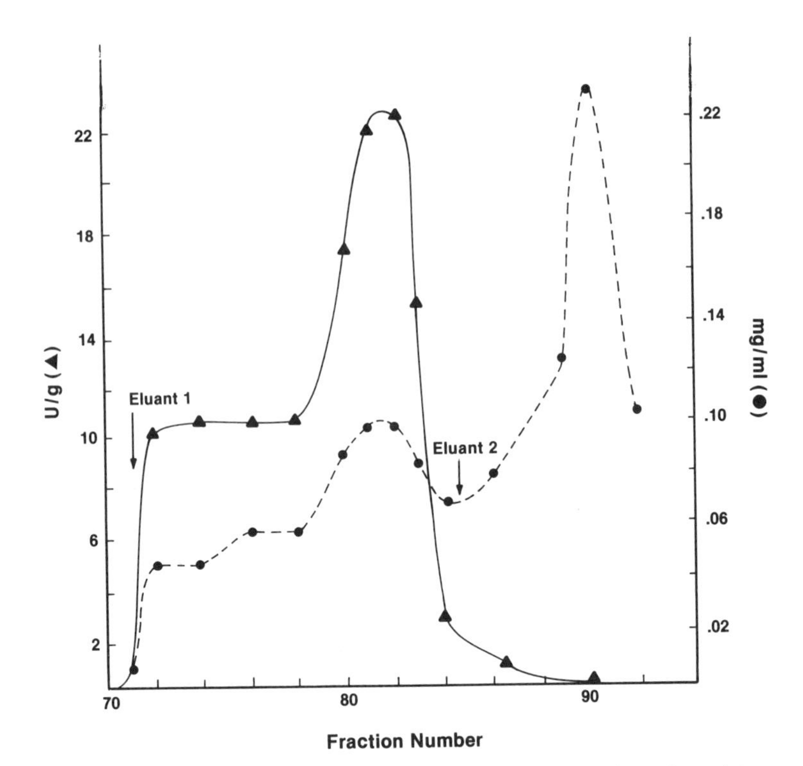

Figure 4. Elution pattern obtained for modified hexokinase containing glucose isomerase activity on a Cellobiose-Sepharose 4B column. Eluent 1 is 0.02M potassium phosphate buffer, pH 8.0 and Eluent 2 is 0.035M potassium phosphate buffer, pH 8.0. The glucose isomerase activity is shown (▲) as well as the protein concentration (●).

assayed by the sorbitol dehydrogenase method. When this modified hexokinase was assayed in 25% fructose at pH 7.6 and 60°C the activity was 0.95 U/mg. After two months of storage at -50°C, no loss in activity was observed.

When hexokinase was modified in solution without the solid support inhibitor, the starting activities ranged from 7-30 U/g at 60°C. After partial purification on a cellobiose-Sepharose 4B column, the activity was 150 U/g at 40°C for the most active fraction. Figure 4 shows a typical chromatogram of semisynthetic glucose isomerase partially purified on a cellobiose affinity column. Unlike the semisynthetic glucose isomerase prepared from Con-A, this CCMP has no activity at pH 6 but good activity at alkaline pH.

4. Glucose Oxidase

Apoglucose oxidase modified with cellobiose, crosslinked with dimethyl suberimidate and dialyzed extensively against 2 mM potassium phosphate buffer, pH 7.0 had between 3-6 U/g activity in 4.5% fructose at pH 8.0 and 60°C. After lyophylization and gel filtration using a Bio-gel A 1.5 m column, the specific activity was 3.5 U/mg. The pH optimum for this CCMP is 6.8.

5. Conclusions

An enzyme or semisynthetic enzyme that catalyzes the conversion of glucose to fructose could have commercial significance, especially in the United States.[20] The CCMP's that we have produced have two advantages over the enzymes now being used commercially. First, none of the semi-synthetic enzymes require metal ion cofactors. Secondly, by choice of the host protein and conditions of modification, the pH optimum can be located in the acid region (See section 2 above).

FLUOROHYDROLASES

The most recent examples of CCMP are fluorohydrolases that have the ability to catalyze the hydrolysis of organophosphates, such as, diisopropyl fluorophosphate (DFP) and phenylmethylsulfonyl fluoride (PMSF). Nearly all of the initial experiments used RNase, however, just recently we have successfully prepared several fluorohydrolases using bovine serum albumin, bovine casein, egg albumin and hexokinase as starting proteins. The fluorohydrolases are particularly interesting since the modified conformation is retained for sufficient time to measure activity at neutral pH without crosslinking.

To obtain the highest activity, it is necessary to add modifier on the same day that the RNase sample is chromatographed at pH 3.0. After one day of storage at 5°C in 1 mM HCl, the activity of RNase modified with HMPA is less than half that obtained on the first day. By the third day the generated activity is only 20%. The addition of HMPA to the newly chromatographed RNase does not increase the stability.

The stability was increased somewhat by adding HMPA to newly chromatographed RNase and then adding potassium phosphate buffer, pH 7.6. The final concentration of HMPA and phosphate buffer was 0.67M. Using this procedure the activity after three days was 38% of the initial activity.

Table 6. Fluorohydrolase activity of modified RNase.

Modifier	Activity (U/mg) DFP	PMSF
Diisopropyl phosphoric acid	0.25	0.94
Diethylisopropyl phosphonate	0.11	0.31
Ethyldiethylphosphonyl formate	0.06	0.16
Diisopropyl phosphite	0.05	0.74
Diethylcyanomethyl phosphonate	0.05	0.12
Hexamethyl phosphoramide	0.04	0.34
Diethylphenyl phosphate	0.03	--
Diethyl phosphite	0.02	0.14
Trimethyl phosphonoacetate	0.02	0.11
Dimethyl phosphite	--	0.33
Diisopropylmethyl phosphonate	--	0.30

Diethylchloromethyl phosphonate, phenylmethyl sulfonate, pyridoxal-5-phosphate, tributyl phosphate, triethyl phosphate and trimethyl phosphate gave no activity.

We investigated 18 organophosphate compounds as potential modifiers. All modified RNase preparations were made as previously described and assayed for activity using both DFP and PMSF. Table 6 lists the activities of these semisynthetic fluorohydrolases and the modifier used to prepare each semisynthetic catalyst. The greatest activity was obtained for RNase modified with diisopropyl phosphoric acid. With DFP as the substrate the activity was 0.25 U/mg and the catalyst had a molecular activity of 8.6 $min.^{-1}$; however, with PMSF as the substrate the activity was 0.94 U/mg and the molecular activity was 17.4 $min.^{-1}$ measured over 13.5 min.

In addition to preparing semisynthetic fluorohydrolases from RNase, we have also explored the modification of BSA, casein, egg albumin and hexokinase using the same procedure described above. Semisynthetic fluorohydrolases prepared from BSA possessed activity as high as 2.6 U/mg with DFP as the substrate and 7.5 U/mg with PMSF as the substrate. The BSA concentration in the assay was 0.01 mg/mL.

Semisynthetic catalysts prepared from casein also had good fluorohydrolase activity with a mean activity of 12 U/mg with PMSF as the substrate and 1.4 U/mg with DFP. The fluorohydrolases prepared from egg albumin and hexokinase had mean activities of 0.67 and 0.32 U/mg with PMSF as the substrate, respectively. However, neither of these modified proteins had activity towards DFP. Table 7 shows the mean activity obtained for each of these modified proteins.

Initial investigations using semisynthetic fluorohydrolases prepared from modified casein and crosslinked with dimethyl adipimidate also had good activity with a mean activity of 4.0 U/mg using PMSF and about 0.40 U/mg with DFP as the substrate. After storage at 5°C for 5 days there was no loss of activity.

Table 7. Fluorohydrolase activity of various proteins modified with HMPA.

Protein	Substrate	Activity (U/mg)
Chromatographed in 1 mM HCl		
BSA	DFP	1.03 ± 0.30
Casein	PMSF	12.67 ± 1.33
Egg albumin	DFP	-
	PMSF	0.49 ± 0.18
RNASE	DFP	0.08 ± 0.03
	PMSF	0.44 ± 0.10
Chromatographed in 5 mM phosphoric acid		
BSA	DFP	2.23 ± 0.87
	PMSF	4.13 ± 0.86
Casein	DFP	1.40 ± 0.96
	PMSF	3.17 ± 0.77
Hexokinase	DFP	-
	PMSF	0.32 ± 0.06
RNASE	PMSF	0.75 ± 0.13

DISCUSSION

Amino acid esterases were the first example of semisynthetic enzymes prepared by the conformational modification method. At least two amino acid esterases are formed having specific activities of approximately 0.5 U/mg after partial purification. The initial activity was very low and the crude material precipitated upon standing. Later work showed that the initial activity could be increased by selecting a more water soluble modifier. No doubt the geometry and functional groups on the modifier is important, but the modifier must be sufficiently soluble to interact with the perturbed protein. The problem with precipitation is probably related to the use of glutaraldehyde which often causes protein precipitation when used as a crosslinking agent.

Many examples of generation and purification of glucose isomerase type CCMP have been explored. Some have been purified until the specific activity is 2 or 3 U/mg. None of them appear to have a metal ion requirement as does the natural enzyme. All but one of them have pH optima in the neutral to slightly basic range. However, CCMP prepared from Con A and crosslinked with glutaraldehyde has a pH optimum at about 5. Initial activities are the same or somewhat higher than the amino acid esterases.

Unlike the other CCMP, the fluorohydrolases generally possess a much higher initial activity (See Table 6). In addition, the activity can be measured without crosslinking the protein. As would be expected when the uncrosslinked material is left at neutral pH overnight the activity is lost. The ability to measure an initial activity before crosslinking is a great advantage because it allows us to easily survey a number of modifiers and proteins without the complexity of crosslinking. We explored the use of 18 potential modifiers with RNase and found activity with eleven of them. Of the proteins tested, three resulted in activity towards DFP and PMSF while two generated CCMP having activity toward PMSF only (Table 7).

Fluorohydrolases are not found abundantly in nature and when located usually have low specific activity. Table 8 compares the activities of natural enzymes with the crude CCMP. Notice that the generated activities in the crude reaction mixtures are as much as two to three orders of magnitude higher than their unpurified natural counterparts. The activity generated from BSA is in fact approximately the same as the purified enzyme from Hog Kidney. These results indicate the power of this simple chemical modification for the generation of catalytic activity.

Table 8. Natural and semisynthetic DFPases.

Source	Activity (U/mg) Crude	Purified
Squid head ganglion[21]	0.057	110
Rabbit plasma[22]	0.0013	-
E. coli[23]	0.00127	-
Hog kidney[24]	0.025	2.5
CCMP (casein)	1.40	-
CCMP (RNase)	0.083	-
CCMP (BSA)	2.23	-

1. Modifiers

Early in the development of CCMP, the authors anticipated that the choice of modifier would be critical to the generation of activity. The modifier, it was thought, would have to be a template which closely mimics the transition state of the substrate. Our experience to date indicates that the choice of modifier is much more forgiving than initially anticipated. For the fluorohydrolases, eleven of the eighteen organophosphate compounds tested resulted in activity. These compounds were merely the organophosphates that are commercially available. For glucose isomerase activity, many sugars were found to be acceptable in the generation of activity.

These results suggest that some groups necessary for catalysis may be neutral as far as the binding of the modifier. The binding site of the modifier is obviously not a "designed" site as in the case of antibodies or other natural binding sites, but simply an accommodation of the perturbed protein. The binding constant is probably quite low in many cases requiring a concentration of modifier of 1 mM or more. For amino acid esterases, increasing the concentration of the modifier above 1mM did, in fact, increase the initial activity.

The ease of preparation of CCMP does allow one to survey a number of modifiers, as well as proteins, to obtain the properties and activity acceptable for a given application. Rapid surveying is particularly easy when the transient activity without crosslinking can be monitored as is the case for fluorohydrolases.

2. Crosslinking Conditions and Reagents

The pH dependence for the generation of esterase activity showed that the highest activity is obtained at approximately 3 (Figure 2). These data were obtained using glutaraldehyde as the crosslinking reagent which forms a stable crosslinked product both in the alkaline and acid region. Initial experiments with fluorohydrolases which can be monitored for activity before crosslinking, showed that crosslinking with most reagents resulted in considerable loss of activity. These results indicate the limitations of commercial crosslinking reagents, most of which react with proteins only in the neutral or basic region. We are presently exploring other reagents and conditions to optimize the crosslinking reaction to the preparation of CCMP.

ACKNOWLEDGMENTS

The authors wish to thank the following organizations for their financial support: (1) Department of the Energy - Office of Basic Energy Science Contract No. DE-AC02-81ER12003 (Esterases), (2) Army Research Office Contract No. DAAL03-86-C-0021 (Fluorohydrolases), (3) Owens-Illinois, Inc., and (4) Anatrace, Inc.

REFERENCES

1. W. V. Shaw, Biochem. J., **246**, 1-17 (1987).
2. T. L. Blundell, B. L. Sibanda, M. J. E. Sternberg, and J. M. Thornton, Nature, **326**, 347-352 (1987).

3. M. Mutter, K.-H. Altmann, G. Tuchscherer, and S. Vuilleumier, Tetrahedron, **44**, 771-785 (1988).
4. R. A. Lerner and A. Tramontano, TIBS, **12**, 427-430 (1987).
5. S. J. Pollack, J. W. Jacobs, and P. G. Schultz, Science, **234**, 1570-1573 (1986).
6. J. Jacobs and P. G. Schultz, J. Am. Chem. Soc., **109**, 2174-2176 (1987).
7. M. H. Keyes, "*Biotec 1,*" C. P. Hollenberg and H. Sahm, Eds., VCH Publishers, Inc., New York, 1987, pp. 137-142.
8. M. Keyes, "*Protein Engineering: Current Status, Proceedings Bioexpo 86,*" Butterworth Publishers, Stoneham, MA, 1986, pp. 273-290.
9. M. H. Keyes, D. E. Albert and S. Saraswathi, in: "*Enzyme Engineering 8,*" A. I. Laskin, K. Mosbach, D. Thomas and L. B. Wincard, Jr., Eds, New York Acad. Sci., New York, 1986, 201-204, "Enzyme Semisynthesis by Conformational Modification of Proteins".
10. M. H. Keyes and S. Saraswathi, in: "*Polymeric Materials in Medication,*" C. G. Gebelein and C. E. Carraher, Jr., Eds., Plenum Publishing Corp., New York, 1985, pp. 249-264, "A Semisynthetic Approach to Induce New Enzyme Activities in Protein".
11. S. Saraswathi and M. H. Keyes, Enzyme Microb. Technol., **6**, 98-100 (1984).
12. M. H. Keyes, U.S. Patents Nos. 4,716,116; 4,714,677; 4,714,676; and 4,713,335. "Protein Modification to Provide Enzyme Activity", 1987.
13. M. H. Keyes and S. Vasan, U.S. Patent No. 4,609,625. "Process for the Production of Modified Proteins and Product Thereof", 1986.
14. H. Tsuge and H. Mitsuda, J. Vitaminol., **17**, 24-31 (1971).
15. B. E. P. Swoboda, Biochim. Biophys. Acta., **175**, 365-379 (1969).
16. P. Monsan, G. Puzo and H. Mazarguil, Biochemie, **57**, 1281-1292 (1975).
17. R. Lubig, P. Kusch, K. Roper and H. Zahn, Monatshifte Fur Chemie, **112**, 1313-1323 (1981).
18. R. Koelsch, M. Fusek, Z. Hostomska, J. Larch, and J. Turkova, Biotechnology Letters, **8**, 283-286 (1986).
19. M. H. Keyes and D. Albert, in: "*Encyclopedia of Polymer Science and Engineering,*" Vol.6, Mark, Bikales, Overberger & Manges, Eds., John Wiley & Sons, Inc., New York, 1986, 189-209, "Enzymes, Immobilized".
20. C. Verbanic, Chemical Business, January, 1986, 29-31.
21. J. M. Garden, et.al., Comp. Biochem. Physiol., **52C.**, 95-98 (1975).
22. K. B. Augustinsson and G. Heinburger, Acta Chem. Scand., **8.**, 1533 (1954).
23. F. C. G. Hoskin, Biochem. Pharm., **34**, 2069-2072 (1985).
24. Estimated from: A. Mazur, "*Methods in Enzymology,*" **1**, 1955, 651-656.

HEPARIN, HEPARINOIDS AND HEPARIN OLIGOSACCHARIDES: STRUCTURE AND BIOLOGICAL ACTIVITIES

Robert J. Linhardt and Duraikkannu Loganathan

Division of Medicinal and Natural Products Chemistry
College of Pharmacy
University of Iowa
Iowa City, Iowa 52242 USA

Heparin is a polydisperse, highly sulfated polysaccharide which has been in widespread clinical use over the past half century. This chapter describes our current understanding of heparin's structure and its biosynthesis. In addition to heparin's usefullness as an anticoagulant, it has a wide range of additional activities. These include its antiatherosclerotic activity, complement inhibitory activity, angiogenic activity, and other additional, recently discovered activities. The rationale for developing biomimetic polymers to replace the natural product, heparin, is discussed. Synthetic polymers, semi-synthetic sulfated polysaccharides, fully synthetic heparin oligosaccharides, fractionated heparins, low molecular weight heparins, and enzymatically prepared heparin oligosaccharides all have been used as heparin substitutes. This chapter examines heparin's structure-activity relationships with a focus on the potential utility of these biomimetic polymers as new therapeutic agents.

HEPARIN: ITS STRUCTURE AND SIGNIFICANCE

Heparin is a polydisperse, highly sulfated, linear polysaccharide comprised of repeating 1→4 linked uronic acid and glucosamine residues.[1,2] Although heparin has been used clinically as an anticoagulant for the past 50 years, its precise structure remains unknown.[1,2] The failure to understand completely heparin's structure is not the result of a lack of effort but rather is due to its extremely complex structure. Heparin is biosynthesized as a proteoglycan (PG, MW is approximately 1 million) consisting of a central core protein from which approximately eleven long linear polysaccharide chains extend (Figure 1).[3] The drug heparin is recovered from porcine intestinal mucosa or bovine lung as a glycosaminoglycan (GAG) or simply a linear polysaccharide chain without any associated protein.[1,2] Originally, it was believed that the harsh commercial processing of tissues from which the drug heparin was prepared resulted in the characteristic complexity of GAG heparin.[4] This processing included treatment with alkali, proteases, bleaching agents.[4] Proteases and β-glucuronidases present in the tissue itself are actually

Biomimetic Polymers
Edited by C. G. Gebelein
Plenum Press, New York, 1990

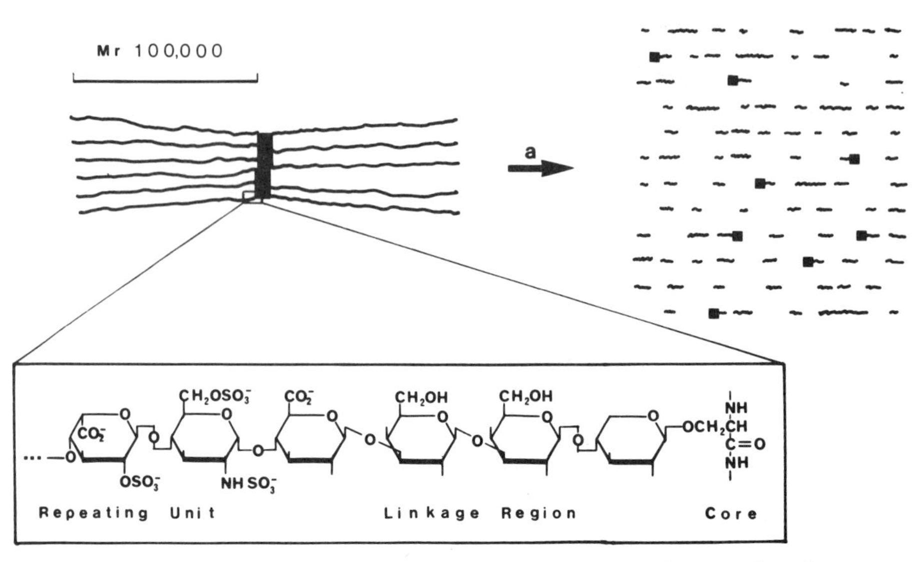

Figure 1. Structure of proteoglycan and glycosaminoglycan heparins. Proteoglycan heparin is processed by endogenous protease and β-endoglucuronidase (a) resulting in glycosaminoglycan heparin

responsible for much of the processing of PG heparin to GAG heparin.[5]

The structural complexity of heparin can be considered at several levels. At the PG level the number, position and nature of the polysaccharide chains attached to the protein core can be examined. Once freed from their core protein by the action of tissue proteases, peptidoglycan is formed (ie., a small peptide to which a single long polysaccharide chain, MW 100,000, is attached.)[3] This peptidoglycan is short-lived as it is immediately processed by a β-endoglucuronidase to a number of smaller polysaccharide chains (only one, corresponding to the original site of attachment to the core protein, would contain peptide material) called GAG heparin. At the level of GAG heparin some of the structural complexity results from its polydispersity. GAG heparin has a MW ranging from 5,000-40,000 (degree of polymerization (DP) 10-80) with an MW (average) of 13,000. Even the heparin chain corresponding to the most prevalent DP represents a mere 5 mole% of a typical GAG heparin preparation.[6] GAG heparin has a second level of structural complexity associated with its primary structure or sequence.[7] Unlike nucleic acids which have a four letter alphabet (ATGC) heparin has at least ten different sugar units comprising its alphabet with many additional sugars possible but as yet unobserved.[2]

The structural features of PG heparin have been primarily established by studying its biosynthesis.[8,9,10] The structural features of GAG heparin have relied on biosynthetic, chemical, enzymatic, and spectroscopic techniques. Recent efforts in our laboratory as well as others have sought to use techniques originally developed to sequence nucleic acids and proteins to establish the sequence of heparin.

PG heparin is just one in a number of families of PG macromolecules (Figure 2). Heparin and heparan sulfate are structurally similar macromolecules with polysaccharide side chains comprised of alternating (1→4) linked sulfated and or acetylated glucosamine and sulfated or nonsulfated

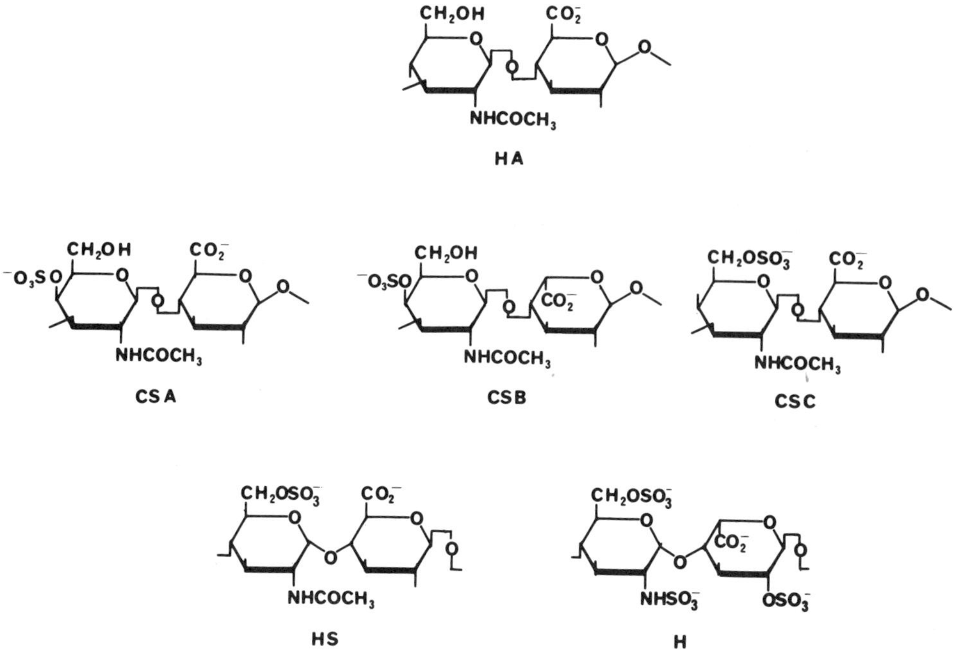

Figure 2. Structure of common glycosaminoglycans. Hyaluronic acid (HA), chondroitin sulfate A (CSA), Chondroitin sulfate B or dermatan sulfate (CSB), Chondroitin sulfate C (CSC), Heparan Monosulfate (HS), and Heparin (H).

iduronic or glucuronic acid residues.[2] The core protein of PG heparin and PG heparan sulfate are different as is the core protein between heparan sulfate found in the extracellular matrix and that found in the membrane.[11,12] Although structurally similar, GAG heparin and GAG heparan sulfates can be distinguished from one another on close examination of their ratio of N-acetylation to O-sulfation.[13] Chondroitin sulfates are another class of PGs which can be broken down into three classes: chondroitin sulfate A, →3) β-D-N-acetylgalactosamine-4 sulfate (1→4) β-D-glucuronic acid (1→; chondroitin sulfate C, →3) β-D-N-acetylgalactosamine-6 sulfate (1→4) β-D-glucuronic acid (1→; and chondroitin sulfate B, →3) β-D-N-acetylgalactosamine-4 sulfate (1→4) α-L-iduronic acid (1→.[14] Keratan sulfate is yet another class of GAGs which has a major disaccharide repeating unit: →3) β-D-galactose(1→4) N-acetylglucosamine-6 sulfate (1→.[14] A final class of glycosaminoglycan, hyaluronic acid contains no sulfate and is not found linked to a core protein.[14] Its structure is →3) N-acetylglucosamine (1→4) glucuronic acid (1→.

Heparin's primary application is as an anticoagulant.[15] It may be more appropriate to consider heparin as a polyelectrolytic drug having a multiplicity of biological activities.[1,16] Heparin is the strongest acid present in the body and thus is present under physiologic conditions as a highly charged polyanion.[16] Virtually any cationic protein (pI>7) is capable of binding to heparin under physiological conditions and the activity of such a protein is more often than not affected by heparin. Even anionic proteins are capable of interacting with heparin, in fact ATIII, the most studied heparin binding protein, has an isoelectric point of 4.9-5.3.[17] Within the past decade a growing number of biological activities[18] have been demonstrated to be regulated by heparin ranging from its effect on angiogenesis[19] to the regulation of the immune response.[20] The discovery of these new activities has brought about an urgency to understand heparin's structure and to elucidate its structure-activity relationship (SAR). Only limited success has been made along these lines primarily due to the efforts of two research groups, headed by Ulf Lindahl and Robert Rosenberg, to understand the SAR of heparin's anticoagulant activity. Their success has spawned additional research aimed at understanding heparin's other biological activities including the true physiological role of endogenous proteoglycan heparin.

A review of heparin would be incomplete if it did not discuss new trends in improving heparin's anticoagulant/antithrombotic activity. A new class of drugs, low molecular weight heparins and heparinoids, have been undergoing clinical trials as anticoagulant/antithrombotic agents[21] for use in a wide variety of disease states ranging from deep vein thrombosis[22,23] to non-hemmorrhagic stroke.[24] These agents include both natural products chemically or enzymatically derived from heparin as well as fully synthetic oligosaccharides and polysaccharides. These new drugs represent the first generation of biomimetic heparins. In addition to being used systemically as soluble agents, recent advances in heparin chemistry has permitted their immobilization onto supports.[25,26] Stable, active, heparinized biomaterials can now be prepared.[27,28] These may someday replace soluble heparins in devices such as the artificial kidney and heart lung machine.[27,28]

PROTEOGLYCAN HEPARIN

1. Biosynthesis

The biosynthesis of heparin has been well studied in a mastocytoma cell culture system (Figure 3).[9,10] The core protein, which contains a high number of serine-glycine repeats is first synthesized in the rough endoplasmic recticulum.[29] To this core, a linkage region (Figure 1) consisting of three neutral sugars is attached to serine through the action of a xylosyl transferase.[29] Onto this neutral sugar linkage region, a repeating copolymer of 1→4 linked glucuronic acid and N-acetylglucosamine is assembled through the stepwise addition of UDP-sugars.[30] The linear polysaccharide chain is extended by approximately 300 sugar units before its synthesis terminates.[9] The chain is then partially de-N-acetylated and sequentially N and O-sulfated.[31] Finally a unique ATIII binding site is introduced by the action of a 3-O-sulfotransferase.[32] The structural variability in the heparin polymer is primarily the result of the incomplete nature of these postpolymerization modifications.[31,32] The sequence heterogeneity in the polysaccharides chains results in a family of PG heparin macromolecules.

2. Biological Functions

PG heparin is primarily found in the granules of mast cells.[33] The precise function of these molecules is not yet understood but they may have a role in: (1) the packing of histamine within the mast cell;[33] (2) the stabilization, inhibition and binding of proteases which are present at very high concentrations within mast cells;[33] and (3) the regulation of complement activation following mast cell degranulation.[34] When mast cells degranulate[34] heparin is released, but as GAG heparin - the result of processing by proteases and endo-β-glucuronidases.[33] Although the GAG

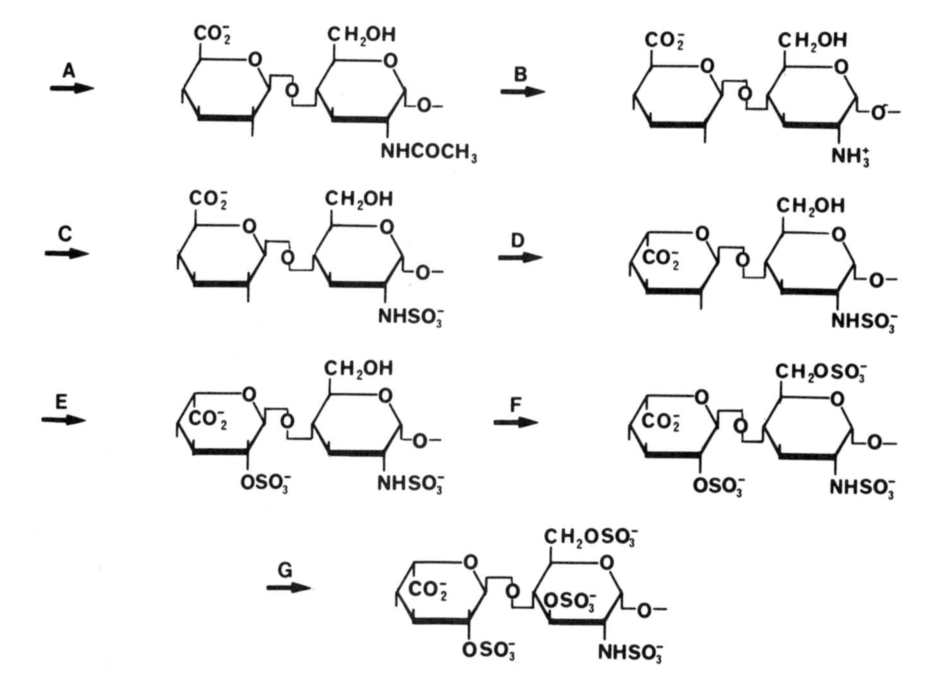

Figure 3. Proposed pathway for the biosynthesis of heparin.[9] (A) Polymerization of UDP sugars; (B) N-deacetylase; (C) N-sulfotransferase; (D) Hexuronosyl C5-epimerase; (E) 2-O-sulfotransferase; (F) 6-O-sulfotransferase; (G) 3-O-sulfotransferase.

heparin released on mast cell degranulation demonstrates anticoagulant activity, the role of this activity is unclear. There is no direct evidence that endogenous mast cell heparin plays a role in maintaining blood flow through the vasculature,[35] even though this is the primary application for exogenously administered GAG heparin.[15]

The true biological function of heparin still remains hotly contested. The structurally related PG, heparan sulfate, has been isolated from the endothelial surface and has "heparin-like" sequences which permit it to bind ATIII, and thus inactivate plasma serine proteases responsible for blood coagulation.[36,37] PG heparan sulfate, found in membranes of a wide variety of cells and in the extracellular matrix, appears to have an active role in cell to cell communication through the binding of protein growth factors,[38] hormones,[39] and various regulators.[19,20,38,40,41] Although the biological roles of endogenous PG heparin, PG heparan monosulfate and the related PG chondroitin sulfates are not completely understood, this has not precluded the use of GAGs derived from these natural products as well as GAG fractions, fragments and synthetic analogues, for a variety of medical applications.

GLYCOSAMINOGLYCAN HEPARIN

Glycosaminoglycan (GAG) heparin (commercial or the drug heparin) has been prepared and used as a clinical anticoagulant since 1939.[42] Nearly all the studies on heparin structure and activity have been performed on this form of heparin.[43]

1. Preparation

GAG heparin is prepared from animal tissues that are rich in mast cells, such as porcine intestinal mucosal and bovine lung.[44] Other species of mammals,[44,45] as well as birds,[45] and even invertebrates, such as lobster[46] and clams,[47,48,49] which do not have a blood coagulation system, also contain heparin.[50] The method of commercial processing of heparin varies between manufacturers and is generally regarded as trade secrets.[4] The basic approach involves the collection of the appropriate tissue followed by proteolytic treatment, extraction and complexing with ion pairing reagents such as cetylpyridinium chloride, followed by fractional precipitation. Treatment with base to remove residual protein by β-eliminative cleavage of xylose O-serine glycoside and/or bleaching with oxidizing agents is commonly used to prepare the pure white drug form of heparin. The major criteria for purity is a high specific activity expressed as USP units per milligram.[4]

The extensive processing of commercial heparin preparations has often lead to speculation that its structure is extensively modified and might in some ways be different than GAG heparin stored in mast cell granules. Heparin has been prepared by researchers using very mild conditions at low temperatures, at neutral pH and without the use of oxidants and bleaches. Such mild preparation methods resulted in a heparin preparation indistinguishable from the commercial one.[4] Studies on conditions required for desulfation of heparin have confirmed heparin's stability under conditions used in its processing.[51,52] Recent work by our research group, using a new oligosaccharide mapping technique involving the enzymatic depolymerization of heparin followed by the analysis of the oligosaccharide products by either strong anion exchange high pressure liquid chromatography[43,53] or gradient polyacrylamide gel electrophoresis,[54] have demonstrated remarkable similarities between heparins prepared from the same tissue by different manufacturers and to heparins prepared under mild conditions (i.e., without bleaching) (Figure 4).

2. Properties

GAG heparin is a polydisperse preparation with a MW ranging from 5,000-40,000 having an MW (average) of 13,000 (20,55). The average molecular weight of porcine mucosal and bovine lung heparin preparations are similar.[56] The primary repeating unit is →4) β-D-glucosamine 2,6-disulfate (1→4) α-L-iduronic acid 2-sulfate (1→. This repeating structure accounts for 85 wt% of porcine and 90 wt% of bovine heparin. The remaining 10-15 wt% of the heparin polysaccharide is comprised of a small number of rarer disaccharide sequences shown in Table 1. These minor disaccharide sequences are the result of incomplete biosynthesis and have a reduced degree of O-sulfation, N-sulfation (containing N-acetyl residues instead) and less iduronic acid (containing glucuronic instead). The disaccharide sequences in Table 1 do not account for all of heparin's mass. Other minor disaccharide sequences have been reported but their presence has not been independently confirmed and some may be artifacts formed in commercial processing or in depolymerization.

The disaccharide sequences are the building blocks of the heparin polymer and represent an alphabet through which information can be stored in the heparin polymer. We will see how this is the case for the antithrombin binding site, which is defined by a particularly rare saccharide sequence, later in this chapter. Until recently, there was little information on the sequence of the heparin polymer or, indeed, whether or not

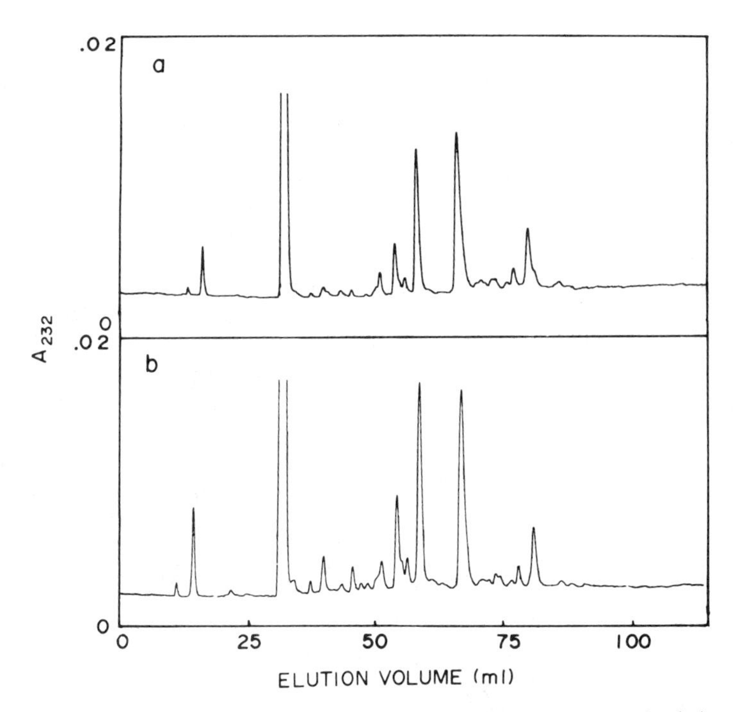

Figure 4. Strong anion exchange-HPLC of unbleached (a) and bleached (b) glycosaminoglycan heparin. Absorbance at 232 nm is plotted against elution volume. For conditions see ref. (43).

there was even a sequence. Early researchers viewed heparin as a polymer having a random sequence where the only restrictions were the type of saccharide units present, the (1→4) linkage configuration, and the required alternating substituted uronic acid and glucosamine residues. Biosynthetic studies, combined with the frequent occurrence of a rare

Table 1. Structure and frequency of disaccharide sequences found in heparin.

Disaccharide	Sequence[a]	Mole%[b]
a	→4)IDU2S(1→4)GN2S,6S(1→	71
b	→4)IDU2S(1→4)GN2S(1→	6.8
c	→4) IDU(1→4)GNAc,6S(1→	4.4
d	→4)GA(1→4)GN2S,6S(1→	9.3
e	→4)IDU(1→4)GN2S,6S(1→	(c)[57]
f	→4)GA(1→4)GN2S,3S,6S(1→	4.4

(a) IDU, α-L-idopyranosyluronic acid; GA, β-D-glucopyranosyluronic acid; GN, 2-deoxy,2-amino-α-D-glucopyranose; GNAc, 2-deoxy-2-acetamido-α-D-glucopyranose; S, sulfate.
(b) Mole% found in a typical commercial porcine mucosal heparin (heparin R).[43]
(c) Not determined.

sequence associated with the antithrombin binding site, has changed this perception.

One approach to understanding order (sequence or information storage) in the heparin polymer has used computer or mathematical simulation.[6,7] For example, the major repeating unit in the heparin polymer, a trisulfated disaccharide (Table 1), could be: (1) clustered in one region of the heparin polymer; (2) distributed uniformly through the polymer; (3) distributed randomly throughout heparin; or (4) occur in unique, well defined positions within the polymer. By computer or mathematical simulation, number chains representing the heparin polymer are constructed following these site distribution patterns. After preparing an ensemble of number chains having the correct polydispersity, the computer simulates their breakdown by cutting the number chains through these sites. The size distribution of the smaller number chains formed are then calculated. These simulated distributions are matched against experimentally measured product distributions obtained using an enzyme, heparinase, which cuts heparin through these sites. The results of this study demonstrated that the major disaccharide repeating sequence was either randomly distributed through the heparin polymer or distributed in a unique sequence that was indistinguishable from the random solution.[6]

A more sophisticated computer simulation examined whether a random distribution of the monosaccharide units, in the experimentally determined quantities found within the heparin polymer, could give rise to the oligosaccharides found in oligosaccharide mapping experiments.[7,43] The results of this study demonstrated that the individual sugar units found in heparin could not be randomly arranged. Further studies along these lines look at the kinetics of heparinase depolymerization of heparin and the distributions of products formed throughout the reaction (not just

Table 2. Compositional and sequence analysis of transient oligosaccharides.

Transient Oligosaccharide	Composition[a]	Sequence[b,c]
HEXASACCHARIDES		
octasulfated-1	2(a),1(b)	a'ab
octasulfated-2	2(a),1(e)	a'ae
nonasulfated	3(a)	a'aa
OCTASACCHARIDES		
undecasulfated-1	3(a),1(b)	nd
undecasulfated-2	3(a),1(e)	a'aae
decasulfated	2(a),1(c),1(f)	a'acf
dodecasulfated	4(a)	a'aaa
DECASACCHARIDES		
tetradecasulfated-1	4(a),1(b)	nd
tetradecasulfated-2	4(a),1(e)	a'aaae
tridecasulfated	3(a),1(c),1(f)	nd
pentadecasulfated	5(a)	a'aaaa

(a) The structure of disaccharide units a, b, c, d, and f are given in Table 1.
(b) a' is ΔIDU2S(1→4)GN2S6S where ΔIDU2S is 4-deoxy-α-L-threo-hex-4-enopyranosyluronic acid.
(c) nd, not determined.

the final product distribution). These simulation studies represent a novel way of understanding sequence of polydisperse polymers such as heparin.

Classical sequencing methods have also been applied to studying the primary structure of the heparin polymer. GAG heparin can be partially depolymerized (either enzymatically or chemically) to prepare transient oligosaccharides (i.e., oligosaccharides not found in the final products because they still contain cleavable sites). These oligosaccharides can then be purified to homogeneity. If a transient oligosaccharide is then exhaustively depolymerized to smaller, previously characterized, final oligosaccharides their composition can be determined.[41] By introducing a tag, such as a radiolabel, at one end of the transient oligosaccharide a sequence can be determined.[58] For more complex transient oligosaccharides, a series of partial depolymerizations can be used and the mixture of these run out on a gel. This represents the polysaccharide equivalent of Maxam-Gilbert sequencing of nucleic acids.[59] The composition and sequence of a number of transient heparin-derived oligosaccharides, which have been determined in this way, are given in Table 2.

New methods of sequencing heparin are now being proposed which involve sequencing the whole intact proteoglycan. These methods depend on the introduction of a reading frame from which the sequence can be determined. One such reading frame is the reducing terminus of the polysaccharide chain where it attaches to the core protein. If the polysaccharide chains can be removed and a label introduced into the same position where each was attached to the core, then a sequence might be read, from this label, down the polysaccharide chain. This method is attractive as it might lead to the total sequence of heparin. A potential pitfall of this approach is that the sequence from the reducing terminus might be interrupted by a series of random sequences that might result in a loss of the reading frame.

3. Medical Applications of Heparin

Systemic Anticoagulation. Heparin is the most commonly used clinical anticoagulant. Over six metric tons of heparin are manufactured worldwide each year, representing over 50 million doses.[16] Heparin is administered intravenously during most extracorporeal procedures (where blood is removed from the body and passed through a device) such as kidney dialysis and membrane oxygenation, used in heart by-pass procedures.[60] Systemic heparinization is also widely used in treating deep vein thrombosis[61] and a variety of other surgical procedures.[61] Heparin is also given by the subcutaneous route and, despite its reduced bioavailability, a low level of heparinization can often be maintained this way for prolonged periods of time.[62] Although it would be highly desirable to prepare an orally active heparin which could be administered outside a hospital setting, no such drug derivative or formulation is currently available.[63,64] Heparin's major side-effect, hemorrhagic complications,[65] is closely linked to its anticoagulant activity.

Thus, while heparin is widely used, it has been cited as the drug most responsible for death in otherwise healthy patients.[66] After administering heparin, it is often necessary to neutralize its anticoagulant activity and this is usually done by administering a cationic polypeptide based drug called protamine.[67] Protamine is not without its own set of side-effects and the resulting protamine heparin complex has been associated with an immune related loss of platelets.[68,69]

Regional Heparinization. The heparin required to maintain flow in an extracorporeal circuit is needed only in the blood outside the body. Once the blood is returned to body and its vascular container it is naturally anticoagulated. In systemic heparinization, heparin is returned to the body along with the blood and this is where it demonstrates many of its side-effects.[70] One approach directed at solving this problem is to infuse heparin as the blood leaves the body and enters the device, and then to remove the heparin from the blood as it leaves the device and re-enters the body.[60] This approach makes use of the ability of immobilized microbial heparinase to destroy heparin's anticoagulant activity catalytically,[60,71] forming inactive and non-toxic[60,72] oligosaccharide products which can be cleared from the body without metabolism.[73] Animal studies in dogs and sheep have demonstrated that regional heparinization is possible, but the need for human clinical trials remains.[74]

Blood Compatible Polymers. The preparation of blood compatible polymers has been undertaken to reduce the undesirable side-effects associated with systemic heparinization during extracorporeal therapy. One approach has been the design and preparation of heparin-coated and heparin-bonded surfaces.[25,27] The rationale behind preparing such surfaces is their similarity to the natural luminal surface of blood vessels which are lined with a heparin-like proteoglycan heparan sulfate. Recently, antithrombin binding sites common to heparin and responsible for its anticoagulant activity have been identified in endothelial heparan sulfate.[36,37] Early efforts at preparing heparinized surfaces have been aimed at the non-covalent attachment of heparin by entrapment,[25] adsorption,[25] or ionic interaction.[25] These approaches had certain limitations the most serious of which was the leaching of the weakly bonded heparin from the device's surface resulting in gradual loss in blood compatibility.[25] In addition a number of different polymers (some porous, some flexible. some rigid, etc.) are commonly used in the construction of an extracorporeal device such as a membrane oxygenator or a hollow fiber kidney dialyzer. Not all of these material types are equally amenable to heparin bonding nor are they equally stable.[25]

Recent advances in heparin chemistry have resulted in a better understanding in the covalent coupling of heparin to polymers.[25] This stronger covalent coupling has resulted in several spectacular advances in fully heparinized extracorporeal devices.[27] The simplest of these is the heparinized venous lines that are now becoming widely used in hospitals, as these involve only a single material type.[75] More complicated devices including fully heparinized membrane oxygenators are just entering clinical trials following very successful animal studies.[27,28] In hind-leg profusions in dogs and sheep membrane oxgenators remained open to flow for up to 24 hours without the use of external pumps.[27,28] Progress in this area depends on understanding which type of binding is the best preventing both leaching and the stripping of surface heparin by the bodies naturally occurring heparinases.[76] The limited effect of these surfaces on formed blood components such as platelets is also of primary importance.[27,28] Other questions which need to be addressed is what type of heparin should be immobilized and whether low molecular weight heparins or synthetic heparinoids would be better ligands for immobilization.

HEPARIN AND ITS ANTITHROMBOTIC ACTIVITY

1. Blood Coagulation

Hemostasis, "the spontaneous arrest of bleeding from ruptured blood vessels",[77] is a broad physiological process of which the blood coagulation system is just one part. *In vivo* hemostasis involves plasma coagulation factors, platelets, monocytes, and endothelial cells which line the blood vessels. The coagulation cascade consists of a sequence of reactions in which protease precussors (apoenzymes) are transformed from enzymatically inactive to enzymatically active forms.[78,79] In the final stages of the coagulation cascade, fibrinogen is transformed by thrombin (factor IIa) into the spontaneously polymerizable fibrin monomer.[80] Polymerization and subsequent crosslinking of fibrin monomers produces gelatinous fibers which emesh platelets forming a primary hemostatic plug.

The coagulation cascade is divided into two pathways, the extrinsic pathway and the intrinsic pathway. The extrinsic pathway is activated by tissue damage which releases thromboplastin, calcium and phospholipid from the cell membrane. These factors activate factor VII, which transforms factor X into its proteolytically active form, factor Xa. The second pathway, the intrinsic pathway, is activated by blood contacting non-endothelial surfaces followed by the transformation of several circulating coagulation factors finally leading to the formation of factor Xa. The remaining steps, referred to the final common pathway,[81] results from factor Xa catalyzing the conversion of prothrombin to thrombin which cleaves fibrinogen to form a fibrin monomer, which produces a clot.[80] The coagulation cascade appears to be autocatalytic and self-limiting and thrombin (factor IIa) plays a central role. The generation of active coagulation factors is explosive and is initiated by a local injury at a vessel wall, while inhibitors of these proteases are present throughout the entire vascular system. The active coagulation proteases, therefore, only exist momentarily at the site of injury where their rate of formation is more rapid than their rate of inactivation.

2. Serine Protease Inhibitors

The blood coaguation enzymes are serine proteases[82] with trypsin-like specificity for arginyl linkages. Unlike trypsin, they have a higher degree of specificity for the linkages they cleave[79,83] and often require cofactors to accelerate their rate of reaction.[84] The entire coagulation process is under the control of a group of glycoprotein serine protease inhibitors, the most important of which is antithrombin III.[85]

Antithrombin III. ATIII is a single chain anionic (pI 4.9-5.3)[17] glycoprotein of molecular weight 58,000.[86] All the coagulation proteases (except for VIIa) are inhibited by ATIII which forms an equimolar covalent complex with these enzymes. ATIII has a high affinity for these enzymes (low K_m) but a slow turnover rate (low K_{cat}). The reaction affords an intact protease covalently linked to the amino terminal portion of ATIII. The ATIII which is released is proteolytically modified at a single site near the carboxy terminus.[87] Heparin binds to thrombin and ATIII in a ternary complex accelerating the rate of thrombin inhibition by ATIII by 2000-fold.[88]

Heparin Cofactor II. Heparin cofactor II is structurally similar to ATIII with a molecular weight of 65,000[89] and a pI of 4.9-5.3,[17] having a similar carboxy terminal sequence but a distinctly different amino terminal sequence.[90] The physiological role of HCII might be as a reserve of thrombin inhibitor when the plasma concentration of ATIII becomes abnormally low.[91] Unlike ATIII, HCII can inhibit thrombin but no other coagulation proteases.[88,92,93] In addition to this unusual specificity, HCII also can be potentiated by GAGs other than heparin including dermatan sulfate and heparan sulfate both of which are found lining the luminal surface of the endothelium.[94] The mechanism by which HCII inhibits thrombin is similar to that proposed for ATIII.[88]

3. Other Antithrombotic Actions of Heparin

Heparin is known to activate platelets, which represent an important component of thrombosis.[95] Once a clot is formed, heparin can act to accelerate finbrinolysis or the dissolution of the clot.[96] A new generation of fibrinolytic drugs including streptokinase,[97] urokinase[98] and tissue plasminogen activator,[99] are particularly promising for the treatment of coronary clots formed in heart attacks.[99] These agents have been tested in conjunction with heparin to accelerate clot dissolution and to prevent clot reformation.[100] The most serious side-effect with such therapy is bleeding complications, but these might be eliminated by using new, low molecular weight heparins and heparinoids with reduced hemmorrhagic side-effects.[101]

HEPARIN'S OTHER BIOLOGICAL ACTIVITIES

1. Antiatherosclerotic Activity

Effect on Lipoprotein Lipase. When heparin is administered intravenously, it causes the release of lipoprotein lipase (LPL) from the endothelium.[102-106] This may result in increased triglyceride lipolysis occurring in the blood stream, thereby lowering the concentration of cholesterol-rich remnant particles in contact with the arterial wall.[102-106] The effect of heparin on the release and activation of LPL has been studied in a number of different animal models including humans.[102-106] Two major problems stand in the way of the application of heparin as an antiatherosclerotic agent. These are its primary activity as an anticoagulant[16] and heparin's low bioavailability when administered orally.[64] To circumvent these problems, low molecular weight heparins, heparin-oligosaccharides,[105] and heparinoids[106] are being studied.

Effect on Smooth Muscle Proliferation. The proliferation of smooth muscle cells following damage to the endothelium is an important part of atherogenesis.[107,108] Both anticoagulant and non-anticoagulant heparin have demonstrated the ability to inhibit the proliferation of smooth muscle cells.[107,108] A heparin-like GAG, present on the luminal surface of the endothelium, probably plays a physiological role in the regulation of smooth muscle proliferation during atherogenesis.[104] A synthetic pentasaccharide containing an ATIII binding site demonstrates anti-proliferative activity comparable to heparin's, as demonstrated in rat aortic smooth muscle cell culture.[108] A similar pentasaccharide, missing the 3-O-sulfate required for ATIII binding, showed a markedly reduced activity, as did disaccharide and tetrasaccharide samples.[108]

Heparins and low molecular weight heparins, with low affinity for ATIII, also demonstrated antiproliferative activity comparable to commercial heparins thus indicating, at least for higher oligosaccharides, that the presence of an ATIII binding sequence is unnecessary.[108] It is likely, therefore, that the anticoagulant and antiproliferative activities are separable. Smooth muscle proliferative activity is highly source dependent as heparins from different commercial suppliers show 2 to 3-fold differences in activity.[108] Full structure-activity studies will be required to exploit heparin as an antiproliferative agent. Heparin's effect on smooth muscle proliferation continues to be an active area of research interest.[109-116]

2. Ability to Inhibit Complement Activation

Heparin's principle location in man is the granules of tissue mast cells and basophils. Because heparin's primary location is so closely linked to the immune response, its ability to regulate complement has become an active area of interest. Heparin acts at multiple sites in both the classical[115-118] and the alternative amplification pathways[20,67,119,123] of complement. Heparin, heparin-oligosaccharides and other polyanions can inhibit the cell-bound alternative amplification pathway C3 convertases, C3b,B, and C3b,Bb,P[119] as well as fluid-phase consumption of B by D in the presence of C3b, suggesting a direct action on C3b.[119] Heparin's anticoagulant activity is primarily associated with a specific pentasaccharide sequence at which ATIII binds.[120,121] Heparin's structure-activity relationship on the complement system, however, is still poorly understood. Although studies indicate the importance of O-sulfation[117,118] and a minimum molecular size,[20] a specific binding site for a complement factor such as C3b (similar to the ATIII binding site) has not yet been implicated in heparin's ability to inhibit complement activity.[41]

3. Angiogenic and Antiangiogenic Activities

Angiogenesis or neovascularization is defined as the formation of new blood vessels. This process, originally identified with vascularization in the placenta,[120] has been extended to other capillary growth including tumor angiogenesis.[121] After a tumor takes hold it grows slowly, remaining quite small (1-2 mm^3) until new capillaries come to within 150-200 μm (the distance which oxygen can diffuse).[122] Once vascularized, the tumor cells multiply at a rapid rate resulting in serious damage to the tissue surrounding the growing tumor. The process of angiogenesis, or the ingrowth of blood vessels, requires the enzymatic degradation of the extracellular matrix underlying tissue (basement membrane), the movement of endothelial cells into this space and their replication.[123] Heparin may play a variety of roles in angiogenesis. Immediately before capillary ingrowth, mast cells, containing heparin, congregate. The heparin from these mast cells can stimulate endothelial cell migration and this activity can be blocked by the addition of protamine.[124]

Heparin can increase the activity, stability or binding of growth factors which stimulate angiogenesis, such as fibroblast growth factor (FGF), endothelial cell growth factor (ECGF).[125-127] Heparin and heparin oligosaccharides can inhibit angiogenesis[19] in the presence of angiostatic steroids, such as the naturally occurring metabolites of cortisone, which were previously thought inactive.[19,128] Specific sequences in the heparin polymer are capable of very tight chelation of copper,[129] long implicated as an angiogenesis modulator.[130] Folkman and Klagsbrun proposed that heparin-like glycosaminoglycans, lining the endothelium, in the presence of other soluble factors, such as steroids, restrain capillary growth.[38] This quiescent microvasculature[131] can rapidly respond to heparin modulated growth factors produced during ovulation, by wounds,[132] or inflammation (as occurs in stroke where the damaged blood brain barrier requires repair),[133] in immune reactions and tumors.[134]

To understand fully the role of heparin in the process of angiogenesis, specific saccharide sequences within the heparin polymer which bind growth factors and growth modulators such as copper need to be elucidated. Once the necessary sequences have been determined heparin-oligosaccharides might be prepared either enzymatically or synthetically. These biomimetic oligomers would potentially represent an important new class of drugs capable of regulating angiogenesis.

4. Additional Activities

Interaction of Heparin with Enzymes. Heparin has been found to bind to, inhibit or activate a large number of enzymes (Table 3). Heparin inhibits leukocyte elastase while having no effect on the related porcine pancreatic elastase or *Pseudomonas aeruginosa* elastases.[143,144] The inhibition was established to be tight-binding (Ki 40nM-100 µM depending on chain length) and hyperbolic non-competitive.[144] Over-sulfation of the heparin chain enhances its activity. The therapeutic application of this activity might be in treating disease states where the role of leukocyte elastase has been well established, such as emphysema or rheumatoid arthritis. Heparin can activate protein kinases found in both liver and skeletal muscle.[161,162] Heparin inhibits the activity of acetylcholinesterase.[163] Heparin's ability to inhibit topoisomerase I, important in cell replication, has been demonstrated to be independent of its anticoagulant activity and primarily the result of its highly charged nature.[145] Heparin accelerates the antiprotease inhibition of protease nexin, a protease found predominantly in the brain and one whose activity is associated with Alzheimer's disease.[164] Heparin effects the proteases involved in fertilization of the egg by the sperm in mammals.[165,166] Heparin is a potent inhibitor of phospholipase A2 which is found elevated in uremia.[167]

Interaction of Heparin and Proteins. Heparin has been shown to bind, to stimulate, or to protect a number of different growth factors including platelet,[168] epithelial,[169] epidermal,[170] fibroblast,[171,172] insulin-like,[173,174] and other growth factors.[111,175-179]

Table 3. Enzymes which interact with heparin.
BINDING
RNA polymerase,[135] THF synthetase,[136] Streptokinase,[137] Tryptase,[138] Adenylate cyclase,[139-141] Complement Convertases,[41] Thrombin.[142]
INHIBITION
Neutraphil elastase,[143,144] Topoisomerase I,[145] Topoisomerase II,[146] Esterase,[147] Lipase,[147] RNase,[148] Glycogen synthase Kinase 1,[149] Diamine oxidase,[150,151] Casein Kinase II,[152] Lysyl oxidase,[153] Sialyl transferase,[154] Protein C.[155]
ACTIVATION
Kinases,[156] Phophatases,[157] RNA-dependent protein Kinase,[158,159] Lipoprotein lipase and Hepatic triglyceride lipase.[160]

Heparin interacts with blood components,[180-184] blood proteins,[100,184-188] and extracellular matrix proteins such as fibronectin,[189] laminin,[190-192] and vitronectin.[193,194] Heparin also binds to low density lipoproteins (LDLs).[104]

Heparin's Other Biological Activities. An endogenous heparin-like anticoagulant has been found in the blood of an AIDS patient and has been associated with coagulation abnormalities.[195] Exogenous heparin is being clinically tested in the treatment of AIDS and has shown itself effective (as has dextran sulfate) in blocking human immunodeficiency virus (HIV) replication *in vivo*.[195-199]

Heparin and other proteoglycans have a role in neoplasia.[200] Heparin binding sites have been found on B16 melanoma cells.[201] Heparin's effect on angiogenesis has been exploited to inhibit tumor growth.[18,19,133] Heparin-like polysaccharides have recently been found in the neuritic plaque formed in Alzheimer's disease.[202] Low concentrations of heparin inhibit the binding of inositol phosphate to its receptor thus effecting calcium release from cells.[203-208]

RATIONALE FOR DEVELOPING BIOMIMETIC POLYMERS TO REPLACE HEPARIN

1. Structural Requirements for Active Biomimetic Polymers

Heparin's biological activities are primarily mediated through its binding to proteins and its regulation of their activities. To substitute for heparin, biomimetic polymers must bind to these proteins and regulate these same activities. Heparin's binding is primarily through electrostatic interactions and depends on its high charge density. Most of these biomimetic polymers are highly sulfated polyanions, prepared enzymatically or chemically, and can substitute for heparin by the positioning of these charged groups. Ideally these heparin substitutes should exhibit tighter binding to these proteins in order to have higher potency. Only the heparin ATIII binding site has been sufficiently studied to develop a well defined structure activity relationship (SAR). The synthetic heparin pentasaccharide binding site and heparin have comparable ATIII binding avidity (heparin-ATIII, $K_{diss} = 10^{-7}$, synthetic pentasaccharide-ATIII, $K_{diss} = 10^{-7}$)[120,121] as well as comparable activity towards the inhibition of ATIII. Structure modification studies are currently underway in an effort to prepare analogs with even higher binding affinities.[209,210] Unfortunately, heparin's interaction with ATIII is the only heparin-protein interaction that has been demonstrated to be restricted to a specific saccharide sequence. Even in this thoroughly studied system, it is uncertain which amino acids within ATIII are interacting with the required sulfate groups in heparin. To systematically develop heparin biomimetic polymers, an increased understanding of heparin's interaction with the proteins it binds to and regulates will be required.

2. Combating Heparin's Side-Effects

Although heparin's major activity as an anticoagulant is widely exploited, it is also associated with its most common side-effect. Bleeding complications can range from mild mucosal oozing to intracranial hemorrhaging. Low molecular weight (LMW) heparins represent a new class of therapeutic agents called antithrombotics.[211] These LMW heparins offer several advantages over the anticoagulant heparin. These include a reduced effect on platelets and increased specificity of action. LMW heparins catalyze the ATIII mediated inhibition of Factor Xa to a greater extent than the inhibition of Factor IIa. This greater specificity towards an early step in the coagulation cascade may be partially responsible for the reduced hemorrhagic side-effects of LMW heparin. Heparin's hemorrhagic side-effect have been reduced by modifying its molecular weight. In the future more subtle structural modifications may be used to prepare antithrombotic heparins which are free from hemorrhagic side-effects.

Heparin-induced osteoporosis (decalcification of the bones) can be reduced by either increasing heparin's anticoagulant potency, thereby reducing the dose of the agent required, or by decreasing the length of time a patient is heparinized. Many of heparin's other secondary side-effects can also be reduced by this approach.

3. Exploiting Heparin's Side-Effects

Not all of heparin's side-effects are undesirable. Some of these side-effects could be exploited and new pharmacologically active agents prepared if only these activities could be enhanced. For example, heparin releases and activates lipoprotein lipase (LPL), but does so only at concentrations which fully anticoagulate.[105] If a heparin could be prepared which was devoid of anticoagulant activity with high LPL releasing activity, it might represent a useful agent in the treatment of atherosclerosis.[105] Heparin inhibits complement activation, but only at concentrations ten-times those required for full anticoagulation.[41] Recent results in our laboratory have demonstrated that it is possible to prepare heparin-oligosaccharides equipotent with heparin (on a weight basis) towards complement, but without anticoagulant activity.[41] Such a drug, or one which contained equal potency as an anticoagulant and as a complement activation inhibitor, might be useful in preventing both coagulation and complement activation in extracorporeal therapy. There are scores of other heparin side-effects which might be usefully exploited resulting in the preparation of new classes of drugs.

4. Controlling Pharmacokinetics

One major problem with heparin is its poor bioavailability when administered by certain routes.[62] Heparin is primarily administered intravenously, and acts systemically as an anticoagulant.[15] It has also has been administered by the subcutaneous route.[62,211] By this route, however, only a fraction of heparin ever makes it to the circulation, and then only after a long delay.[62,211] This delayed release of low levels of heparin into the circulation is often desirable.[62,211] Low-dose heparinization has been effectively used in the treatment of deep vein thrombosis.[22,23] Both intravenous and subcutaneous routes require hospitalization of the patient being treated. If an orally active heparin could be prepared which was safe, it might permit the treatment of outpatients.[63,64] This type of therapy might be useful for disorders which do not generally call for hospitalization, such as the reduction of VLDLs and LDLs in the prevention of atherosclerosis.[105]

A major second problem associated with the use of heparin has been the lack of understanding of its metabolism and its erratic and sometimes unpredictable rate of clearance.[15] Many of these problems are due to the absence of a sufficiently sensitive assay for the measurement of chemical heparin in the plasma.[212] Currently the only methods having the required sensitivity are bioassays, which are also effected by subtle changes in the levels of coagulation proteins in the plasma and are subject to numerous additional interferences. The preparation of homogeneous heparins, which are pure single entities, would go a long way towards solving these problems in facilitating the development of sensitive chemical assays.

5. Preparing Stable Blood Compatible Surfaces

Part of the problem, associated with the preparation of blood compatible surfaces, is our lack of a complete understanding of coagulation and thrombosis. Ideally, one would like to mimic blood's natural container, the vessel lined with endothelial cells, as closely as possible. In addition, the flow characteristics of the device being designed must also closely resemble those observed in the vasculature.[213] Lastly, the surface must be stable and survive enzymatic and chemical attack from the components present in the circulation. It is beyond the scope of this review to consider the first two requirements except for saying that the natural endothelium is lined with heparin-like molecules. The last requirement, the production of a stable antithrombotic surface, is not as simple as it might appear. Simple adsorption of heparin onto a polymer produces a blood compatible surface which only lasts a short period of time until the heparin leaches from the surface.[25,213]

Covalent immobilization offers an alternative in that the linkage is chemically stable, but these surfaces also have a short lifetime, possibly due to the enzymatic stripping of heparin from the surface. Enzymes which act on heparin are present in the circulation such as exo and endoglycuronidases.[73,214] The precise nature of the surface-heparin linkage as well as control on the enzyme accessible regions of the immobilized heparin chain are required to form a stable heparinized surface. Our

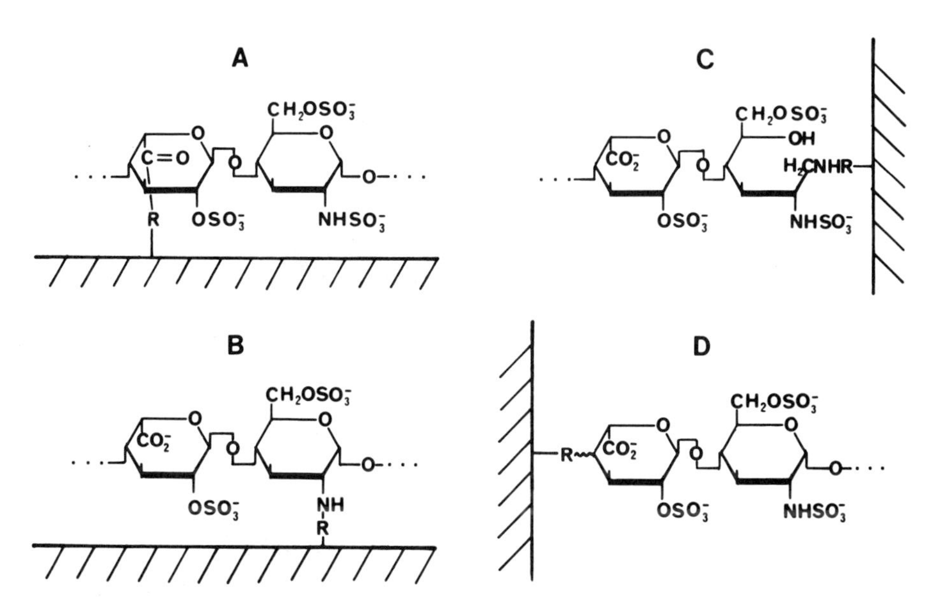

Figure 5. Immobilization of heparin to surfaces, (A) heparin coupled through carboxyl group; (B) heparin coupled through amino group; (C) heparin coupled through its reducing end; (D) heparin coupled through its non-reducing end (R = linkage arm).

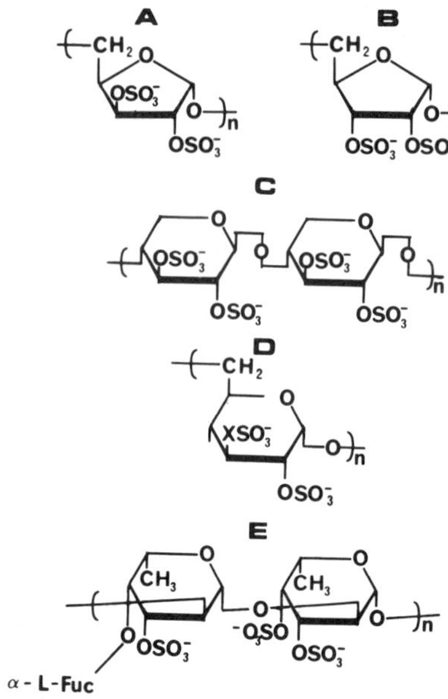

Figure 6. Structure of heparinoid natural products and their derivatives, (A) Xylan sulfate; (B) Ribose sulfate; (C). Pentosan sulfate; (D) Dextran sulfate (X = O or NH); and (E) Fucoidan (n = degree of polymerization).

laboratory is currently examining the question of how the orientation of a heparin chain immobilized to a surface (i.e., coupled through either its reducing-end or its non-reducing end) effects its linkage stability when exposed to enzymes in the circulation (Figure 5). Work along these lines is required to design blood compatible surfaces intelligently.

SYNTHETIC SULFATED POLYSACCHARIDES

Polymers that are structurally related to heparin and which possess certain of its biological properties, such as its anticoagulant activity, are commonly called heparinoids. These heparinoids have been prepared by modification of naturally occurring polysaccharides (Figure 6), by the total synthesis of heparin-like polymers, and most recently by the synthesis of small sulfated heparin-like oligomers.

1. Sulfation of Natural Polysaccharides

Chitin, a polymer of 2-acetamido-2-deoxy-D-glucose, is the major organic component of the exoskeleton found in insects, crabs, etc.[14] Chitin can be de-N-acetylated to prepare chitosan, which on chemical sulfation and/or carboxymethylation results in a polymer having certain structural similarities to heparin.[215] These chitosan derivatives show anticoagulant activity related to their degree of sulfation. Sulfated, carboxymethylated chitosan inhibits thrombin activity through ATIII to almost the same degree as heparin.

Pentosan, a (1→4)-β-linked xylopyranose with a single laterally positioned 4-O-methyl-α-D-glucuronic acid, is extracted from the bark of the beech tree, *Fagus sylvantica*.[216] When fully sulfated by chemical methods, it has been demostrated to be an anticoagulant with one-tenth of heparin's activity on a weight basis.[217] Its primary anti-IIa activity has been proposed to be HCII mediated.[218] This heparinoid also has a demonstrated antiheparin activity in that it can compete with heparin for antithrombin III binding reducing the rate of heparin catalyzed ATIII mediated inhibition of thrombin.[219]

Dextran, a (1→4)-β-D-glucan(1→3)-α-D-branched polymer[14] can be chemically sulfated to prepare dextran sulfate.[220] Both dextran (a plasma extender) and dextran sulfate have been used as pharmaceuticals.[221] Dextran sulfate has low anticoagulant activity with high LPL releasing activity.[222] This has permitted the exploitation of this agent as an anti-atherosclerotic in Japan.[222] Dextran sulfate has been used as a heparin replacement in anticoagulation and has recently been immobilized on plastic tubes to prepare non-thrombogenic surfaces.[223] A clinical study on the use of dextran sulfate as an HIV inhibitor in the treatment of AIDS has recently begun.

2. Fully Synthetic Sulfated Polymers

Synthetic polymers such as poly(vinyl sulfate) and poly(anethole sulfonate) are highly charged heparin-like polyanions (Figure 7) which exhibit anticoagulant activity.[224] This activity has only been exploited *in vitro* in preventing coagulation of collected blood or plasma samples during storage for assay. The use of these agents *in vivo* has largely been precluded by their high toxicity. These synthetic polymers are resistant to metabolism and thus remain in the body for extended periods during which they can display severe side-effects.

3. Small Sulfated Oligomers

Homogeneous, structurally defined, heparin oligosaccharides have been prepared by both enzymatic and synthetic means.[225-229] These possess both anticoagulant activity as well as other heparin associated activities. Often these heparin-oligosaccharides are specific displaying only a single activity.

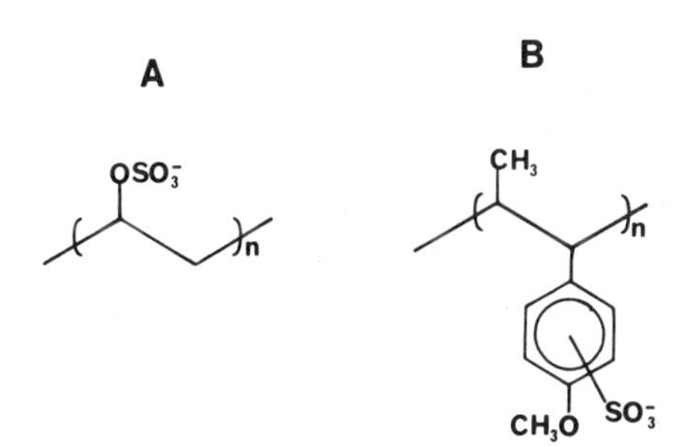

Figure 7. Structure of synthetic heparinoids. (A) Poly(vinyl sulfate) and (B) Poly(anethole sulfonate) where n = degree of polymerization.

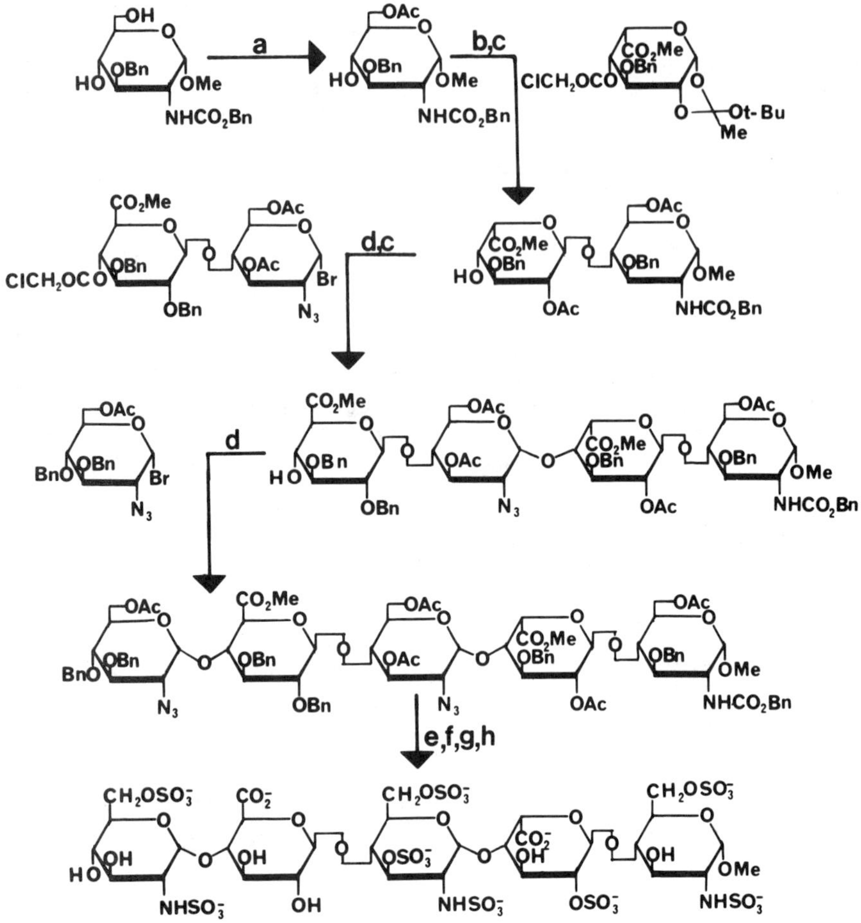

Figure 8. Synthesis of synthetic pentasaccharide comprising heparin's ATIII binding site.[227] (a), N-Acetylimidazole, $ClCH_2CH_2Cl$; (b), Chlorobenzene, 2,6-dimethylpyridinium perchlorate, heat; (c), CH_3OH, CH_2Cl_2, 2,6-dimethylpyridine, AcOH, hydrazinedithiocarbonate; (d), $ClCH_2CH_2Cl$, 2,4,6-trimethylpyridine, -20°C, silver triflate; (e), $CHCl_3$-CH_3OH-H_2O, NaOH; (f), N,N-dimethylformamide, trimethylamine-sulfur trioxide, 50°C; (g), CH_3OH-H_2O, 5% Pd/C; (h), Pyridine-sulfur trioxide.

The synthetic 3-O-sulfated pentasaccharide, representing heparin's ATIII binding site was first prepared by Choay *et al.*, in a multi-step synthesis in less than 5% yield.[226] This synthesis has been optimized by Choay *et al.*,[227] (Figure 8) and repeated by other groups.[228,229] Activity studies demonstrate that this agent is a potent accelerator of ATIII mediated inhibition of Factor Xa.[230] Clinical studies on this pentasaccharide as an antithrombotic agent demonstrated that it was not as effective as heparin itself.[211] The cost of synthesis may also preclude its use as a therapeutic agent.

Heparin-oligosaccharides of defined structure have been prepared from heparin in our laboratory using enzymatic methods.[225] These heparin oligosaccharides possess a number of important biological activities, including their capacity to inhibit complement activation *in vitro*, with nearly equipotency to heparin on a weight basis.[41] Further studies will be required to demonstrate this activity *in vivo* as well as to develop large-scale inexpensive methods to prepare these heparin oligosaccharides.[225]

Synthetic, highly sulfated, lactobionic acids have recently been demonstrated to have high activity in the potentiation of HCII inhibition of thrombin.[231] The regulation of the coagulation cascade at this step resembles dermatan sulfate's activity and might represent an interesting new class of antithrombotic agents.

Synthetic analogs of heparin have a great advantage in that they can permit access to unusual structures which do not occur in nature and thus are not found in natural products.[210,232] On the other hand, carbohydrate synthesis is extremely complex and tedious, involving many blocking and de-blocking steps required to protect sensitive functionality. Thus, it is difficult to predict when structurally complex pentasaccharide to decasaccharide sized structures will be preparable in a cost effective manner by the synthetic chemist. A natural product, such as heparin, provides a relatively inexpensive source of the desired oligosaccharide sequence, thus its enzymatic recovery from the polymer probably represents the best method for its preparation. Ultimately, the biotechnologist might be able to displace both the synthetic and the natural product routes if the desired sequence can be prepared using recombinant genetics. This option is still far off as glycosaminoglycan biosynthesis requires many enzymes to act in concert in a highly compartmentalized cell. Prokaryotes, in which most recombinant genetics is currently carried out, probably are not sufficiently compartmentalized to conduct such synthesis. Thus the preparation of glycosaminoglycans by recombinant genetics awaits the development of improved eukaryotic vectors.

FRACTIONATED HEPARINS

1. On the Basis of Size or Charge

Heparin can be fractionated on the basis of size using gel permeation chromatography (GPC).[233] Both low pressure and high pressure GPC has been used to obtain heparin fractions of average molecular weight between 5,000 and 40,0000.[233] Detection methods include continuous measurement of absorbance at 206 nm,[53] colorimetric assay of fractions using metachromatic dyes such as azure A,[212] or uronic acid assay using carbazole.[234] Because of its high charge density, heparin exists as an extended helical rod and thus has a greater molecular size than does a globular protein of similar molecular mass,[235] as estimated by GPC. For example, a heparin of MW (average) 13,000 has a K_{avg} comparable to a protein of MW 60,000 when sized on a Sephadex G150 column.[20] GPC is primarily used as an analytical technique but has been used for the small scale preparation of heparin fractions[20,55] Some activities of heparin which vary according to chain size are shown in Table 4.

Being a sulfated polysaccharide (with an average of 2.7 sulfate groups per repeating disaccharide unit), heparin can be fractionated on the basis of its charge using anion exchange chromatography. Both low and high pressure strong anion exchange (SAX) chromatography of heparin has been reported.[43,53] Typically, heparin is bound to an anion exchanger at low ionic strength and released by either a stepwise or gradient elution with solute of increasing ionic strength.[43,53] Low pressure supports such

Table 4. Heparin activities as a function of polymer chain length.

	Approximate MW	Anti IIa activity[236] (U/mg) (b) mediated by ATIII	HCII	Relative complement activation inhibition (b)[41]	Relative lipase releasing activity[105]
Tetrasaccharides[a]	1200	<1	<1	4	8
Hexasaccharides	1900	<1	<1	5	10
Octasaccharides	2400	<1	<1	10	9
Decasaccharides	2900	<1	<1	25	30
Dodecasaccharides	3500	5	2	35	c
Tetradecasaccharides	4100	13	16	45	-
Hexadecasaccharides	4700	17	16	55	-
Octadecasaccharides	5300	26	18	-	-
Eicosasaccharides	5900	24	14	-	-
Heparin fraction 1	4,000	6	23	100	32
Heparin fraction 2	5,700	26	44	100	-
Heparin fraction 3	10,700	38	115	100	-
Heparin fraction 4	14,500	196	271	-	-
Heparin fraction 5	16,900	130	331	-	-
Heparin fraction 6	25,300	92	356	100	-
Unfractionated heparin	14,000	171	225	100	100

(a) All heparin and heparin oligosaccharides are from porcine intestinal mucosa.
(b) Oligosaccharides prepared using heparin lyase are mixtures of those shown in Table 2.
(c) Not determined.

as DEAE or QAE cellulose, Sephadex or Sephacel bind heparin at ionic strengths of <0.2M and release heparin at 0.5-1.5 M concentrations of salts such as sodium chloride.[236] Heparin fractions can be prepared with a degree of sulfation ranging from 2 to 3 sulfate groups per disaccharide repeating unit.[236] Some of heparin's biological activities vary with degree of sulfation making this a useful technique to separate or enrich specific heparin based activities (Table 5).

2. Affinity Fractionation

A major breakthrough in heparin research came with the observations that there were two populations of heparin chains, having different affinities to ATIII, which could be separated by affinity chromatography.[237] The different ATIII binding affinities of these heparin fractions translated into major differences in ATIII-mediated activities.[120,121] For example, unfractionated heparin having an ATIII-mediated anti-thrombin activity of 150 U/mg can be fractionated into 33 wt% high-affinity heparin (500 U/mg) and 66 wt% low-affinity heparin (50 U/mg).[238] The affinity separation is dependent on tight-binding of the high affinity fraction ($K_{diss} = 10^{-9}$) and weak-binding (through non-specific interactions) of the low-affinity fraction ($K_{diss} = 10^{-4}$). A

Table 5. Heparin activity as a function of degree of sulfation or charge.

	Anti IIa Activity (U/mg) mediated by[236]	
	ATIII	HCII
Heparin fraction 1[a]	17	90
Heparin fraction 2	108	261
Heparin fraction 3	125	336
Heparin fraction 4	227	468
Heparin fraction 5	540	448
Unfractionated heparin	171	225

(a) Heparin fractionated on the basis of charge using DEAE Sephadex with salt gradient elution. The degree of sulfation of heparins 1 to 5 ranged from 2.73 - 2.83 sulfates/disaccharide repeating unit.

minimum binding affinity of $K_{diss} = 10^{-5}$ to 10^{-6} is required for efficient separations by affinity chromatography. Heparin has been fractionated by its affinity towards other proteins (Table 6), and although many of these demonstrate enhanced biological activities, none have resulted in the level of enrichment seen in ATIII affinity fractionation, nor have any resulted in the isolation and characterization of specific sequences through which binding and activity is expressed.

Immunoaffinity chromatography has been successfully used to fractionate a wide variety of proteins. The application of this technique to fractionating heparin, based on the presence of specific saccharide sequences which could be recognized as individual epitopes, would represent an important advance. The problem with this approach is that heparin is remarkably non-immunogenic. No true (IgE-mediated) allergic reactions to heparin have ever been demonstrated in patients, (only one unconfirmed report of a human antibody to heparin is reported in the scientific literature.)[245] despite heparin's widespread clinical use for nearly 60 years.[42] Attempts to generate either polyclonal or monoclonal antibodies

Table 6. Proteins which have been used to affinity fractionate heparin.

Protein	References
Antithrombin III	239
Heparin Cofactor II	236
Thrombin	240
Poly(lysine)	241
Platelet Factor IV	242
Fibronectin	(a)
C3b	(a)
Lipoprotein Lipase	105
Concanavallin A	243
Protamine	244

(a) Unpublished data from our laboratory

to unmodified heparin chains in animals using a variety of approaches including adjuvants, immobilized heparins, a variety of heparin (hapten) carriers, etc., have failed. Recently, our group has prepared polyclonal antibodies to heparin-oligosaccharides having a modified unsaturated uronic acid residue at the chains nonreducing terminus. A recent report of antibodies has also appeared which again uses heparin-oligosaccharides modified at the reducing terminus, containing a ring contracted anhydromannose produced by deaminative cleavage.[50]

LOW MOLECULAR WEIGHT HEPARINS

1. Preparation

Low molecular weight (LMW) heparins are generally prepared through the controlled, partial, chemical or enzymatic depolymerization of commercial GAG heparin. Early attempts at acid or base catalyzed hydrolysis of heparin failed primarily due to the resistance of the glycosidic linkages in uronic acid to hydrolysis, coupled with the sensitivity of other functional groups to the strong hydrolysis conditions required.[58] A convenient method to cleave heparin involves its treatment under relatively mild conditions with nitrous acid.[246] Heparin is cleaved at glucosamine residues containing an N-sulfate group.[246] Not all N-sulfated glucosamines are equally susceptible to deaminative cleavage, the resistance of 3,6-O-sulfated glucosamine N-sulfate (found in the center of the ATIII binding site) was a fortunate accident which lead to the isolation of ATIII binding oligosaccharides following deaminative cleavage of heparin.[121] This method of preparing LMW-heparins introduces an artifact into the oligosaccharide's reducing end, a ring-contracted anhydromannose residue.[246,252] There is currently no easy way of removing this residue from LMW-heparins prepared by this method. Because of the selectivity of nitrous acid for residues outside the ATIII binding site, an enrichment in this site is obtained in these preparations. Thus depolymerization by nitrous acid treatment is seldom run to completion to preserve these resistant sulfated glucosamine residues and to retain ATIII binding activity.[120,121]

Peroxidative cleavage of sugars by Smith degradation permits the cleavage of heparin. Heparin contains unsulfated glucuronic acid residues which have vicinal diol functionality particularly susceptible to certain oxidants. The cleavage of the glucuronic results in an acyclic residue which is sensitive to hydrolysis under relatively mild conditions. The ATIII binding site contains an unsulfated uronic acid and thus can also be cleaved by this method. Again controlled, partial oxidation followed by hydrolysis is required to retain activity. It is possible that some acyclic, oxidized sugar residues may remain in LMW-heparin prepared using this method, however further studies will be required to establish this point.[45,252] Heparin can be eliminatively cleaved at its iduronic acid residue (this residue contains the appropriate anti-relationship between the 5-position proton and the 4-position glycoside anion. This eliminative cleavage can be further facilitated by first forming the benzyl ester of uronic acid before treatment with base. Only the glucuronic acid residues survive such eliminative cleavage and these can then be debenzylated.[247] The resulting oligosaccharides have an unsaturated uronic acid residue at their non-reducing end. This residue may prolong the *in vivo* half-life of these oligosacharides by blocking their biotransformation by exoglucuronidases.[73] This unsaturated sugar, however can be easily removed by treatment with ozone followed by acidic work-up. More selective eliminative depolymerization of heparin can be accomplished enzymatically

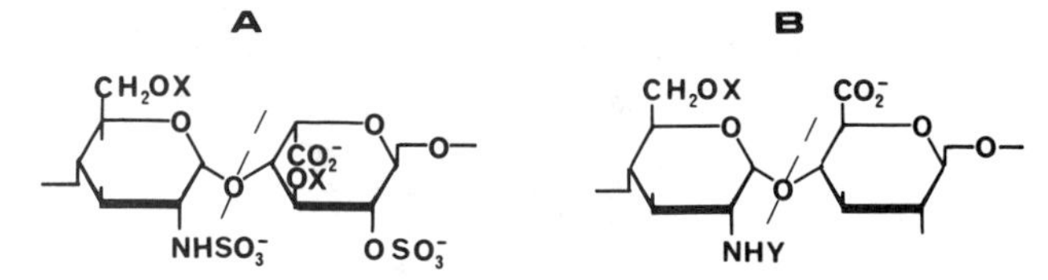

Figure 9. Site-specificity of heparin lyase and heparan monosulfate lyase, (A), heparin lyase site and (B), heparan monosulfate lyase site, where X = SO_3^- and Y = SO_3^- or $COCH_3$.

using heparin and heparan sulfate lyases (Figure 9).[248]

Heparin lyase cleaves heparin's most common likage, 2,6-disulfated-glucosamine-(1→4)-2-sulfated iduronic acid.[248] This enzyme will cleave this linkage even if the 6-sulfate is missing or in the presence of a 3-O-sulfate in the glucosamine residue, found in heparin's ATIII binding site.[249] In fact, it now appears that heparin lyase preferentially cleaves heparin's ATIII binding site possibly because of its high degree of sulfation. This makes heparin lyase a more useful tool in preparing oligosaccharides with low ATIII mediated anticoagulant activity (for applications such as inhibition of complement activation).[41] Heparan sulfate lyase can also be used to depolymerize heparin and, although it cuts at fewer linkages, one of these is contained within the ATIII binding site as well.[249] Thus, both lyases must be used in controlled partial depolymerizations to prepare LMW heparins with anticoagulant activity.[250] As with the LMW-heparin prepared by chemical eliminative cleavage, the unsaturated sugar can either be left on each chain to block biotransformation, or can be removed by ozonolysis.[251]

2. Chemical and Physical Properties

Although each LMW heparin is different (ie, each possess artifacts generated in their preparation) they all have several features in common.[252] These LMW heparins are still polydisperse with average molecular weights of approximately 5000 (Table 7), and having molecular weights ranging from 2000-8000.[250,253] These LMW heparins have been mapped using SAX-hplc and gradient PAGE techniques (Figures 10 and 11). As with the standard commercial heparins which have been mapped, LMW heparins are primarily comprised of six oligosaccharides (Figures 10 and 11).[252] In addition to these major oligosaccharides these maps show uncharacterized oligosaccharide components which are probably introduced in the depolymerization step.[121,255,256] More study of the structure of these LMW heparin preparations will be required to understand and to rationalize the differences in their biological activities.

3. Biological Activities

The anticoagulant activities of several common LMW heparins are presented in Table 7. A ratio of ATIII mediated anti-Xa to ATIII mediated anti-IIa activities of 5-25 is observed in LMW heparins resulting in their inhibition of coagulation early in the coagulation cascade.[238] presumably reducing their hemmorrhagic side-effects.[21] These LMW heparins also exhibit a decreased effect on platelets reducing both their activation and aggregation.[21] Increases in bioavailability following subcuta-

Table 7. Properties of commercial low molecular weight heparins.

Agent	Source	Origin	Preparation[a]	Wt Avg MW[b]	Anti Xa Potency[c]
CY216 (Fraxiparin)	Choay(FR)	PM	EtOH FRC, NA	5506	95
CY222	Choay(FR)	PM	Prolonged NA	3410	60
Kabi 2165 (Fragmin)	Kabi(SW)	PM	NA, GPC	6370	142
LMWH/DHE	Sandoz(SZ)	PM	AN	6322	-
RD118885	Hepar/ Wyeth(US)	PM	OX	6221	62
OP2123	Opocrin(IT)	BM	OX	6511	83
PK10169 (Enoxaparin/ Lovinox)	Pharmuka(FR)	PM	Benzylation/β-E	3789	96
LHN-1 (Logiprain)	Novo(DN)	PM	Enzymatic β-E	4850	87
StdLMWH 85/600	NIBSC(UK)	PM	NA, FRC	-	100
Heparin	Hepar(US)	PM		14,000	150

(a) NA, nitrous acid; GPC, gel permeation chromatography; AN, isoamylnitrate; OX, peroxidative; β-E, β-elimination, FRC, fractionation.
(b) By GPC.[253]
(c) Relative Anti-Xa potency against NIBSC standard LMWH.[254]

neous administration and increased half-life of LMW heparin preparations have also been reported.[211] These improved biological activities have paved the way for clinical trials of these new drugs.

4. Clinical Studies

Studies using LMW heparins performed in healthy volunteers (Phase 1) have demonstrated their safety at doses that are generally considered effective.[24] Phase 2 clinical trials have begun to study both the safety and efficacy of LMW heparins in the treatment of a variety of disease

Table 8. Clinical studies on low molecular weight heparins.

Study	References
Long term Prophylaxis of thromboembolism	257
Non-hemorrhagic stroke	24
Deep Vein Thrombosis	258
Membrane Oxygenation	259
Kidney Dialysis	260
With Fibrinolytics	101

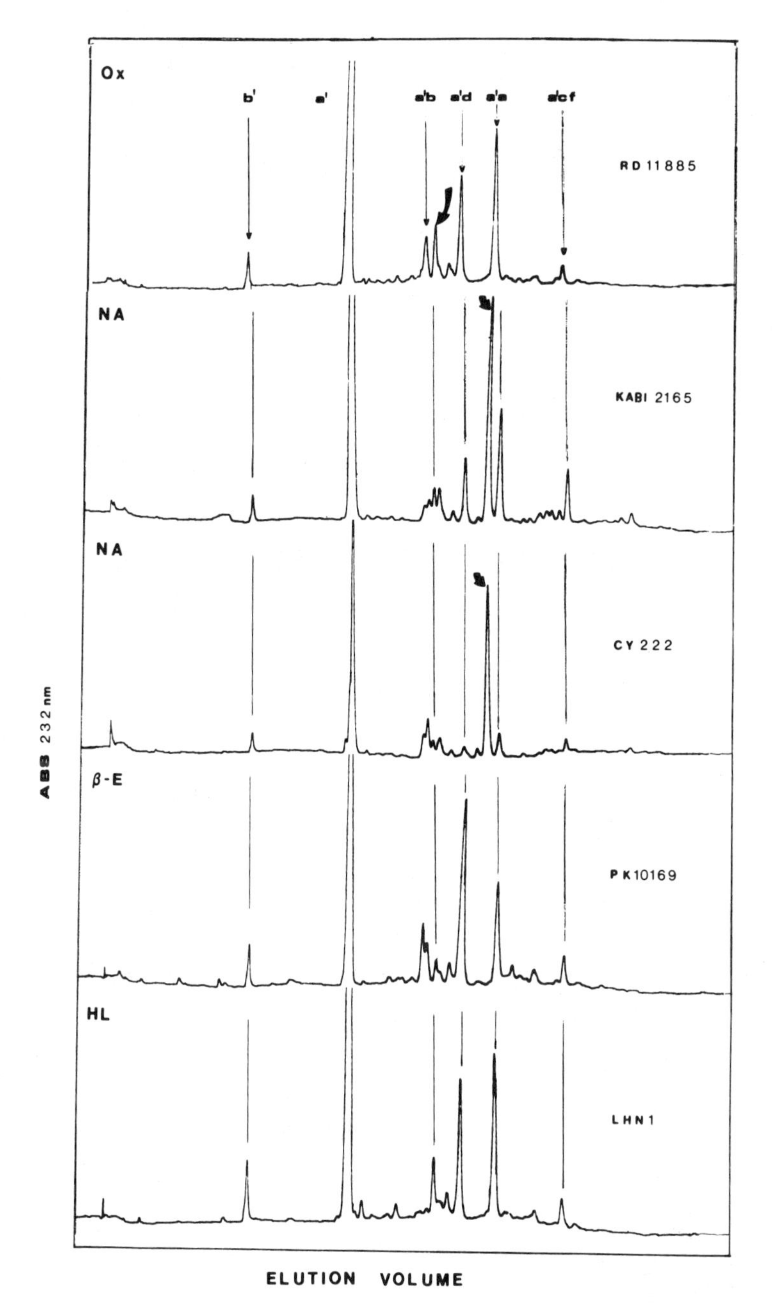

Figure 10. Strong anion exchange-hplc oligosaccharide map of low molecular weight heparins. These LMW heparins (see Table 7) were prepared oxidatively (OX) by nitrous acid depolymerization (NA), by chemical β-elimination (β-E) and using heparin lyase. Absorbance at 232 nm is plotted against elution volume and chromatography is performed using conditions given in ref. 43. The position at which standard oligosaccharides elute are marked using vertical lines and abbreviations of their chemical structure are described in Tables 1 and 2.

states (Table 8). The results of these studies are promising, suggesting that they are safer than heparin, generally resulting in less hemmorrhagic complications, and have greater bioavailability and longer half-lives resulting in better dose control.[24] One potential problem of LMW heparins is the failure of protamine sulfate to reverse their action effectively (as it does for heparin),[261] particularly when the longer half-life of LMW-heparin is considered.[211] This may stimulate research into new method of LMW-heparin reversal and the preparation of protamine sulfate replacements.

HEPARIN OLIGOSACCHARIDES

Chemically defined, homogeneous heparin oligosaccharides have several advantages over the polydisperse microheterogeneous preparations (including LMW heparins) currently in use.[21] First, it should be possible to prepare extremely selective agents, i.e, one which regulates complement activation but not coagulation or one which targets a specific step in the coagulation cascade. Second, the concentration of homogeneous drug entities are easier to monitor *in vivo* and hence better dose control may be possible, reducing the side-effects common in heparinization.[65] Third, homogeneous heparins make metabolism studies easier to perform as well as studies of heparin SAR, potentially resulting in the faster development of second and third generation heparin-like drugs with a variety of pharmacological activities.

1. Chemical Synthesis

The first total synthesis of a heparin-oligosaccharide was accomplished in 1984 resulting in the preparation of a pentasaccharide ATIII binding site.[226] This synthesis both confirmed the reported structure and activity described for the natural product and also permitted the unequivocal elucidation of the functionality required for the expression of ATIII mediated activity.[262] The initial synthetic scheme affording ATIII binding pentasaccharide required multiple steps resulting in a less than 5% overall yield.[226] Subsequent synthesis reported by a number of groups have reduced the required steps and increased the yield.[227-229] Chemical synthesis of the ATIII binding pentasaccharide possesses two major advantages over its preparation from the natural product, heparin. The first, is that it results in independent confirmation of structure and eliminates any doubts that the identified structure (as opposed to a minor contaminant) is responsible for biological activity. The second advantage of chemical synthesis is that, if the synthetic strategy is designed well, a large number of derivatives that can not be obtained from the natural product, can be synthesized and examined for biological activity.

2. Enzymatic Preparation

Heparin oligosaccharides can be prepared from the natural product in a single enzymatic step followed by fractionation and purification.[225] The simplicity of this approach and the low cost of the heparin starting material certainly makes it competitive with classical synthesis. One major failure, however, is that often the desired sequence is either not present in the starting heparin or it cannot be enzymatically removed from the polymer. One approach to these problems is to first use enzymes such as heparin lyase and heparan sulfate lyase to prepare a collection

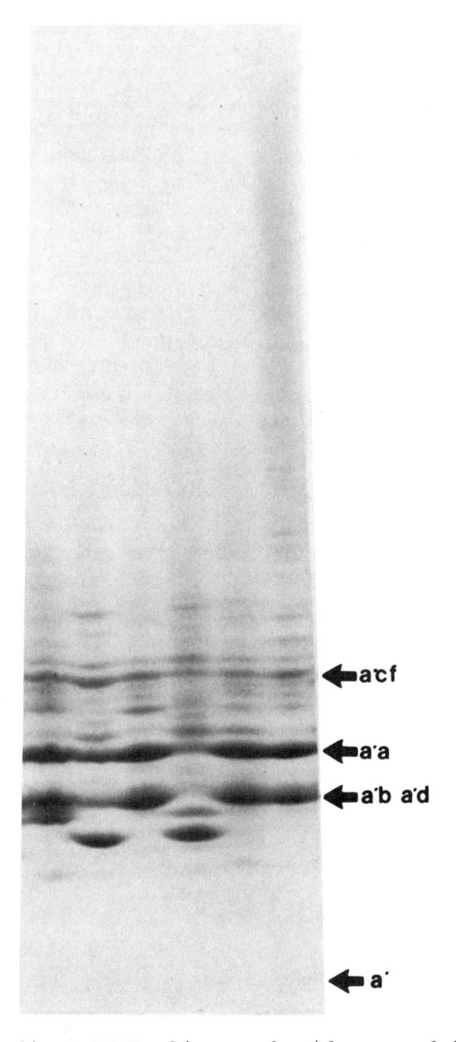

Figure 11. Gradient-PAGE oligosaccharide map of low molecular weight heparins. Lanes: (1) LMWH, β-elimination (PK 10169); (2) LMWH, Nitrous Acid (Kabi 2165); (3) LMWH, Heparinase (LHN-1); (4) LMWH, Nitrous Acid (Cy 222); (5) LMWH, Peroxidative (RD 11885); (6) Porcine heparin, (Hepar PM 18085). Standards are indicated with arrows (see Tables 1 and 2 for abbreviations).

of small oligosaccharide building blocks. These building blocks might then be combined either chemically or enzymatically in a number of proportions or sequences to prepare a wide variety of heparin sequences. Initial studies on the use of heparin lyase to synthesize large oligosaccharides from smaller ones by reverse enzymic catalysis have been unsuccessful. An approach currently under study in our laboratory involves the treatment of a partial ATIII binding site a'cf (see Tables 1 and 2 for chemical structure) with heparin lyase in the presence of ATIII as a trap for a completely intact ATIII binding site arising from reverse catalysis. Further research along these lines will be necessary.

3. Chemical and Physical Properties

The chemical and physical properties of structurally defined oligosaccharides are essentially the same as those of polydisperse preparations. Structurally defined, homogeneous, oligosaccharides, however, offer certain distinct advantages. For example, recent studies in our laboratory using ^{13}C NMR over a range of pH has for the first time permitted the measurement of the pKa of the various types of carboxyl groups found in the heparin polymer.[263] Such a measurement would not be possible by examining polydisperse mixtures. A second advantage of using homogeneous oligosaccharides of defined structure is that complete ^{13}C and ^{1}H NMR spectral assignment is often possible. Definitive and complete assignment is a prerequisite for analyzing the secondary structure of these oligosaccharides using advanced NMR techniques including 2D COSY and 2D NOESY.[264] These permit distance, which can be used to deduce the average conformation of these molecules. Since these measurements are performed in aqueous solutions under near physiological conditions, the data collected may be more physiologically relevant than what might be obtained from X-ray crystal structure (this approach is also only possible using homogeneous preparations). Even a simple property such as molecular weight is complicated by polydispersity, particularly when the MW distribution of species is not symmetrical. Such distributions result in number average, weight average, and Z average molecular weights which can be quite different.

USING HEPARIN OLIGOSACCHARIDES TO UNDERSTAND HEPARIN SAR

1. Mapping the ATIII Binding Site

The story behind the determination of heparin's ATIII binding site provides a useful lesson in understanding the structures behind heparin's other biological activities. The first step is the demonstration of high affinity binding to a protein and that not all heparin molecules are capable of such binding (selectivity). The second step is a reductionist approach resulting in the isolation of the smallest oligosaccharide capable of high affinity binding. The third step is preparation of that oligosaccharide in its pure form from the natural product and determination of its structure. The fourth step is the independent synthesis of this high affinity oligosaccharide and the demonstration that the synthetic product has activity equivalent to that produced from the natural product. Fifth, the determination of the molecule's SAR by chemically or enzymatically altering specific functional groups or by the denovo synthesis of a variety of structurally related derivatives.

2. Other Activities

Screening a library of structurally defined oligosaccharides with a battery of bioassays might result in separation of two biological activities, as it has for complement regulatory and anticoagulant activities.[41] Oligosaccharide samples of defined structure have been sent by our laboratory around the world for screening by groups for a variety of biological activities including anti-gout, immunomodulatory, topoisomerase, antifertility,[265] and osteoporosis.[266] The use of pure, structurally characterized oligosaccharides can avoid misleading and often unreproducible results frequently obtained when mixtures are examined.

3. Metabolism

Here the use of defined homogeneous oligosaccharides is self evident. Without such samples it is not possible to study the biotransformation and clearance of these drugs. The absence of such defined oligosaccharides in the past has made heparin the drug with one of the most poorly understood pharmacokinetics. The knowledge of heparin pharmacokinetics will undoubtedly improve with the introduction of these defined oligosaccharides.[73] This may also lead to a better understanding of a variety of genetic diseases that result in the incomplete metabolism of GAGs resulting in their accumulation and ultimately causing mental retardation and death.[14]

HEPARIN OLIGOSACCHARIDES POTENTIAL AS NEW THERAPEUTIC AGENTS

Whether or not structurally defined, homogeneous heparin-oligosaccharides will ever be used as therapeutic agents depends on several factors. The first would be improved anticoagulant properties including better bioavailability, pharmacokinetics, dose control and dose monitoring, reduced side-effects, easier reversal and higher specificity. The second would be the therapeutic exploitation of other activities such as complement regulatory activity, angiogenesis regulatory activity, smooth muscle proliferative regulation, lipoprotein lipase release and activation activities, antiviral activity, and heparin's many additional observed biological activities. Before these activities can be exploited they must be separated and the absence of side-effects as well as high potency (a high therapeutic index) must be demonstrated. The third potential is one based on understanding how heparin acts and what its natural physiological roles are. Once known, this may result in previously unforeseen applications for heparin oligosaccharides. Finally, by understanding heparins SAR and physiological roles, analogues with greatly improved properties and potency, the "next generation" of heparin development, might be possible.

REFERENCES

1. B. Casu, Adv. Carbohydr. Chem. Biochem., **43**, 51 (1985).
2. W. D. Comper, "*Heparin and Related Polysaccharides,*" Polymer Monographs, Volume 7, Ed. M. B. Huglin, Gordan and Breach Science Publ., New York, 1981.
3. A. A. Horner & E. Young, J. Biol. Chem. **257**, 8749 (1982).
4. E. Coyne, Fed. Proc., **36**, 32 (1977).
5. K. G. Jacobsson & U. Lindahl, Biochem. J., **246**, 409 (1987).
6. R. J. Linhardt, Z. M. Merchant, K. G. Rice, Y. S. Kim, G. L. Fitzgerald, A. C. Grant & R. Langer, Biochemistry, **24**, 7805 (1985).
7. R. J. Linhardt, D. M. Cohen & K. G. Rice, Biochemistry, **28**, 2888 (1989).
8. I. Jacobsson & U. Lindahl, J. Biol. Chem., **255**, 5094 (1980).
9. U. Lindahl, D. S. Feingold & L. Roden, **TIBS**, 221 (1986).
10. U. Lindahl & L. Kjellen in: "*The Biology of the Extracellular Matrix Proteoglycans,*" T. N. Wight & R. Mecham, Eds., Academic Press, New York, 1987, p. 59.
11. B. M. Vertel, W. B. Upholt & A. Dorfman, Biochem. J., **217**, 259 (1984).
12. R. K. Chopra, C. H. Pearson, G. A. Pringle, D. S. Frackre & P. G. Scott, Biochem. J., **232**, 277 (1985).

13. J. T. Gallagher & M. Walker, Biochem. J., **230**, 665 (1985).
14. J. F. Kennedy & C. A. White, "*Bioactive Carbohydrates*," Ellis Horwood Limited, New York, 1983.
15. W. W. Coon. J. Clin. Pharmacol. **19**, 337 (1979).
16. L. B. Jaques, Science, **206**, 528 (1979).
17. Y. S. Kim, K. B. Lee & R. J. Linhardt, Thromb. Res. **51**, 97 (1988).
18. J. Folkman, Biochem. Pharmacol., **34**, 905 (1985).
19. J. Folkman, R. Langer, R. J. Linhardt, C. Haundenschild & S. Taylor, Science, **221**, 719 (1983).
20. M. D. Sharath, J. M. Weiler, Z. M. Merchant, Y. S. Kim, K. G. Rice & R. J. Linhardt, Immunopharmacol., **9**, 73 (1985).
21. J. Fareed, Sem. Thromb. Hemosts., **11**, 227 (1985).
22. V. V. Kakkar, B. Djazaeri, J. Fok, M. Fletcher, M. F. Scully & J. Westwick, British Medical J., **284**, 375 (1982).
23. G. Bratt, E. Tornebohm, S. Granquist, W. Aberg & D. Lockney, Thromb. Haemosts., **54**, 813 (1985).
24. J. Biller, E. W. Massay, H. P. Adams, J. N. Davis, J. R. Marler, A. Bruno, R. A. Henriksen, R. J. Linhardt, L. Goldstein, M. Alberts, C. Kisker, G. Toffol, C. Greenberg, K. Banwart, C. Bertels, M. Walker, & H. Maganani, Neurology, **39**, 262 (1989).
25. N. A. Plate' & L. I. Valuev, Adv. Polym. Sci., **79**, 95 (1986).
26. A. Farroqui & A. Horrocks, Adv. Chromatogr., **23**, 127 (1983).
27. D. P. Kapelanski, personal communication.
28. J. M. Toomasian, L. C. Hsu, R. B. Hirschl, K. F. Heiss, K. A. Hultquist & R. H. Bartlett, ASAIO Trans., **34**, 410 (1988).
29. H. P. Hoffman, N. B. Schwartz, L. Roden & D. J. Prockap, Connect. Tissue Res., **12**, 151 (1984).
30. K. Lidholt, J. Riesenfeld, K-G. Jacobsson, D. S. Feingold & U. Lindahl, Biochem. J., **254**, 571 (1988).
31. J. Riesenfeld, M. Hook & U. Lindahl, J. Biol. Chem., **254**, 421 (1982).
32. M. Kusche, G. Backstrom, J. Riesenfeld, M. Petitou, J. Choay & U. Lindahl, XIVth International Carbohydrate Symposium, Stockholm, Sweden, August, 1988, c 22.
33. R. L. Stevens, C. C. Fox, L. M. Lichtenstein & K. F. Austen, Proc. Natl. Acad. Sci., U.S.A., **85**, 2284 (1988).
34. R. D. Guhmann, Ed., "*Immunology*," The Upjohn Company, Kalamazoo, MI, 1981.
35. S. Vanucchi, M. Ruggiero & V. Chiarugi, Biochem. J., **227**, 57 (1985).
36. J. A. Marcum & R. D. Rosenberg, Semin. Thromb. Haemostasis, **13**, 464 (1987).
37. G. Pejler, G. Backstrom, U. Lindahl, M. Paulsson, M. Dziadek, S. Fujiwara & R. Timps, J. Biol. Chem., **262**, 5036 (1987).
38. J. Folkman & M. Klagsburn, Science, **235**, 442, (1987).
39. K. M. Krianciunas, F. Grigorescu & C. R. Kahn, Diabetes, **36**, 163 (1987).
40. J. T. Gallagher, M. Lyon & W. P. Steward, Biochem. J., **236**, 313 (1986).
41. R. J. Linhardt, K. G. Rice, Y. S. Kim, J. D. Engelken & J. M. Weiler, J. Biol. Chem., **263**, 13090 (1988).
42. L. E. Bottinger, Acta Med. Scand. **222**, 195 (1987).
43. R. J. Linhardt, K. G. Rice, Y. S. Kim, D. L. Lohse, H. M. Wang & D. Loganathan, Biochem. J., **254**, 781 (1988).
44. P. Hovingh, M. Piepkorn & A. Linker, Biochem. J., **237**, 573 (1986).
45. P. Bianchini, personal communication
46. P. Hovingh & A. Linker, J. Biol. Chem., **257**, 9840 (1982).
47. S. L. Burson, Jr., M. J. Fabrenback, L. H. Frommhagen, B. A. Riccardi & R. A. Brown, J. Am. Chem. Soc., **78**, 5874 (1956).
48. L. H. Fromhagen, M. J. Fahrenback, J. A. Brockman, Jr. & E. L. R. Stokstad, Proc. Soc. Exptl. Biol. (Biomed.), **82**, 280 (1953).
49. G. Pejler, A. Danielsson, I. Bjork, U. Lindahl, H. B. Nader & C. P. Dietrich, J. Biol. Chem., **262**, 11413 (1987).

50. G. Pejler, "*Why Does Heparin Bind Antithrombin?*" Ph.D. Thesis, Swedish University of Agricultural Sciences, Uppasala, Sweden, 1988.
51. M. Kosakai & Z. Yosizawa, J. Biochem., **85**, 147 (1979).
52. A. I. Usov, K. S. Adamyants, L. I. Miroshnikova, A. A. Shaposhnkova & N. K. Kochetkov, Carbohydr. Res., **18**, 336 (1981).
53. K. G. Rice, Y. S. Kim, A. C. Grant, Z. M. Merchant & R. J. Linhardt, Anal. Biochem., **150**, 325 (1985).
54. K. G. Rice, M. K. Rottink & R. J. Linhardt, Biochem. J., **244**, 515 (1987); Z. M. Merchant, Y. S. Kim, K. G. Rice & R. J. Linhardt, Biochem. J., **229**, 369 (1985).
55. A. C. Grant, R. J. Linhardt, G. L. Fitzgerald, J. J. Park & R. Langer, Anal. Biochem., **137**, 25 (1984).
56. N. M. McDuffie, Ed., "*Heparin: Structure, Cellular Functions and Clinical Applications,*" Academic Press, New York (1979).
57. Z. M. Merchant, Y. S. Kim, K. G. Rice & R. J. Linhardt, Biochem. J., **229**, 369 (1985).
58. K. G. Rice, "*A Sequencing Strategy for Heparin,*" Ph.D. Thesis, University of Iowa, Iowa City, USA, 1987.
59. A. M. Maxam & W. Gilbert, Proc. Natl. Acad. Sci., (USA), **74**, 560 (1977).
60. R. Langer, R. J. Linhardt, C. L. Cooney, M. Klein, D. Tapper, S. M. Hoffberg & A. Larsen, Science, **217**, 261 (1982).
61. J. R. Fletcher, A. E. McKee, M. Mills, K. C. Snyder & C. M. Herman, Surgery, **80**, 214 (1976).
62. H. F. Schran, D. W. Bitz, F. J. DiSerio & J. Hirsh, Thromb. Res. **31**, 51 (1983).
63. S. E Lasker & M. L. Chium, U.S. Patent No. 3,766,167 (1973).
64. T. K. Sue, L. B. Jaques & E. Yeun, Can. J. Physiol. Pharmacol., **54**, 613 (1976).
65. A. S. Gervin, Surg. Gyn. Obstet., **140**, 789 (1975).
66. J. Porter & H. Fick, J.A.M.A., **237**, 879 (1987).
67. J. M. Weiler, Immunopharmacol., **6**, 245 (1983).
68. J. M. Weiler, P. Freiman, M. D. Sharath, W. J. Metzger, J. M. Smith, H. B. Richerson, Z. K. Ballas, P. C. Halverson, D. J. Shulan, S. Matsuo, R. L. Wilson, J. Allergy Clin. Immunol., **75**, 297 (1985).
69. M. D. Sharath, W. J. Metzger, H. B. Richerson, R. K. Seupharm, R. L. Meng, B. H. Ginsberg & J. M. Weiler, J. Thorac. Cardiovasc. Surg., **90**, 86 (1985).
70. W. R. Bell & R. M. Royall, N. Eng. J. Med., **303**, 902 (1980).
71. R. J. Linhardt & R. Langer, Biomat., Art. Cells, Art. Org., **15**, 91 (1987).
72. A. K. Larsen, R. J. Linhardt, M. Klein, D. Tapper & R. Langer, Artificial Organs, **8**, 198 (1984).
73. A. K. Larsen, R. J. Linhardt, K. G. Rice, W. Wogan & R. Langer, J. Biol. Chem., **264**, 1570 (1989).
74. H. Bernstein, M. Randawa, D. Lund, V. Yang & R. Langer, Kidney Int., **32**, 452 (1987).
75. H. W. Collins, P. E. Measells, H. M. Gajewski, J. E. Miripol, U. C. Geissler, W. J. Stith, G. A. Grode & R. A. Williams, Baxter Travenol Laboratories Inc., S. African U.S. Patent No. 3,766,167 (1980).
76. M. Nakajima, T. Irimura & G. L. Nicolson, Anal. Biochem., **157**, 162 (1986).
77. P. N. Walsh in: "*Hemostasis and Thrombosis,*" R. Colman, J. Hirsh, V. J. Marder & E. Salzman, Eds., Lippincott Co., Philadelphia, 1981, p. 404.
78. M. E. Silva & C. P. Dietrich, J. Biol. Chem., **251**, 6841 (1975).
79. E. W. Davie & K. Fujikawa, Ann. Rev. Biochem., **44**, 799 (1975).
80. A. Henschen & J. McDonagh, New Compre. Biochem., **13**, 171 (1986).
81. B. N. Bouma & J. H. Griffin, New Compre. Bichem., **13**, 103 (1986).
82. E. W. Davie, K. Fujikawa, K. Kurachi & W. Kisiel, Adv. Enzymol. **48**, 277 (1979).

83. C. M. Jackson & Y. Nemerson, Ann. Rev. Biochem., **49**, 765 (1980).
84. G. Tans & J. Rosing, New Compre. Biochem., **13**, 59 (1986).
85. M. Guillin & A. Bezeaud, Path. Biol., **33**, 917 (1985).
86. K. Kwachi, G. Schmer, M. A. Hernodson, D. C. Teller & E. W. Davie, Biochem., **15**, 368 (1976).
87. M. O. Longas & T. H. Finlay, Biochem. J., **189**, 481 (1980).
88. R. D. Rosenberg & P. S. Damus, J. Biol. Chem., **248**, 6490 (1973).
89. D. M. Tollefson & M. K. Blank, J. Biol. Chem., **257**, 2162 (1982).
90. M. J. Griffith, C. M. Noyes & F. C. Church, J. Biol. Chem., **260**, 2218 (1985).
91. T. H. Tran, B. Zbiden & F. Duckert, Thromb. Haemost., **11**, 342 (1985).
92. P. Wunderwald, W. G. Schrenk & H. Post, Thromb. Res., **25**, 177 (1982).
93. D. M. Tollefsen, Nouv., Rev. Fr. Hematol., **26**, 233 (1984).
94. V. Buonassisi, Exp. Cell Res., **76**, 363 (1973).
95. A. J. Quick, J. N. Shanberge & M. Stefani, J. Lab. Med., **33**, 1424 (1948)
96. J. Verstraete, Klin. Wochenschr., **66**, 5 (1988): Chem. Abstr. **109**, 47617 (1988).
97. M. Verstraete, M. Bory, D. Collen, R. Erbel, R. J. lennane, D. Mathey, H. R. Michels, M. Schartl, R. Bebis, R. Bernard, et. al., Klin. Wochenschr., **66**, 77 (1988).
98. L. J. Lewis, Abbott Laboratories, S. African U.S. Patent No. 020876 (1976).
99. S. J. Crabbe & C. C. Cloninger, Clin. Pharm., **6**, 373 (1987).
100. T. Susawa, Y. Yui, R. Hattori, M. Takahashi, T. Aoyama, Y. Takatsu, K. Sakaguchi, N. Yui & C. Kawai, Jpn. Circ. J., **51**, 431 (1987).
101. J. M. Stassen, I. Juhan-Vague, M. C. Alessi, F. DeCock & D. Collen, Thromb. Haemostasis, **58**, 947 (1987).
102. H. Engelberg, Sem. Thromb. Hemostas., **11**, 48 (1985).
103. R. D. Rosenberg, L. M. S. Fritze, J. J. Castellot & M. J. Karnovsky, Nouv. Rev. Fr. Hematol., **26**, 255 (1984).
104. H. Engelberg, Pharmacol. Rev., **36**, 91 (1984).
105. Z. M. Merchant, E. E. Erbe, W. P. Eddy, D. Patel & R. J. Linhardt, Atherosclerosis, **62**, 151 (1986).
106. T. W. Barrowcliffe, Thromb. Res., **42**, 583 (1986).
107. J. J. Castellot, Jr., D. L. Beeler, R. D. Rosenberg & M. J. Karnovsky, J. Cell. Physiol., **120**, 315 (1984).
108. J. J. Castellot, Jr., J. Choay, J.-C. Lormeau, M. Petitou, E. Sache & M. J. Karnovsky, J. Cell. Biol., **102**, 1979 (1986).
109. H. P. Ehrlich, T. R. Griswold & J. B. Rajaratnam, Exp. Cell. Res., **164**, 154 (1986).
110. W. E. Benitz, D. S. Lessler, J. D. Coulson & M. Bernfield, J. Cell. Physiol, **127**, 1 (1986).
111. J. A. Winkles, R. Friesel, W. H. Burgess, R. Howk, T. Mehlman, R. Weinstein & T. Maciag, Proc. Natl. Acad. Sci., **84**, 7124 (1987).
112. M. W. Majesky, S. M. Schwartz, M. M. Clowes & A. W. Clowes, Circ. Res., **61**, 296 (1987).
113. A. W. Clowes & M. M. Clowes, Int. Angiol., **6**, 45 (1987).
114. M. F. Graham, D. E. Drucker, H. A. Perr, R. F. Diegelmann & H. P Ehrlich, Gastroenterology, **93**, 801 (1987).
115. M. Loos, J. E. Volanakis & R. M. Stroud, Immunochem., **13**, 789 (1976).
116. E. Raepple, H.-V. Hill & M. Loos, Immunochem., **13**, 251 (1976).
117. M. Loos, J. E. Volanakis & R. M. Stroud, Immunochem., **13**, 257 (1976).
118. G. B. Caughman, R. J. Boackle & J. Vesely, Mol. Immunol., **19**, 287 (1982).
119. J. M. Weiler, R. W. Yurt, D. T. Fearon & K. F. Austen, J. Exp. Med., **147**, 409 (1978).
120. D. H. Atha, A. W. Stephens, A. Rimon & R. D. Rosenberg, Biochem., 23, 5801 (1984).

121. U. Lindahl, L. Thurnberg, G. Backstrom, J. Reisenfeld, K. Nordling & I. Bjork, J. Biol. Chem., **259**, 12368 (1984).
122. M. D. Kazatchkine, D. T. Fearon, D. D. Metcalfe, R. D. Rosenberg & K. F. Austen, J. Clin. Invest., **67**, 223 (1981).
123. E. Cofrancesco, F. Radaelli, E. Pogliani, N. Amici, G. G. Torri & B. Casu, Thromb. Res., **14**, 179 (1979).
124. R. G. Azizkhan, J. C. Azizkhan, B. R. Zetter & J. Folkman, J. Exp. Med., **152**, 931 (1980).
125. S. C. Thornton, S. N. Mueller & E. M. Levine, Science, **222**, 623 (1983).
126. D. Giospodarowicz & J. Cheng, J. Cell. Physiol., **128**, 475 (1986).
127. A. B. Schreiber, J. Kenney, W. J. Kowalski, R. Friesel, T. Mehlman & T. Maciag, Proc. Natl. Acad. Sci., **82**, 6138 (1985).
128. R. Orem, S. Szabo & J. Folkman, Science, **230**, 1375 (1985).
129. E. Grushka & A. S. Cohen, Analyt. Lett., **15**, 1277 (1982).
130. W. F. Long & F. B. Williamson, Med. Hypoth., **13**, 147 (1980).
131. J. Denekamp in: "*Progress in Applied Microcirculation*," F. Hammerson & O. Hudlicka, Eds., Karger, Basel, 1984.
132. D. R. Knighton, T. K. Hunt, H. Scheuenstuhl, B. J. Halliday, Z. Werb & M. J. Banda, Science, **221**, 1283 (1983).
133. D. W. Beck, J. J. Olson & R. J. Linhardt, J. Neuropathology & Exptl. Neurology, **45**, 503 (1986).
134. J. Folkman, Important Adv. Oncol., **42**, (1985); K. Jaschonek, C. Faul, W. Daiss & H. Weisenberger, Prostaglandins Leukotrienes Med., **24**, 199 (1986).
135. Z. Y. Zhang-Keck & M. R. Stallcup, J. Biol. Chem., **263**, 3513 (1988).
136. C. Staben, T. R. Whitehead & J. C. Rabinowitz, Anal. Biochem., **162**, 257 (1987).
137 V. N. Nikandrov, V. I. Votiakov, S. A. Naumovich, G. V. Vorob'eva, & S. G. Tsymanovich, Vopr. Med. Khim., **33**, 54 (1986).
138. S. C. Alter, D. D. Metcalf, T. R. Bradford & L. B. Schwartz, Biochem., J., **248**, 821 (1987).
139. H. Mioh & J.K. Chen, Biochem. Biophys. Res. Commun., **146**, 771 (1987).
140. B. Willuweit & K. Aktories, Biochem. J., **249**, 857 (1988).
141. K. Jaschonek, C. Faul, W. Daiss & H. Weisenberger, Prostaglandins Leukotrienes Med., **24**, 199 (1986).
142. M. J. Griffith, J. Biol. Chem., **257**, 7360 (1982).
143. R. E. Jordan, J. Kilpatrick & R. M. Nelson, Science, **237**, 777 (1987).
144. F. Redini, J.-M. Tixier, M. Petitou, J. Choay, L. Robert & W. Hornerbeck, Biochem. J., **252**, 515 (1988).
145. K. Ishii, S. Futaki, H. Uchiyama, K. Nagasawa & T. Andoh, Biochem. J., **241**, 111 (1987).
146. P. Ackerman, C. V. Glover, & N. Osheroff, J. Biol. Chem., **263**, 12653 (1988).
147. I. Posner & J. Desanctis, Arch. Biochem. Biophys., **253**, 475 (1987).
148. D. Gauthier & M. R. Ven Murthy, Neurochem. Res., **12**, 335 (1987).
149. T. J. Singh, Arch. Biochem. Biophys., **160**, 661 (1988).
150. G. R. Corassa, A. Falasca, A. Strocchi, C. A. Rossi & G. Gasbarrini, Dig. Dis. Sci., **33**, 956 (1988).
151. J. S. Thompson, D. A. Burnett, R. A. Cormier & W. P. Vaughan, Dis. Colon Rectum., **31**, 529 (1988).
152. L. Leiva, D. Carrasco, A. Taymor, M. V'eliz, C. Gonzalez, C. C. Allende & J. E. Allende, Biochem. Int., **14**, 707 (1987).
153. P. Gavriel & H. M. Kagan, Biochem., **27**, 2811 (1988).
154. G. A. Schwarting, A. Gajewski, P. Carroll & W. C. DeWolf, Arch. Biochem. Biophys., **256**, 69 (1987).
155. M. Gaiger, M. J. Heeb, B. R. Binder & J. H. Griffin, FASEB J., **2**, 2263 (1988).
156. M. Pfaff & F. A. Anderer, Biochim. Biophys. Acta, **969**, 100 (1988).

157. A. D. Deana & L. A. Pinna, Biochim. Biophys. Acta, **968**, 179 (1988).
158. A. G. Hovanessian & J. Galabru, Eur. J. Biochem., **167**, 467 (1987).
159. R. K. Sihag, A. Y. Jeng & R. A. Nixon, FEBS Lett. **233**, 181 (1988).
160. E. Persson, Acta Med. Scand., **724**, 1 (1988); K. Jaschonek, C. Faul, W. Daiss & H. Weisenberger, Prostaglandins Leukotrienes Med., **24**, 199 (1986).
161. F. Erdodi, P. Gergely & G. Bot, Int. J. Biochem., **16**, 1391 (1984).
162. Z. Ahmad, A. A. DePaoli-Roach & P. J. Roach, FEBS Lett., **179**, 96 (1985).
163. E. Brandan & N. C. Inestrosa, Biochem. J., **221**, 415 (1984).
164. R. W. Scott, B. L. Bergman, A. Bajpai, R. T. Hersh, H. Rodriguez, B. N. Jones, C. Barreda, S. Watts & J. B. Baker, J. Biol. Chem., **260**, 7029 (1985).
165. S. P. Boyers, B. C. Tarlatsiz, J. N. Stronk & A. H. DeCherney, **48**, 628 (1987).
166. J. J. Parrish, J. Susko-Parrish, M. A. Winer & N. L. First, **38**, 1171 (1988).
167. R. Fransen, P. Patriarca & P. Elsbach, J. Lipid. Res., **15**, 380 (1974).
168. G. Fager, G. K. Hansson, P. Ottosson, B. Dahllof & G. Bondjers, Exp. Cell Res., **176**, 319 (1988).
169. W. L. McKeehan & P. S. Adams, In Vitro Cell Dev. Biol., **24**, 243 (1988).
170. C. F. Reilly, L. M. Fritze & R. D. Rosenberg, J. Cell Physiol., **136**, 23 (1988).
171. M. Seno, R. Sasada, M. Iwane, K. Sudo, T. Kurokawa, K. Ito & K. Igarashi, Biochem. Biophys. Res. Commun. , **151**, 701 (1988).
172. Y. Shing, J. Biol. Chem., **263**, 9059 (1988).
173. B. A. Kudriashov, I. M. Ammosova, L. A. Liapina & A. M. Ulianov, Farmakol-Toksikol, **50**, 49 (1987).
174. S. Mohan, J. C. Jennings, T. A. Linkhart & D. J. Baylink, Biochim. Biophys. Acta, **966**, 44 (1988).
175. B. A. Kudrajhov, F. B. Shapiro & A. M. Ulyanov, Acta Physiol. Hung., **69**, 197 (1987).
176. H. Hoshi, M. Kan, H. Mioh, J.-K. Chen & W. L McKeehan, FASEB. J., **2**, 2797 (1988).
177. E. Dupuy, P. S. Rohrlich & G. Tobelem, Cell Biol. Int. Rep., **12**, 17 (1988).
178. J. W. Crabb, L. G. Armes, C. M. Johnson & W. L. McKeehan, Biochem., Biophys. Res. Comun., **136**, 1155 (1986).
179. T. K. Rosengart, W. V. Johnson, R. Friesel, R. Clark & T. Maciag, Biochem. Biophys. Res. Commun., **152**, 432 (1988).
180. E. R. Lazarowski, J. A. Santome', N. H. Behrens & J. C. Sanchez Avalos, Thromb. Res., **41**, 437 (1986).
181 J. N. Moore, E. A. Mahaffey & M. Zboran, Am. J. Vet. Res., **48**, 68 (1987).
182. B. H. Chong & P. A. Castaldi, Aust. NZ. J. Med., **16**, 715 (1986).
183. M. Sobel & B. Adelman, Thromb. Res., **50**, 815 (1988).
184. M. Silverberg & S. V. Diehl, Biochem. J., **248**, 715 (1987).
185. H. Fukui, Nippon Ketsueki Gakkai Zasshi, **50**, 15108 (1987).
186. K. T. Preissner, U. Delvos & G. Muller-Berghaus, Biochem., **26**, 2521 (1987).
187. L. L. Dumenco, B. Everson, L. A. Culp & O. D. Ratnoff, J. Lab. Clin. Med., **112**, 394 (1988).
188. C. Doutremepuich, F. Toulemonde, F. Doutremepuich, O. de Seze, B. Bayrou & F. Pereira, Thromb. Res., **50**, 335 (1988).
189. M. Y. Khan, N. S. Jaikaria, D. A. Frenz, G. Villanueva & S. A. Newman, J. Biol. Chem., **163**, 11314 (1988).
190. A. P. N. Skubitz, J. B. McCarthy, A. S. Charonis & L. T. Furcht, J. Biol. Chem., **263**, 4861 (1988).
191. J. B. McCarthy, A. P. Skubitz, S. L. Palm & L. T. Furcht, J. Natl. Canc. Inst., **80**, 108 (1988).

192. A. S. Charonis, A. P. N. Skubitz, G. G. Koliakos, L. A. Reger, J. Dege, A. M. Vogel, R. Wohlueter & L. T. Furcht, J. Cell Biol., **107**, 1253 (1988).
193. T. Akama, K. M. Yamada, N. Seno, I. Matsumoto, I. Kono, K. Kashiwagi, T. Funaki & M. Hayaski, J. Biochem. (Tokyo), **100**, 1343 (1986).
194. J. Tschopp, D. Masson, S. Schafer, M. Peitsch & K. T. Preissner, Biochem., **27**, 4103 (1988).
195. D. de Prost, C. Katlama, G. Pialoux, F. Karsenty-Mathonnet & M. Wolf, Thromb. Haemost., **57**, 239 (1987).
196. M. E. Gonzalez & L. Carrasco, Biochem. Biophys. Res. Commun., **146**, 1303 (1987).
197. T. Nagumo & H. Hoshino, Jpn. J. Cancer Res., **79**, 9 (1988).
198. M. Baba, M. Nakajima, D. Schols, R. Pauwels, J. Balzarini & E. D. Clercq, Antiviral Res., **9**, 335 (1988).
199. M. Baba, R. Pauwels, J. Balzarini, J. Arnout, J. Desmytr & E. De Clerq, Proc. Natl. Acad. Sci., **85**, 6132 (1988).
200. R. V. Iozzo, Cancer Metastasis Rev., **7**, 39 (1988).
201. C. Biswas, J. Cell. Physiol., **136**, 147 (1988).
202. A. D. Snow, H. Mar, D. Nochlin, K. Kimata, M. Kato, S. Suzuki, J. Hassell & T. N. Wight, Am. J. Phathol., **133**, 456 (1988).
203. D. L. Cochran, Calcif. Tissue Inst., **41**, 79 (1987).
204. T. D. Hill, P.-O. Berggren & A. L. Boynton, Biochem. Biophys. Res. Commun., **149**, 897 (1987).
205. P. J. Cullen, J. G. Comerford & A. P. Dawson, FEBS Lett., **228**, 57 (1988).
206. T. Nilsson, J. Zwiller, A. L. Boynton & P.-O. Berggren, FEBS Lett., **229**, 211 (1988).
207. T. K. Ghosh, P. S. Eis, J. M. Mullaney, C. L. Ebert & D. L. Gill, J. Biol. Chem., **263**, 11075 (1988).
208. S. Kobayashi, A. V. Somlyo & A. P. Somlyo, Biochem., Biophys. Res. Commun., **153**, 625 (1988).
209. C. A. A. van Boeckel, S. F. van Aelst, T. Beetz & H. Lucas, XIVth Intl. Carbohydr. Symposium, Stockholm, Sweden, August 14-19, 1988, B-88.
210. C. A. A. van Boeckel, J. E. M. Basten, H. Lucas & S. F. Van Aelst, ibid., B-89.
211. H. K. Breddin, J. Fareed & N. Bender, Eds., "Low-Molecular-Weight Heparins," 4th Congress on Thrombosis and Haemostasis Research, in Haemostasis, **18**, 1-87 (1988).
212. M. Klein, R. A. Drongowski, R. J. Linhardt & R. Langer, Anal. Biochem., **124**, 59 (1982).
213. C. G. Gebelein & D. Murphy, Polym. Sci. Technol. (Plenum), **35**, 277 (1987).
214. A. K. Larsen, "*Toxicological Aspects of an Enzymatic System for Removing Heparin in Extracorporeal Therapy,*" Ph.D. Thesis, MIT, 1984.
215. Nishimura, Carbohydr. Res., **156**, 286 (1986).
216. G. O. Aspinall, Adv. Carbohydr. Chem., **14**, 429 (1983).
217. M. F. Scully, K. M. Weerasinghe, V. Ellis, B. Djazaeri & V. V. Kakkar, Thrombosis Res., **31**, 87 (1983).
218. F. Dol, P. Sie, D. Dupouy & B. Boneu, Thromb. Haemostas, **56**, 295 (1986).
219. M. F. Scully & V. V. Kakkar, Biochem. J., **218**, 657 (1984).
220. Rickets & Walton, U.S. Patent No. 2715091, 1955.
221. A. Osol., Ed., "*16th Remington's Pharmaceutical Sciences,*" Mack Publishing Co., Easton, PA, 1980, pp. 760-761.
222. M. Windholz, "*The Merck Index,*" 9th Edition, Merck Publishing Co., Rahway, NJ, 1976, p. 2911.
223. G. Oshima, Thromb. Res., **49**, 353 (1988).

224. R. Langer, R. J. Linhardt, M. Klein, P. M. Galliher, C. L. Cooney & M. M. Flanagan, in "*Biomaterials: Interfacial Phenomenon and Applications*", Advances in Chemistry Symposium Series, Chapter 31, p. 493., S. Cooper, A. Hoffman, N. Pepas & B. Ratner (Eds.), Washington, DC, 1982.
225. K. G. Rice & R. J. Linhardt, Carbohydr. Res., **190**, 219 (1989).
226. P. Sinay, J.-C. Jacquinet, M. Petitou, P. Duchaussoy, I. Lederman, J. Choay & G. Torri, Carbohydr. Res., **132**, c5 (1984).
227. M. Petitou, P. Duchaussoy, I. Lederman, J. Choay, J.-C. Jacquinet, P. Sinay & G. Torri, **ibid**, **167**, 67 (1987).
228. C. A. A. Van Boeckel, T. Beetz, J. N. Vos, A. J. M. De Jong, S. F. Van Aelst, R. H. Van Den Bosch, J. M. R. Mertens & F. A. Van der Vlught, J. Carbohydr. Chem., **4**, 293 (1985).
229. Y. Ichikawa, R. Monden & H. Kuzuhara, Carbohydr. Res., **172**, 37 (1988).
230. J. Choay, M. Petitou, J.-C. Lormeau, P. Sinay, B. Casu & G. Torri, Biochem. Biophys. Res. Commun., **116**, 492 (1983).
231. J. Walenga, J. Fareed & H. Schumacher, The 30th Annual Meeting of the American Society of Haemotology, August, 1988.
232. C. A. A. Boeckel, S. F. van Aelst, J.-R. Mellema & G. Wagnenaors, Proceedings of the 9th International Symposium on Glycoconjugates, Lille, France, July 6-11, 1987, B5.
233. T. Laurent, A. Tengblad, L. Thunberg, M. Hook & U. Lindahl, Biochem. J., **175**, 691 (1978).
234. T. Bitter & H. M. Muir, Anal. Biochem., **4**, 330 (1962).
235. B. A. Khorramian & S. S. Stivala, Arch. Biochem. Biophys., **247**, 384 (1986).
236. Y. S. Kim & R. J. Linhardt, Thromb. Res., **53**, 55 (1989).
237. M. Hook, I. Bjork, J. Hopwood & U. Lindahl, FEBS Lett., **66**, 90 (1976).
238. Y. S. Kim, "*Interaction of Heparin with Antithrombin III and Heparin Cofactor II,*" Ph.D. Thesis, University of Iowa, 1988.
239. J. Denton, W. E. Lewis, I. A. Nieduszynski & C. F. Phelps, Anal. Biochem., **118**, 388 (1981).
240. B. Nordenman & I. Bjork, Thromb. Res., **19**, (1980).
241. S. F. Mohammad, H. Y. K. Chuang & R. G. Mason, Thromb. Res., **20**, 599 (1980).
242. J. Denton, D. A. Lane, L. Thunberg, A. M. Slater & U. Lindahl, Biochem. J., **209**, 455 (1983).
243. M. E. Fernandez deRecondo, C. Legorburu, J. C. Monge & E. F. Recondo, Biochem. Biophys. Res. Commun., **155**, 216 (1988).
244. C. L. Teng, J. S. Kim, F. K. Port, T. W. Wakefield, G. O. Till & V. C. Yang, ASAIO-Trans., **34**, 743 (1988).
245. E. Szondy & J. L. Beaumont, Immunobiol., **157**, 407 (1980).
246. J. E. Shivley & H. E. Conrad, Biochemistry, **15**, 3932 (1976).
247. J. Mardiguian, U.S. Patent, 4,440,926, April 3, 1984.
248. R. J. Linhardt, C. L. Cooney, P. M. Galliher, Appl. Biochem. & Biotechnol., **12**, 135 (1986).
249. R. J. Linhardt, J. E. Turnbull, H. M. Wang, D. Loganathan & J. T. Gallagher, Biochemistry, in press.
250. J. Fareed, J. M. Walenga, A. Racanelli, D. Hoppensteadt, X. Huan & H. L. Messmore, Hemostasis, **18**, 33 (1988).
251. L. H. Mallis, H. M. Wang, D. Loganathan & R. J. Linhardt, Anal. Chem., **61**, 1453 (1989).
252. R. J. Linhardt, D. Loganathan, A. Al-Hakim, H. M. Wang, J. M. Walenga, D. Hoppensteadt & J. Fareed, J. Med. Chem., 1990, in press.
253. J. Fareed, J. M. Walenga, D. Hoppensteadt, X. Huan & A. Racanelli, Haemostasis, **18**, 3 (1988).
254. J. Fareed, J. M. Walenga, A. Racanelli, D. Hoppensteadt, X. Huan & H. L. Messmore, Haemostasis, 1988, in press.
255. M. J. Bienkowski & H. E. Conrad, J. Biol. Chem., **260**, 356 (1985).

256. F. Fussi, U.S. Patent, 4,281,108, July 28, 1981.
257. S. Haas & G. Blumel, Haemostasis, **18**, 82 (1988).
258. D. Bergqvist, Acta Chir. Scand. Suppl., **543**, 87 (1988).
259. G. Reber, A. Schweizer, P. de Moerloose, M. E. Sinclair, C. A. Bouvier & J. P. Gardaz, Thromb. Res., **49**, 157 (1988).
260. J. Schrader, W. Stibbe, V. W. Armstrong, M. Kandt, R. Muche, H. Kostering, D. Seidel, F. Scheler, Kidney Int., **33**, 890 (1988).
261. J. Fareed in: "Symposium on Molecular Heterogeneity in Low Molecular Weight Heparins: Implications in Standardization and Clinical Use," Strich School of Medicine, Loyola University Medical Center, Maywood, IL, November, 1988.
262. M. Petitou, P. Duchaussoy, I. Lederman, J. Choay & P. Sinay, Carbohydr. Res., **179**, 163 (1988).
263. H. M. Wang, D. Loganathan & R. J. Linhardt, manuscript in preparation.
264. D. Loganathan, H. M. Wang, L. M. Mallis & R. J. Linhardt, Biochemistry, 1990, in press.
265. N. M. Delgado, R. Reyes, J. Mora-Galindo & A. Rosado, Life Sciences, **42**, 2177 (1988).
266. T. Maetzsch, D. Bergqrist, U. Hender & B. Nilsson, Thromb. Haemosts., **56**, 293 (1986).

AN IMMOBILIZED PROTAMINE SYSTEM FOR REMOVING HEPARIN IN EXTRACORPOREAL BLOOD CIRCULATION

Victor C. Yang* and Ching-Leou C. Teng

College of Pharmacy
The University of Michigan
Ann Arbor, Michigan 48109-1065

Extracorporeal blood circulation (ECBC) has become one of the most frequently employed medical procedures in recent years. Approximately twenty million such procedures are performed each year. It requires heparin to provide blood compatibility. However, heparin employed in this procedure often leads to a high incidence of hemorrhagic complications. To avoid such complications, it is generally accepted in clinical practice to administer protamine, a heparin antagonist, at the conclusion of the ECBC procedure to neutralize the anticoagulant activity of heparin. Unfortunately, intravenous administration of protamine can cause adverse responses such as hypotension and shock. To date, there has been no real alternatives to control the bleeding risks associated with systemic use of heparin and the adverse effects resulting from heparin neutralization with protamine. We suggest a novel approach which would potentially control both heparin and protamine induced complications. The approach consists of placing a filter device containing immobilized protamine at the distal end of an ECBC procedure. Such a filter would remove heparin before heparin is returned to the patient. Meanwhile, the filter would allow for an external protamine treatment. Since protamine toxicity results from direct contact of protamine with certain cells present in the liver, lungs and other organ tissues, the use of an external protamine treatment would minimize the toxic effects of protamine. Protamine was immobilized onto a cellulosic hollow fiber bundle obtained from a clinically used hemodialyzer. The bundle was accessed to the vascular system of a dog by femoral artery and vein cannulation. *In vivo* experiments show that the protamine-bound fiber bundle not only removes more than 80% of heparin from the blood circuit, but also causes no clinically significant hemodynamic changes in the dog. In addition, the protamine bundle also significantly reduces the thrombocytopenic response normally associated with the administration of protamine.

* To whom correspondence should be addressed.

Biomimetic Polymers
Edited by C. G. Gebelein
Plenum Press, New York, 1990

INTRODUCTION

Extracorporeal blood circulation (ECBC) has been widely employed in many clinical situations such as kidney dialysis, open heart operation, organ transplantation, plasmaphoresis, and blood oxygenation. Kidney dialysis has been used during the past two decades to treat over 200,000 kidney disease patients worldwide.[1] Open heart operations employing cardiopulmonary bypass on the heart-lung machine are an everyday occurrence at most medical centers. Functional renal transplants have given new lives to nearly 37,000 patients.[2] In 1984, an estimated 90,000 therapeutic plasmaphoresis procedures were carried out in the U.S. alone.[1] The artificial lung (i.e. blood oxygenator) has become an essential life-saving device used to treat patients suffering from acute respiratory failure and infants with diaphragmatic hemia.[3] Less often employed are the artificial liver and the artificial heart, which are still in a conceptual stage.[4] Altogether, it is estimated that approximately twenty million ECBC procedures are performed each year.

In all these applications, blood is drawn from the patient and passed through an extracorporeal device. Upon contact with the synthetic materials forming the surfaces of such extracorporeal devices, the blood's precisely regulated hemostasis is disturbed and it tends to clot within the device.[5] The thrombi formed occlude the perfusion channels in the machine. To prevent occlusion and maintain the fluidity of blood in the circulation, heparin, the most widely used anticoagulant, is systemically administered to the patient and the device prior to treatment. However, the high level of heparin required for this purpose poses a considerable hemorrhagic hazard to the patients. Gervin, for instance, reports a 8-30% incidence of hemorrhagic complications occurs during heparinization.[6] Similarly, Fletcher and co-workers report that 6-10% of patients develop coagulation abnormalities with excessive bleeding following open heart surgery.[7] Swartz also estimates that nearly 25% of all patients suffering from acute renal failure are subject to increased bleeding risk during and immediately following dialysis.[8] The intent of these complications is enhanced for elderly patients, diabetic patients, patients with ulcers or other multiple traumata, and patients with current cardiac or vascular surgery.[9,10] In fact, the Boston Collaborative Drug Surveillance Program has cited "heparin continues to be the drug responsible for a majority drug deaths in patients who are reasonably healthy."[11]

Because of the life-threatening nature of the hemorrhage associated with systemic heparinization, considerable amounts of effort have been directed at solving this problem. The approaches taken include: (1) the administration of anti-heparin compounds (e.g. protamine) to neutralize the anticoagulant activity of heparin,[4] (2) the development of heparin substitutes (e.g. prostacyclin),[12] and new antithrombotic agents (e.g. low molecular weight heparins),[13] (3) the use of low-dose heparinization or regional heparin anticoagulation (i.e. by infusion of heparin into the blood entering the dialyzer,[8] and neutralization by infusion of protamine into the heparinized blood as it returns to the patient),[14] (4) the use of regional citrate anticoagulation (this approach is similar to that of the regional heparin anticoagulation as described in (3), except for that citrate is used as the anticoagulant agent and calcium is used as the neutralizing agent),[15] (5) the development of new blood compatible materials such as materials with surface-bound heparin for construction of the extracorporeal devices,[16] and (6) the development of an immobilized heparinase filter to degrade heparin at the termination of the ECBC procedure.[17-19]

Although these approaches have led to some improvement, they have not met much clinical success. For instance, prostacyclin infusion is associated with the unpredictable occurrences of sudden hypotension.[12] The use of low molecular weight heparins as new antithrombotic agents has been restricted due to the lack of neutralizing agents for these compounds.[20] The use of low-dose heparinization or regional heparin anticoagulation proves unsuccessful in preventing bleeding associated with dialysis.[8,14] In addition, regional heparin anticoagulation does not obviate the need of protamine as the neutralizing agent. Regional citrate anticoagulation is rarely used because of the technical difficulties in performing the procedures and in determining the adequate amount of calcium required for citrate neutralization. The use of heparin-bound materials for the construction of extracorporeal devices is associated with a major problem that unless the entire extracorporeal unit is made of these materials, clotting will still occur in the devices. The heparinase-based heparin removal system under development looks most promising because it obviates the need of any heparin neutralizing agents. However, it is still in the experimental stage. Moreover, the approach is also associated with concerns such as the accumulation of heparin degradation products in the patient's body may pose some toxic effects, and the use of the microbial heparinase may trigger some immunological responses. To date, the administration of protamine at the conclusion of an ECBC procedure to reverse the anticoagulant activity of heparin still remains as the only, and most accepted, approach in clinical practice.

Protamine is a low molecular weight protein rich in lysine, arginine and other basic amino acids. It binds heparin through the electrostatic interaction, and neutralizes the anticoagulant activity of heparin. One milligram of protamine is clinically used to neutralize 100 USP units of heparin.[21] Although protamine has been approved by the Food and Drug Administration (FDA) for such clinical use, it is nevertheless a toxic compound. Protamine induced toxicity ranges from mild hypotension[22,23] to severe cardiovascular collapse.[24,25] In a recent survey involving 1400 perfusionists the most frequently cited perfusion accident in cardiopulmonary bypass operations was the "protamine reaction."[26] However, as long as there are no alternatives to control the more serious and life-threatening bleeding problems associated with the systemic use of heparin, protamine will remain in use in conjuncture with heparin therapy despite its toxic effects.

We suggest herein a novel approach which would potentially control both heparin and protamine induced complications. The approach consists of placing a blood compatible filter device containing immobilized protamine (defined as a protamine filter) at the distal end of the ECBC circuit.[27] Such a protamine filter would bind and selectively remove heparin after heparin serves its anticoagulant purpose in the extracorporeal device and before it is returned to the patient. Meanwhile, the filter would restrict the protamine therapy to an external "spot" treatment. Since protamine toxicity results from the direct interaction of protamine with certain cells present in the lungs, liver and other tissues,[21] the use of an external protamine treatment would theoretically minimize the adverse effects of protamine.

In this paper, we present some preliminary data demonstrating the feasibility of the approach. Protamine was first immobilized onto 8% agarose beads activated with cyanogen bromide. An investigation of the parameters affecting protamine immobilization of the agarose beads was conducted in order to establish a protocol so that the immobilization process can be quantitatively controlled. Based upon this protocol, we have immobilized desired amounts of protamine onto a blood compatible cellulosic hollow fiber bundle which is obtained from a clinically used hemodialyzer. *In vivo* experiments show that this protamine-bound fiber bundle (i.e., the protamine filter) is capable of removing more than 80% of the clinically used amounts of heparin. In addition, the protamine filter produces no protamine-linked adverse effects in dogs, as revealed by the relatively insignificant hemodynamic changes in the dog after using the filter. Moreover, the filter has also markedly reduced the thrombocytopenic response which is generally associated with the administration of protamine.

EXPERIMENTAL MATERIALS

1. Materials

Heparin solid (sodium salt from porcine intestinal mucosa, 120 USP units/mg) was purchased from Hepar Industries (Franklin, OH). Heparin injection (1000 USP units/mL) was purchased from Elkins-Sinn (Cherry Hill, NJ). Protamine sulfate injections (10 mg/mL) was obtained from Eli Lily & Co. (Indianapolis, IN). Azure A dye and cyanogen bromide (CNBr) were obtained from Fisher Scientific Co. (Fair Lawn, NJ). Bio-Rad Dye reagent and Biogel A1.5 (8% agarose) was purchased from Bio-Rad Laboratories (Richmond, CA). Actin Activated Cephaloplastin Reagent for the APTT assay was from Dade (Miami, FL). Freshly frozen citrated human plasma was obtained from the American Red Cross (Detroit, MI).Cellulose hollow fibers were obtained from a Travenol Model 1500 CF Dialyzer. All other chemicals were reagent grade and water was distilled and deionized.

2. Methods

2A. Assays. Protamine concentration was assayed according to the method of Bradford.[28] This method has been found very sensitive for measuring protamine concentration in the range of 1-5 µg/mL. Heparin concentration in solution was determined by the Azure A metachromasia assay.[29] This method is based upon the fact that heparin causes a shift in the absorption maximum of the Azure dye from 620 nm to 530 nm. The degree of shift is linearly related to the heparin concentration in the range of 1-10 µg/mL. By measuring the absorbance at 620 nm (or 530 nm), one can precisely determine the heparin concentration in the sample from a standard curve. The biological activity of heparin in plasma was measured by the APTT (activated partial thromboplastin time) clotting assay.[20,30] For the APTT assay, a standard curve involving the APTT and the plasma heparin level was constructed. Samples outside the linear range were diluted with normal plasma prior to the measurements. The degree of activation of the heparin was determined by measuring the cyanate ester concentrations on the activated polymers, according to the method of Kohn and Wilchek.[31]

2B. Protamine Immobilization. Protamine was successfully immobilized onto 8% agarose beads and cellulose hollow fibers, according to a modified cyanogen bromide activation method[32] of March, et al.[33] A general description of the method using the agarose beads as an example is discussed below. One hundred milliliters of 8% agarose beads were washed with chilled distilled water and then suspended in 100 mL of chilled distilled water and 200 mL of chilled 2 M sodium carbonate. Forty milliliters of CNBr in acetonitrile (1 g of CNBr per mL of acetonitrile) was added to the suspension, and the mixture was vigorously stirred in a fume hood for 5 minutes. The reaction mixture was quickly filtered and washed with chilled distilled water (1000 mL) and 1 mM HCl solution (500 mL). The gel was then washed with 500 mL of 0.1 M sodium bicarbonate solution (pH 8.3) containing 0.5 M NaCl. The suction-dried beads were immediately transferred into 400 mL of "coupling buffer" (0.1 M $NaHCO_3$, pH 8.3) containing 10-20 mg/mL protamine. The reaction mixture was placed on an orbital shaker, and the coupling procedure was carried out overnight at room temperature. After protamine immobilization, albumin was introduced to the beads in order to block the residual activated groups that were not used during the immobilization process. Polymers coated with albumin have been reported to have improved blood compatibility.[34] Similar procedures were followed for the immobilization of protamine onto the cellulosic hollow fibers.

2C. Fabrication of the Cellulosic Hollow Fiber Bundle. A Travenol Model 1500 CF hemodialyzer was disassembled, and the fibers were cut down to a length of about 15 cm. The fibers were counted into groups of 150 and tied together at each end to form bundles. A piece of silicon tubing (2 cm in length) was served as the exterior mold for the epoxy potting of the fibers. A cotton plug was inserted around the fibers and inside the mold away from the fiber ends. This plug would support the epoxy and give strength to the junction between the fibers and the epoxy. Epoxy was then applied around the fibers and within the mold in such a way to prevent air bubbles. The epoxy was cured in a 37°C oven for 2 hours and then allowed to stand at room temperature for 30 min. When the epoxy was not sticky, it was cleaved perpendicular to the fibers with a razor cut. The bundle was then ready to use. Figure 1 shows a picture of the fabricated bundle.

2D. *In Vivo* Experiments. Nine female mongrel dogs including four experimental dogs and five control dogs were prepared for this study. The dogs were anesthetized with sodium pentobarbital (30 mg/kg) and mechanically ventilated to maintain physiologic blood gases. Hydration was maintained by administration of lactated Ringer's solution (20 mL/kg bolus followed by a 10 mL/kg/hour infusion). The femoral artery and vein of the dog was cannulated and a hollow fiber bundle (diameter: 1.2 cm; length: 15 cm) was attached to allow blood to flow from the artery through the bundle and into the vein. Heparin was given intravenously at a dose of 150 units/kg of body weight. In the experimental dogs the bundles used contained 60-70 mg of immobilized protamine, whereas in the control dogs the bundles used were not treated with protamine. Two types of control experiments, the "control with protamine" and the "control without protamine", were conducted. The control with protamine consisted of three dogs in which the dogs were given protamine intravenously 10 minutes after heparin administration at a dose of 1 mg/100 units of heparin. The

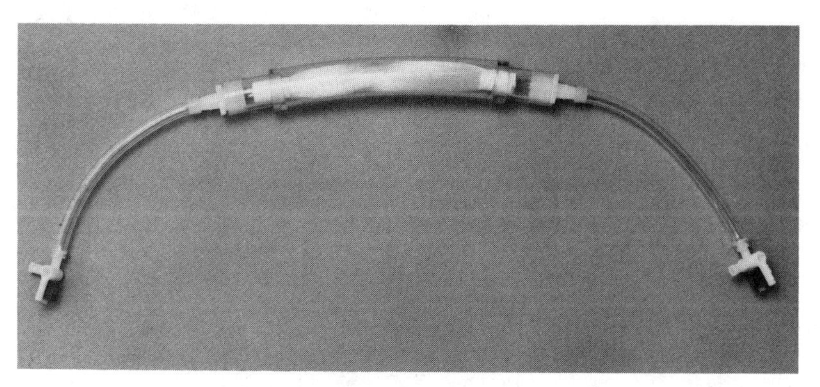

Figure 1. A photograph of the protamine filter used for the *in vivo* experiments.

control without protamine consisted of two dogs in which no protamine was administered. Blood samples were drawn periodically from the two sample valves located at the arterial line and the venous line of the bundle, and were assayed using the APTT heparin assay.

Hemodynamic changes were monitored throughout the entire study. Parameters measured included systemic arterial blood pressure (BP), pulmonary artery systolic (PAS) and diastolic (PAD) pressures, and cardiac output (CO). Systemic arterial blood pressure measurements were made using a catheter in the femoral artery, while pulmonary artery pressures and cardiac output determinations were accomplished via a Swan Ganz catheter threaded from the jugular vein into the pulmonary artery. Cardiac output was determined by thermal dilution technique.

Hematologic parameters measured included hematocrit, red cell counts, white cell counts, platelet counts and total blood hemoglobin. Blood samples were collected for all dogs from the atrial line of the filter at the following intervals: (1) before heparin administration (i.e., baseline), (2) 10 minutes after heparin administration, and (3) 3, 10, 20, 30, 60 minutes after protamine administration (for the control with protamine), after opening the untreated bundle (for the control without protamine), or after opening the protamine filter (for the experimental dogs). White cell counts, red cell counts, total hemoglobin and hematocrit were measured using a Coulter Counter. Platelet counts were performed manually using a light microscope.

RESULTS

Protamine was first immobilized onto a 8% agarose beads activated with cyanogen bromide. Figure 2 shows that the immobilization process consists of two steps: (1) the activation of the agarose gel with CNBr, and (2) the coupling of protamine onto the activated gel. The activation step appears to be controlled by the amount of CNBr added. Doubling the amount of CNBr added nearly doubles the degree of activation, as reflected by the number of activated cyanate ester groups on the gel (data are not shown). The coupling step, on the other hand, is dependent on the degree of activation on the gel as well as the kinetics of protamine coupling. Figure 3 shows that the amount of protamine immobilized on the gel increases with both increasing the cyanate ester concentration on the gel and the protamine concentration in the coupling buffer. The latter is obviously due to an increase in the coupling kinetics resulting from the increase in protamine concentration in the coupling solution. It is important to note that the coupling step has been carried out overnight, a time period that is long enough to complete protamine immobilization. For the immobilization process already reaching completion, the degree of immobilization (i.e. the total amount of protamine immobilized on the gel) should be independent of the protamine concentration rather than dependent of it as seen in Figure 3. In other words, if the immobilization is carried out to completion, the total amount of protamine immobilized on the beads possessing the same degree of activation should be the same, regardless of what protamine concentration is used during the coupling process. The concentration dependency of the degree of immobilization shown in Figure 3 therefore suggests that the activated cyanate ester groups are not stable under the experimental conditions employed for the coupling step, and are continuously inactivated during the coupling process. Due to the continuous inactivation of the cyanate ester groups, the amount of protamine that is immobilized on the gel at a fixed incubation period is governed by both the kinetics of protamine coupling and the inactivation of the cyanate ester groups. Figure 4 shows that the

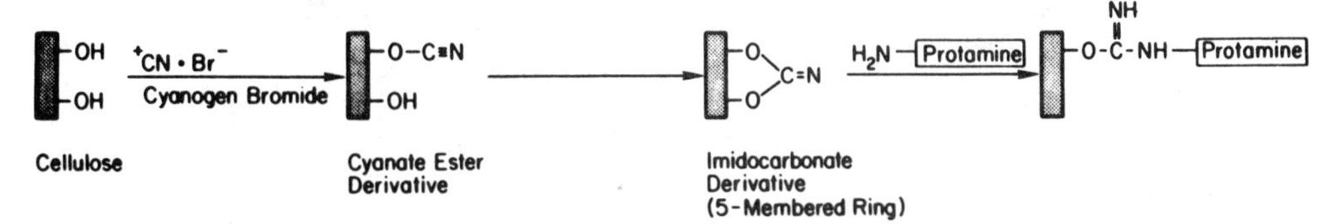

Figure 2. The mechanism describing the activation and coupling of protamine on cellulose hollow fibers.

cyanate ester groups are indeed not stable in the coupling buffer, and undergoing a time-dependent hydrolysis reaction. After one hour of incubation at room temperature in the coupling buffer, nearly 50% of the cyanate ester groups were leached off from the beads due to hydrolysis. Less than 30% of the activated cyanate ester groups on the beads remained after 2 hours of incubation.

Figure 5 shows that the degree of immobilization reaches a plateau after the activated beads were incubated in the protamine solution for 1 hour. This leveling in the degree of immobilization is not due to a shortage of protamine in the coupling solution, since the coupling solution still contains protamine even after the plateau is reached (data not shown). The leveling is in fact due to the lack of available cyanate ester groups on the gel. After one hour of incubation, the portion of cyanate ester groups which survives the hydrolysis (about 50%, as shown in Figure 4) has been completely utilized for protamine coupling.

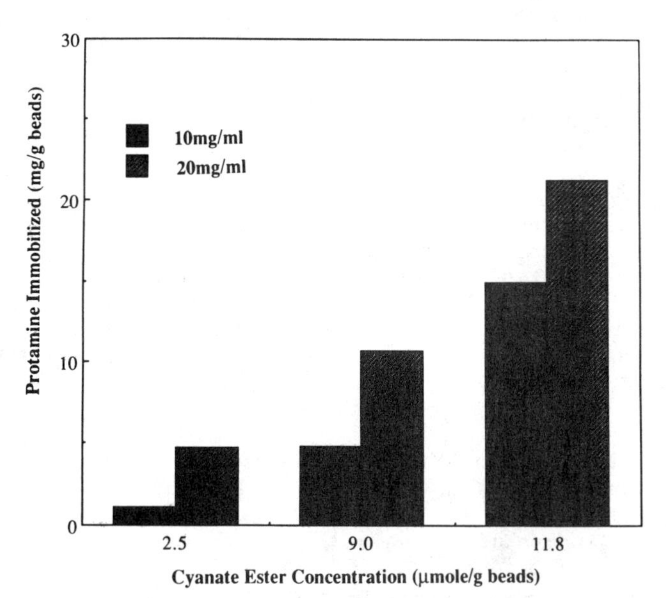

Figure 3. The effect of cyanate ester concentration and protamine concentration on the immobilization of protamine. Two and half grams of suction-dried 8% agarose beads containing different cyanate ester concentrations were mixed with 5 mL of the coupling solution (0.1 M $NaHCO_3$, pH 8.3) containing either 10 or 20 mg/mL of protamine. The mixture was incubated at room temperature overnight. Protamine concentrations before and after the coupling procedure were measured by the method of Bradford.[28]

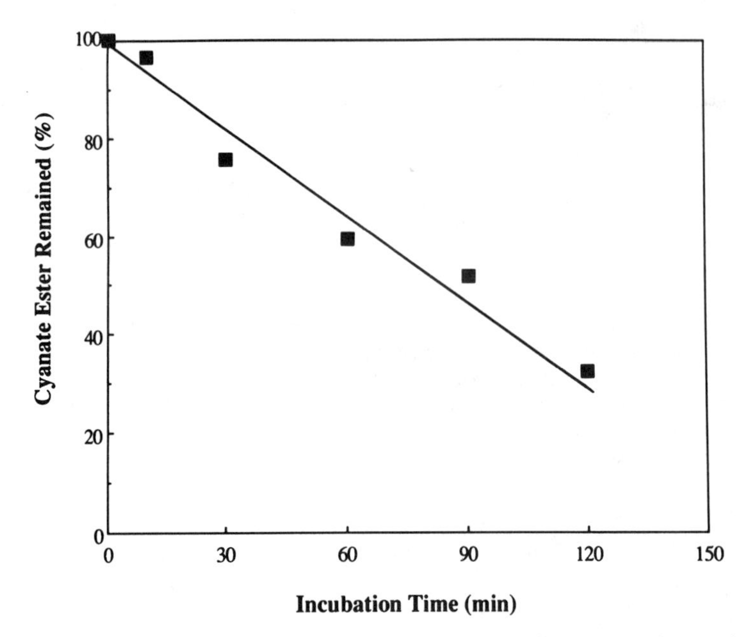

Figure 4. The effect of hydrolysis on the cyanate ester groups. Two and half grams of suction-dried agarose beads containing about 10.5 umoles of cyanate esters per gram of beads were suspended in 5 mL of 0.1 M $NaHCO_3$ buffer (ph 8.3). The suspension was incubated at room temperature and agitated with an orbital rotator. At various times, aliquots containing 1 mL of the suspension were drawn and assayed for the cyanate ester content according to the method of Kohn and Wilchek.[31]

The reason for the presence of higher degree of immobilization at higher protamine concentrations seen in Figure 3 now becomes evident. For two sets of beads possessing the same degree of activation (i.e., containing the same amount of cyanate ester groups on the beads) the rate of hydrolysis of the cyanate ester groups should be the same. Under this condition, the degree of immobilization is controlled only by the kinetics of protamine coupling. The gel incubated at a higher protamine concentration would therefore have a higher rate of coupling and consequently, possesses a higher degree of immobilization.

In view of the above results, the degree of immobilization can be controlled by three independent ways: (1) by changing the amount of activating agent (i.e. CNBr) added, (2) by changing the protamine concentration in the coupling solution, and (3) by changing the incubation time during the coupling procedure. By adopting any of the above methods or a combination of these methods, we were able to quantitatively control the immobilization process and prepare beads containing desired amounts of immobilized protamine.

The protamine-bound agarose beads were tested for their efficiency in removing heparin *in vitro*. One milliliter of suction-dried beads containing 0.2-3.8 mg of immobilized protamine was mixed with 9 mL of citrated normal human plasma containing 500 units of total heparin. The residual heparin concentration in each of the samples was measured by the APTT heparin assay. Figure 6 shows a plot of the amount of heparin removed

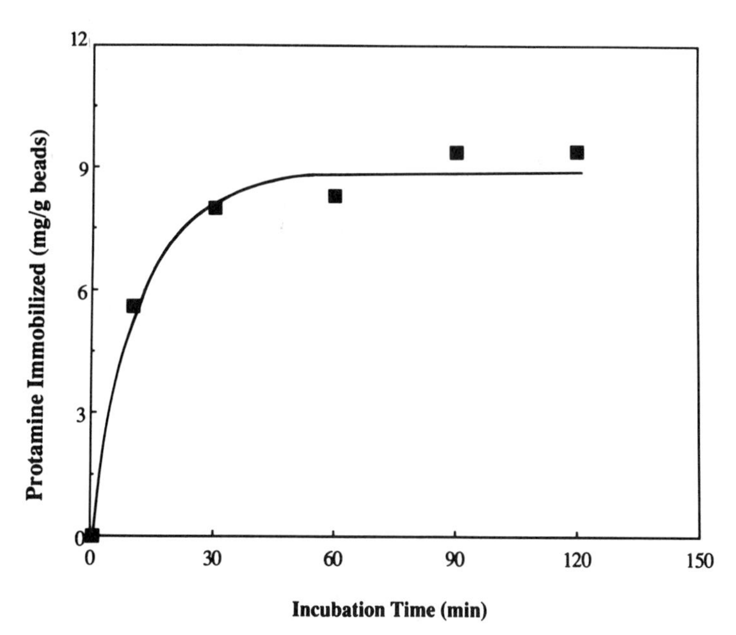

Figure 5. The kinetics of protamine immobilization. Two and half grams of suction-dried agarose beads containing 11.0 umoles of cyanate esters were suspended in 5 mL of coupling buffer (0.1 M $NaHCO_3$, pH 8.3) containing 10 mg/mL of protamine. The suspension was incubated at room temperature and agitated with an orbital rotator. At various times 50 ul aliquots were drawn from the supernatant of the suspension and assayed for their protamine concentrations according to the method of Bradford.[28]

verses the amount of immobilized protamine used. Two linear regions were seen in this plot. In the first region where less than 0.6 mg of immobilized protamine was used, 1 mg of immobilized protamine was capable of neutralizing 120 ± 20 units of heparin. This "equivalent" value is consistent to that reported in the literature for free protamine (One mg of free protamine neutralizes 100-150 units of heparin).[20,21] It therefore suggests that at a very low protamine content immobilized protamine behaves similarly to that of free protamine. However, once the protamine content on the agarose beads is increased over the value of 0.6 mg, the equivalent value in heparin neutralization for the immobilized protamine drops. One mg of immobilized protamine in this case was capable of neutralizing only 75 ± 25 units of heparin. The drop in the equivalent value may result from the shielding effect due to heparin-protamine binding. When the protamine content on the beads is high, the arrangement of protamine molecules on the surface becomes crowded. Binding of heparin to an immobilized protamine molecule could then provide a shielding effect over neighboring protamine molecules, causing them sterically inaccessible to other heparin molecules. This would significantly reduce the equivalent value for the immobilized protamine.

To prepare a protamine filter for *in vivo* use, the selection of the support polymer matrix must be carefully considered. One major prerequisite for the support material is that the material must be blood compatible. Although there is no real definition for blood compatibility, a blood compatible material is generally recognized as the one which will

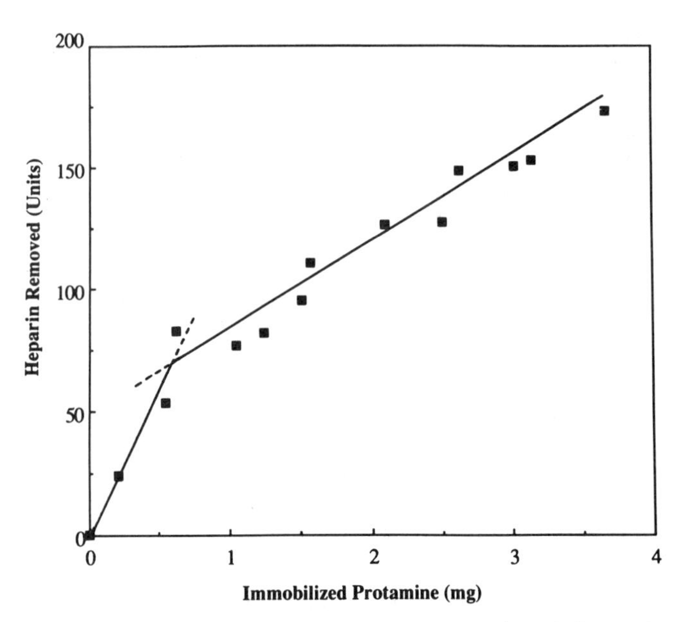

Figure 6. *In vitro* neutralization of heparin with protamine-bound agarose beads. One mL of suction-dried beads containing 0.2 - 3.8 mg of immobilized protamine was mixed with 9 mL of citrated normal plasma containing 500 units of total heparin. The samples were incubated at room temperature for 10 minutes. After the incubation, the residual heparin concentrations in each of the samples were measured by the APTT clotting assay.

not cause hemolysis, undue platelet binding, thrombus formation, and damage of the blood components. In view of these criteria, agarose gel is certainly not quite a blood compatible material. We have therefore attempted to immobilize protamine onto cellulose hollow fibers. This is because that cellulose hollow fibers are the constructing material of the clinically used artificial kidney. They have been extensively screened with cell culture cytotoxicity and hemolysis tests before being approved for clinical use, and should presumably be a blood compatible material. In addition, cellulose hollow fibers also contain abundant hydroxyl groups (like that of the agarose beads) and can be activated with the same CNBr activation method. Using the same immobilization procedures as described above, hollow fiber bundles containing 40-70 mg of immobilized protamine were readily prepared.

The protamine-bound hollow fiber bundle was tested both *in vitro* and *in vivo* for its efficiency in removing heparin. Figure 7A shows *in vitro* results. When 50 mL of the plasma containing 60 units/mL of heparin were passed through the bundle once (the bundle contains 48 mg of immobilized protamine) at a flow rate of 50 mL/min., 45% of the initial heparin in the plasma was removed by the bundle. After circulating the plasma through the bundle for 10 minutes, 60% of the heparin were adsorbed by the bundle. However, the plasma heparin level remained unchanged after 30 minutes of circulation, suggesting that the bundle was saturated with heparin. Based upon these results, it is estimated that one mg of the (hollow) fiber-bound protamine removes 50 ± 15 units of heparin. This equivalent value in heparin neutralization is slightly lower than that obtained for the (agarose) gel-bound protamine. The decrease in the value

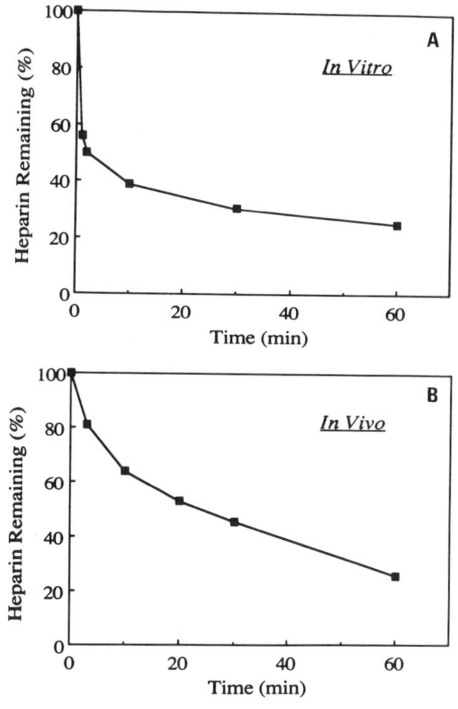

Figure 7. (A) *In vitro* and (B) *in vivo* heparin removal using the protamine-bound hollow fiber bundles. The experimental procedures were described in detail in the text. Time zero represents the opening of the protamine filter.

may be due to the fact that only protamine molecules on the inner surface of the hollow fibers are available for heparin neutralization. Unlike 8% agarose gel, which possess a molecular weight cutoff of about a half-million daltons, the cellulose fiber membrane has a molecular weight cutoff of only 10,000 daltons. Protamine with an average molecular weight of 4,000 daltons can penetrate through the fiber membrane and get immobilized inside the membrane matrix. However, heparin with an average molecular weight of 11,000 daltons will be excluded by the fiber membrane (this is not the case for the agarose gel), and can only interact with protamine molecules located at the surface of the fiber membrane. This would result in a drop in the equivalent value in heparin neutralization for the fiber-bound protamine.

In vivo results (Figure 7B) are consistent with those observed in the *in vitro* experiments. Twenty minutes after opening the protamine filter, heparin level in the animal's arterial line, as measured by the APTT assay, dropped to 54% of the initial value. At a blood flow rate of 100 mL/min. (this was measured by inserting an electromagnetic flow meter in the dog), 20 minutes are equivalent to the time required to circulate the entire blood through the filter once, assuming that the dog has a total blood volume of 2 liters. Heparin was continuously removed by the filter, and after 60 minutes almost 80% of the heparin was removed by the protamine filter. The TCT (i.e., thrombin clotting time) and PTT (i.e., prothrombin time) heparin assays gave similar results. Heparin levels in the samples drawn from the arterial line were consistently higher than those drawn from the venous line (data not shown), presumably due to the mixing of blood in the dog. Blood continued to flow in an unrestricted fashion during the first 40 minutes of the experiment, but slowed down slightly at the end of the experiment. Owing to a slight pressure build up between the two ends of the filter, the experiment was discontinued after 60 minutes. The drop in the blood flow rate and the build up of the filter's interior pressure may be due to partial clotting inside the filter when most heparin was removed.

The toxicological effects of the protamine filter was monitored by the protamine induced hemodynamic changes in the dog. As shown in Table 1, there were important hemodynamic changes in the control group with protamine in which the dogs were given protamine intravenously. A marked drop in blood pressure (-59.7 mmHg) and cardiac output (-1.7 L/min.), and a significant elevation in pulmonary artery systolic (+18.0 mmHg) and diastolic pressure (+16.3 mmHg) were noted for these dogs. Since the control dogs without protamine showed no statistically significant hemodynamic changes, all the changes observed in the control dogs with protamine are believed to be due to protamine administration rather than the hollow fiber bundle.

On the other hand, using the protamine filter has abolished almost all the hemodynamic changes produced by protamine (Table 1). None of the variations seen in the experimental dogs were statistically significant, except for the minor change (-0.5 ± 0.3 L/min.) in the cardiac output. However, even if this change is statistically significant, the depression must result from the bundle rather than the immobilized protamine, since the control group without protamine shows nearly an identical change in the cardiac output.

Table 2 shows that the use of the protamine filter has also significantly attenuated protamine induced thrombocytopenic responses. Peripheral platelet counts were reduced by 67% in the control dogs with protamine, but only 18% in the experimental dogs. It is interesting to note that the protamine filter causes only a slightly larger reduction in platelet counts than that of the untreated bundle (as shown in the control without protamine). At this moment it is not clear how the immobi-

Table 1. Hemodynamic changes in dogs.

	Mean Systemic Blood Pressure (mmHg)	Pulmonary Artery Systolic Pressure (mmHg)	Pulmonary Artery Diastolic Pressure (mmHg)	Cardiac Output (L/min.)
Experimental Dogs (N = 4) (with Protamine filter)	-3.8 ± 4.8	+1.8 ± 1.7	+1.3 ± 1.3	-0.5 ± 0.3
Control Dogs (N = 3) (with protamine)	-59.7 ± 22.5	+18.0 ± 7.6	+16.3 ± 7.8	-1.7 ± 1.1
Control Dogs (N = 2) (without protamine)	-4.0 ± 4.2	+1.6 ± 1.7	+2.0 ± 2.0	-0.4 ± 0.5

The data are presented as the mean ± SD.
The "+" or "-" indicates that an elevation or reduction, respectively, is observed in these parameters.
Only the maximum changes are listed.

Table 2. Platelet Counts.			
	Baseline	After Heparin	After Protamine
Experimental Dogs (N = 4) (with protamine filter)	246,500 ± 75,949	245,500 ± 79,450	203,250 ± 68,300(a)
Control Dogs (N = 3) (with protamine)	361,000 ± 219,228	331,667 ± 252,834	120,333 ± 89,813(b)
Control Dogs (N = 2) (without protamine)	160,500 ± 61,518	142,000 ± 55,154	-

(a). After opening the protamine filter.
(b). After intravenous administration of protamine.
The data are presented as the mean ± SD.
Only the maximum changes are listed.

lized protamine suppresses the thrombocytopenic responses caused by free protamine. Recent studies have suggested a correlation between thrombocytopenia and the development of hypotension.[35] The diminution in the degree of thrombocytopenia may be a direct consequence of the lack of the related hypotensive responses to the protamine filter.

For both the experimental group (i.e., with a protamine filter) and the control group with protamine, the white cell counts dropped within the first 10 minutes of the experiment after which the counts began to rebound. However, the drop was less significant in the experimental dogs than in the control group with protamine. Preliminary data show than an average maximum drop of 18% in white cell counts were observed for the experimental dogs, as compared to a drop of 60% for the control group with protamine.

The red cell counts, the mean hematocrits, and the total blood hemoglobin varied between 100% and 95% over the course of the experiment for both the experimental group and the control group with protamine. None of these values were statistically significant.

DISCUSSION

Heparin, a polydisperse sulfated polysaccharide obtained from mammalian tissues, has been the most widely used anticoagulant since its introduction nearly 50 years ago. The annual production of heparin in the United States totals over 6 metric tons (10^{12} USP units). The major consumption of heparin appears to come from the application in the extracorporeal blood circulation (ECBC) procedure. It is estimated that nearly 20 million such procedures are performed each year. In spite of all the brilliant applications, the use of heparin has always been shadowed by a major setback; that is, the high incidence of bleeding complications associated with the systemic use of heparin. To avoid the risk of these life-threatening bleeding complications, it is greatly accepted in clinical practice to administer protamine at the conclusion of the ECBC procedure in order to counteract the anticoagulant activity of heparin. Unfortunately, the use of protamine for heparin reversal is also associated with numerous side effects. These side effects appear to be due to the action of protamine as a histamine liberator and stimulator in mast cells.[21] Intravenous administration of protamine has been found to produce typical symptoms of histamine release such as hypotension, bradycardia, dyspena, fatigue and pronounced blood pressure depression.[21,36] Such protamine reactions range from mild circulatory changes to severe cardiac arrest.[22-25] Recently, several cases of fatal shock following protamine administration have been reported in the literature.[24,25] To mitigate the protamine induced adverse effects, a number of approaches including the alternation of the route and rate of protamine administration,[22,35,37,38] the pretreatment of the patients with small doses of protamine,[35] and the employment of agents such as antihistamines and steroids[37,40] have been attempted. Although these approaches have led to some improvements, at the present time there still remains no real alternatives to control the bleeding risks associated with heparinization and the toxic effects resulting from heparin neutralization with protamine.

The protamine filter suggested in this paper may prove by far the only means to control simultaneously both heparin and protamine induced complications. It can prevent heparin induced bleeding complications by removing heparin from the extracorporeal blood circuit. It can also prevent protamine induced complications by restricting protamine from entering the patient. The results presented above have clearly demonstrated the feasibility of this approach. The protamine-bound hollow fiber bundle removes more than 80% of the clinically used amounts of heparin both *in vitro* and *in vivo*. The filter also causes no clinically significant protamine responses in dogs. In addition, the filter has markedly reduced the thrombocytopenic response normally associated with the use of protamine.

Heparin-protamine complex has been reported to cause the activation of the blood complement system.[41,42] Rent, et al. indicate that the consumption of complements by the heparin-protamine complex is dependent upon both the incubation time and temperature.[41] Consumption of complements after 30 minutes of incubation at 14°C is about half of that at 37°C. At 4°C the consumption is only 10% of that at 37°C. Reduction of the incubation time results in a considerable reduction of complement consumption at all temperatures. This time and temperature dependent nature of the complement consumption may provide another potential advantage of using a protamine filter to neutralize heparin. The protamine filter eventually will be operated at a very high flow rate (the blood flow rate employed in kidney dialysis is about 250 mL/min., and in open heart operation is about 3-4 L/min.). At these flow rates, the incubation time of blood to heparin-protamine complex should be very short. Under this circumstances, the consumption of complement should presumably be quite limited. In addition, the protamine filter can be operated at a low temperature to reduce complement consumption, and later followed by rewarming the blood to the body temperature before the blood is returned to the patient. Experiments involving these studies are currently being conducted in our laboratory.

The proposed protamine filter for extracorporeal heparin removal has a far reaching significance. Its primary value would be as described above, in diminishment of the bleeding risks associated with systemic heparinization during an ECBC procedure, as well as the toxic effects resulting from heparin neutralization with protamine. This would significantly increase the safety of the ECBC procedure. With nearly 20 million such procedures performed each year, the benefits of the filter could be widespread.

The proposed system also posses two practical advantages. One advantage would be the simplicity and flexibility to operate such a system. The filter can be easily interfaced with the current ECBC procedure without any additional apparatus or invasive procedures. The other advantage of the system would be the potential ease of acceptance for clinical use. Since the filter was established using a clinically accepted blood compatible material (i.e. the cellulose hollow fibers obtained from a clinical hemodialyzer) with an existing FDA (Food and Drug Administration) approved drug (i.e. protamine), and at the same time minimized the toxic effects associated with the use of the drug, its acceptance for clinical use would be anticipated.

In a more general view, the protamine filter may be a prototype for other selective drug removal systems. It may also open up the possibility for the use of immobilized chemicals for extracorporeal blood treatment, particularly when the chemicals are too toxic to be administered safely into the vascular system.

ACKNOWLEDGMENTS

We would like to thank Drs. Thomas W. Wakefield and Christopher Vincent for technical assistance, Drs. Friedrick K. Port and Gerd O. Till for valuable advice. The work was supported by the National Institute of Health, National Heart, Lung and Blood Institute Grant 1-R29-HL38353-01, and the Whitaker Foundation Medical Research Grant.

REFERENCES

1. H. E. Kambic, S. Murabayashi, and Y. Nose, C&EN, April 14, 1986, p. 31.
2. A. J. Wing, F. P. Brunner, H. O. A. Brynger, C. Jacobs & P. Kramer in: "*Replacement of Renal Function by Dialysis*", W. Drukker, F. M. Parson & J. F. Maker, Eds., Martinus Nijhoff, Boston, 1983, p. 850.
3. J. P. Vacanti, R. K. Crone, J. D. Murphy, S. D. Smith, P. K. Blank, L. Reud & W. H. Hendren, J. Pediatr. Surg., **19**, 672 (1984).
4. K. Atsumi, I. Fujimasa, K. Imachi, M. Nakajima, S. Tsukagoshi, K. Mabushi, K. Motoruma, A. Kouno, T. Ono, A. Miyamoto, N. Takido & N. Inou, ASAIO J., **8**, 155 (1985).
5. R. G. Mason, H. Y. K. Chuang & S. F. Mohammad in: "*Replacement of Renal Function by Dialysis*", W. Drukker, F. M. Parson & J. F. Maher, Eds., Martinus Nijhoff, Boston, 1983, p. 186.
6. A. S. Gervin, Surg. Gynecol. Obstet., **140**, 789 (1975).
7. J. R. Fletcher, A. E. McKee, M. Mills, K. C. Suyden & C. M. Herman, Surgery, **80**, 214 (1976).
8. R. D. Swartz & F. K. Port, Kidney Int., **16**, 513 (1979).
9. M. A. Galen, S. M. Steinberg & E. G. Lourie, Ann. Inter. Med., **82**, 359 (1975).
10. N. E. Tolkoff-Rubin, J. Nardini, R. N. Leslie, S. T. Fang & R. H. Rubin, Dialysis Transplant., **15**, 125 (1986).
11. J. Porter & J. Jick, J. Am. Med. Assoc., **237**, 9 (1977).
12. M. C. Smith, D. Kowit, J. W. Crow, A. E. Cato, G. D. Park, A. Hassid & M. J. Dunn, Am. J. Med., **73**, 669 (1982).
13. A. Laned, I. R. McGregor, R. Michalski & V. V. Kakkar, Thromb. Res., **12**, 256 (1978).

14. R. D. Swartz, Nephron., **28**, 65 (1981).
15. R. Pinnik, T. Wiegmann & D. Diederich, N. Engl. J. Med., **308**, 258 (1983).
16. H. Miyama, N. Harumiya, Y. Mori & H. Tanzawa, J. Biomed. Mater. Res., **11**, 251 (1977).
17. R. Langer, R. L. Linhardt, P. M. Galliher, M. M. Flanagan, C. L. Cooney & M. D. Klein, Science, **217**, 261 (1982).
18. H. Berstein, V. C. Yang, D. Lund, M. Randhawa, W. Harmon & R. Langer, Kidney Int., **32**, 452 (1987).
19. V. C. Yang, H. Berstein & R. Langer in: "*Bioreactor Immobilized Enzymes and Cells: Fundamentals and Applications*", Moo-Young Ed., Elsevier, U. K., 1988, p. 83.
20. V. C. Yang, H. Berstein, C. L. Cooney, J. Kadam & R. Langer, Thromb. Res., **44**, 599 (1986).
21. L. B. Jaques, Can. Med. Assoc., J., **108**, 129 (1973).
22. N. M. Katz, Y. D. Kim, R. Siegelman, S. A. Ved, S. W. Ahned & R. B. Wallace, J. Thoracic Cardiovasc. Surg., **94**, 881 (1987).
23. J. K. Kirklin, D. E. Chenoweth, D. C. Naftel, E. H. Balckstone, J. W. Kirklin, D. D. Bitran, J. G. Curd, J. G. Reves & P. N. Samuelson, Ann. Thoracic Surg., **41**, 193 (1986).
24. G. N. Olinger, R. M. Becker & L. I. Bonchek, Ann. Thoracic Surg., **29**, 20 (1980).
25. C. A. Cobb, III & D. L. Fung, Surg. Neurol., **17**, 245 (1982).
26. C. Kurusz, "*6th. Annual Meeting of Pathophysiology and Extracorporeal Technology*", San Diego, CA, 1986.
27. V. C. Yang, Extracorporeal Blood Deheparinization System, patent approved.
28. M. M. Brandford, Anal. Biochem., **72**, 248 (1976).
29. L. B. Jaques, Science, **206**, 528 (1979).
30. J. Walenga, J. Fareed, D. Hoppensteadt & R. M. Emanuele, CRC Crit. Rev. Clin. Lab. Sci., **22**, 361 (1986).
31. J. Kohn & M. Wilcheck, Anal. Biochem., **115**, 375 (1981).
32. H. Bernstein, V. C. Yang & R. Langer, Appl. Biochem. Biotechnol., **16**, 129 (1988).
33. S. C. March, I. Parikh & P. Cuatrecasas, Anal. Biochem., **60**, 149 (1974).
34. S. W. Kim, R. G. Lee, H. Oster, D. Coleman, J. D. Andrade, D. L. Lentz & D. Olsen, Trans. Am. Soc. Artif. Intern. Organs, **20**, 449 (1974).
35. T. W. Wakefield, W. M. Whitehouse, Jr. & J. C. Stanley, J. Vasc. Surg., **1**, 346 (1984).
36. L. B. Jaques, Brit. J. Pharmacol., **4**, 135 (1949).
37. T. W. Wakefield, C. B. Hantler, B. Lindblad, W. M. Whitehouse, Jr. & J. C. Stanley, J. Vasc. Surg., **3**, 885 (1986).
38. A. L. Pauca, J. E. Graham & A. S. Hudspeth, Ann. Thorac. Surg., **35**, 637 (1983).
39. J. M. Weiler, P. Freiman, M. D. Sharath, W. J. Matzger, J. M. Smith, H. B. Richerson, Z. K. Ballas, P. C. Halverson, D. J. Shulan, S. Matsuo & R. L. Wilson, J. Allergy Clin. Immunol., **75**, 297 (1985).
40. J. F. Kelley, R. Patterson, P. Lieberman, D. A. Mathison & D. D. Stevenson, J. Allergy Clin. Immunol., **62**, 181 (1978).
41. R. Rent, N. Ertel, R. Eisenstein & H. Gewurz, J. Immunol., **114**, 120 (1975).
42. B. A. Fiedel, R. Rent, R. Myhrman & H. Gewurz, Immunology, **30**, 161 (1976).

POLY(VINYL ALCOHOL)-POLYELECTROLYTE BLENDED MEMBRANES-BLOOD COMPATIBILITY AND PERMEABILITY PROPERTIES

A. J. Aleyamma and C. P. Sharma

Biosurface Technology Division
Sree Chita Tirunal Institute for Medical Sciences
and Technology
Biomedical Technology Wing
Poojapura, Trivandrum-695012, India

Attempts have been made using poly(vinyl alcohol) to develop hemodialysis membranes with optimum properties for artificial kidney applications. Chemical crosslinking of poly(vinyl alcohol) with paraformaldehyde was developed and standardized to make it suitable for fabricating quality membranes. Further modifications of these membranes have been done with a synthetic heparinoid polyelectrolyte to enhance blood compatibility. Surface properties of these membranes, like platelet adhesion, water contact angle and protein adsorption, and other parameters such as permeability characteristics and mechanical properties were evaluated. The property variations of the membranes due to different sterilization techniques were also studied. The results were compared with that of dialysis grade cellulose acetate membranes.

INTRODUCTION

Extracorporeal hemodialysis has been in clinical use for several years and the search for a nonthrombogenic membrane having high permselectivity continues to be a field of extensive investigation. Different materials have been investigated to overcome certain limitations of the cellulosic membranes. It is indicated that the diffusive permeation of water soluble solutes proceeds only through the water phase in water swollen membranes.[1] Therefore, the water content of the membranes is a significant parameter governing the diffusive permeation of the solutes. On this ground poly(vinyl alcohol) [PVA] hydrogel is known for its permeability characteristics and can be considered as a candidate for artificial kidney applications.

PVA is known as a good film-forming highly hydrophilic polymer with outstanding chemical stability.[2] It is a water soluble polymer and to make it water insoluble, various methods are employed including both chemical crosslinking [3,4] and Co^{60} γ-irradiation techniques.[5,6] We have attempted to standardize a method for the chemical crosslinking of PVA

Biomimetic Polymers
Edited by C. G. Gebelein
Plenum Press, New York, 1990

using paraformaldehyde. This is a simple method and reproducible quality membranes can be fabricated easily.

Heparinization of hydrogels is a strategy of producing nonthrombogenic materials by combining the anticoagulant properties of heparin with the inherent biocompatibility of hydrogels.[7] The low interfacial tension, high permeability to small molecules, and soft and rubbery nature make hydrogels well suited candidate for biomaterials.

The outstanding anticoagulant activity of heparin, the natural anticoagulant responsible for the fluidity of blood, is attributed to the high concentration of sulfate, sulfamate and carboxylic groups and their steric order and conformation along the chain.[8] Recently, the addition reaction of N-chlorosulfonyl isocyanate and cis-1,4-poly(isoprene) has been reported.[9] It was shown that the addition product, upon treatment with alkali, formed a water soluble polyelectrolyte whose structural unit is shown in Figure 1. Beugeling, et al[10] observed good anticoagulant activity with this polyelectrolyte having sulfamate and carboxylic groups which can be used as a substitute for heparin. The synthetic aspects of the CSI-poly(olefin) reaction has been studied recently and insoluble versions of the hydrolyzed poly(isoprene)-CSI adducts have been reported.[11,12]

Recent studies show that polyelectrolyte bound surfaces also provide very good blood compatibility.[13,14] It has been suggested that when an artificial surface comes in contact with the blood, there may exist a simultaneous adhesion of platelets and protein to the surface.[15] Since polyelectrolytes have proven to have both anticoagulant and antiplatelet activity, this may discourage the chance of both ways of thrombus formation. Hence, these were chosen for the present studies.

In this work, an attempt is made to improve the blood compatibility of polyvinyl alcohol. Two approaches were employed, one is the bulk modification of PVA which results by blending with polyelectrolytes. The second approach is the surface modification procedures which have been suggested by different investigators to make the surface more compatible towards blood. Surface properties of these membranes, like water contact angle, protein adsorption using ^{125}I labeled fibrinogen and platelet adhesion, were evaluated to correlate with the blood compatibility. Also other parameters such as permeability characteristics and mechanical properties were studied. The results were compared with those obtained for dialysis grade Cellulose Acetate membranes. Finally the membranes are

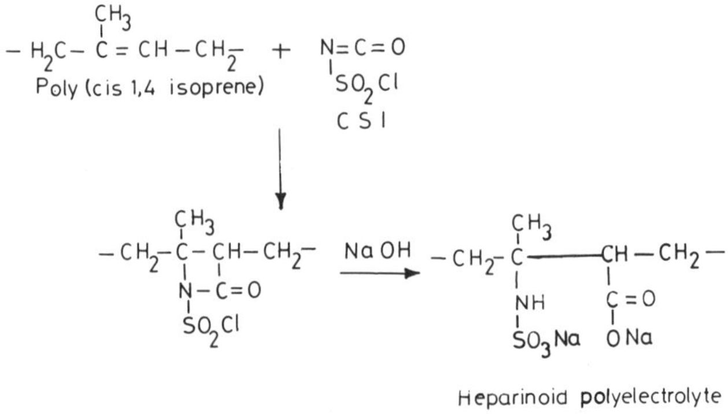

Poly (cis 1,4 isoprene)

CSI

NaOH

Heparinoid polyelectrolyte

Figure 1. Reactions involved in the synthesis of heparinoid polyelectrolytes from natural rubber.

subjected to different sterilization techniques to evaluate the property variations due to different sterilization processes.

EXPERIMENTAL

PVA (MW = 125,000) and paraformaldehyde were obtained from BDH Chemicals Ltd., Poole, England. The polyelectrolyte used was synthesized in our laboratory from natural rubber (*Hevea brasiliensis*) as illustrated below. The protein used for adsorption studies was fibrinogen, human, fraction I, MW = 360,000, Lot No. 83F-9305, approx. 95% clottable (Sigma Co.) and ^{125}I labeled human fibrinogen (Lot 926) obtained from Amersham International, Plc, UK. Dialysis grade cellulose acetate membranes were supplied from Bel-Art Products, Pequan Cock, NJ 07440, USA. The albumin used in the dialysis experiments was bovine Fraction V (Sisco Research Lab. Pvt. Ltd., Bombay). All other chemicals used were of reagent grade.

1. Synthesis of Polyelectrolyte

Natural rubber, poly(cis-1,4-isoprene), upon reaction with chlorosulfonyl isocyanate (CSI) had sulfamate and carboxylate groups introduced, which produced an electronegativity in the product similar to that of heparin. This was prepared by dissolving 5 g of rubber in 700 mL of toluene and adding to a stirred solution of CSI in the same solvent (0.15 mole in 100 mL). The addition reaction was conducted at room temperature under N_2 atmosphere, with stirring, for 2 hours. A white solid adduct of chlorosulfonyl isocyanate and rubber was obtained in 40-65% yield and was washed with diethyl ether.

In the second step, the adduct was boiled with 2.5% NaOH for 1 hour. Initially the mixture was warmed gently in order to avoid excess foaming. After boiling for 1 hour, a clear viscous solution was obtained and the polyelectrolyte was isolated by pouring the viscous solution into dry ethanol. The precipitated product was washed several times with acetone. The entire reaction sequence is shown in Figure 1.

2. Preparation of PVA Membranes

PVA was chemically crosslinked using paraformaldehyde. The paraformaldehyde, on treatment with sodium hydroxide, generated formaldehyde which reacted with the polymer forming a crosslinked gel.

The PVA solution was prepared in distilled water (10% w/v). The 20% paraformaldehyde [PF] solution was prepared in 5% NaOH. The PVA and PF solutions were mixed in the ratio 10:3 (v/v). The well mixed solution was kept for some time to remove air bubbles and was then spread over a clean glass plate and air dried at room temperature for 2 days. The membranes (~0.1 mm thickness) were peeled from the glass plate and washed with a large amount of distilled water.

3. Preparation of PVA:Electrolyte Blended Membranes

Both PVA and the polyelectrolyte are water soluble polymers and crosslinked PVA can be produced from the primary materials simultaneously with the incorporation of the polyelectrolyte [PE]. An aqueous solution of PVA, with different known concentrations of polyelectrolyte, was treated in a manner similar to that described above to obtain polyelectrolyte blended membranes (~0.1 mm thickness). Three types of blended membranes were prepared.

1. 2% polyelectrolyte added (100:2) (PVA + PE-2)
2. 5% polyelectrolyte added (100:5) (PVA + PE-5)
3. 10% polyelectrolyte added (100:10) (PVA + PE-10)

4. Surface Modification of PVA with Polyelectrolyte

For this modification, the bare PVA membranes were exposed to 500 mg% aqueous solution of polyelectrolyte overnight, vacuum dried and γ-irradiated with a dose of 0.275 mrds. from a Co^{60} source while in a nitrogen atmosphere.

5. Characterization of the Membranes

5A. Determination of the degree of hydration. The membrane samples were allowed to soak in beakers of distilled water for 24 hours. These were then removed, blotted dry and weighed. This weight was recorded as the wet weight of the membranes, which were then placed in a vacuum oven to dry and then reweighed. The degree of hydration was determined using Equation 1, where W_1 and W_2 are the wet and dry weights of the membranes, respectively.

$$\text{Degree of hydration} = 100(W_1 - W_2)/W_1 \qquad \text{(Equation 1)}$$

5B. Permeability test for the membranes. A dialysis cell was used for determining the passage of various molecules through the membranes as a function of time. Prior to use, the PVA membranes were soaked in distilled water for about 1 hour, and an overnight swelling period was used for the standard cellulose acetate [CA] membranes. The swollen membrane was clamped between the two compartments of the dialysis cell using suitable supporting and sealing devices. One compartment was filled with 0.1 M phosphate buffer (PH = 7.4) and the other with a mixture of solutes, prepared in the same buffer, containing urea (100 mg%), MW 60), creatinine (10 mg% MW 113), uric acid (10 mg% MW 168) Inulin (25 mg% MW 5,000), albumin (100 mg% MW 69,000) and sodium and potassium ions (140 and 5 meg/liter respectively). The dialysate was analyzed colorimetrically using a spectrophotometer (Spectronic 20, Milton Roy Co, USA). The permeability percentages were calculated.[16] The experiment was run in triplicate to ensure reproducibility within 10% limits.

5C. Platelet adhesion studies with washed platelets. Calf blood platelets were isolated by differential centrifugation within 2 hours after collection, from citrated calf blood, as described elsewhere,[17] and were washed with tyrode solution (0.055 M D-glucose, 0.138 M NaCl, 0.012 M $NaHCO_3$, 0.0018 M $CaCl_2$, 0.0049 M $MgCl_2$,,0.0027 M KCl, 0.00036 M NaH_2PO_4; pH = 7.4) for the adhesion studies.[17,18] Briefly, 10 mL of blood was collected in 1 mL of 3.8% sodium citrate and centrifuged at 700 x g for 10 minutes. The supernatant liquid having platelet rich plasma (PRP) was centrifuged at 1000 x g for 10 minutes and the white blood cell button was removed. The PRP was again centrifuged at 2000 x g for 10 minutes to get the platelets button and washed with tyrode solution three times and then suspended in the same solution.

Platelet suspensions (approximately 1.0 x 10^8 platelets per mL) were exposed to various membrane surfaces for 15 min. at room temperature (~30°C). The surfaces were rinsed with 0.1 M phosphate buffer, pH = 7.4, which fixed the platelets adhering to the substrates. These were then counted using an optical microscope in an identical fashion for all samples.

5D. Mechanical properties. The mechanical properties of the membranes were determined by the ASTM standard test protocol using a Chatillon Universal Test stand model UTSE-2.[19] The membranes were cut in the form of standard dumbbell-shaped test specimens, having a 2.5 cm region between grips, width of 0.5 cm and a cross head speed of 1 inch per minute. The tensile stress and the tensile strain (percentage of elongation) were calculated using Equations 2 and 3. The relative humidity noted at the time of experiment was ~90%, at room temperature.

$$\text{Tensile stress} = \frac{\text{Load at break}}{\text{Minimum cross sectional area}} \quad \text{(Equation 2)}$$

$$\text{Tensile strain} = \frac{\text{Elongation x 100}}{\text{Original length}} \quad \text{(Equation 3)}$$

5E. Protein adsorption studies. The adsorption of fibrinogen from a mixture of solutes was investigated. The adsorption/desorption experiments were carried out as described elsewhere.[20,21] Briefly, the samples of suitable sizes (2 x 1.5 cm) were exposed to the mixture containing urea 100 mg%, uric acid 10 mg%, creatinine 10 mg%, Dextran 25 mg%, NaCl 140 eg/liter, KCl 5 eg/liter, and fibrinogen 20 mg%, with a known amount of ^{125}I labeled fibrinogen in phosphate buffer, pH = 7.4. The exposure was done at 37°C and the experiments were run over a period of 3 hrs. The samples were thoroughly rinsed three times in distilled water with controlled flow. The amount of protein adsorbed was estimated with the help of a γ-counter (EC Corporation India Ltd.) using Equation 4, where λ = the surface concentration (µg/cm^2), C_p = the bulk concentration, R_f = the surface count rate, A = the total surface area, R_s = the count rate per mL of the mixture containing the labeled material.[21] The desorption kinetics were also studied, using the 3 hours preadsorption samples to phosphate buffer pH = 7.4, which were estimated in the same manner.

$$\lambda = C_p R_f / A R_s \quad \text{(Equation 4)}$$

The adsorption of fibrinogen from the same mixture as used above, during dialysis, was also studied. One compartment of the dialysis cell was filled with the mixture containing ^{125}I labeled fibrinogen and the other with a buffer solution at pH = 7.4. After a period of 3 hours the membranes were removed, cut into suitable sizes, and analyzed for the adsorbed protein in the same way as described above.

6. Sterilization of the Membranes

PVA alone, polyelectrolyte blended PVA (PVA + PE-10) and standard cellulose acetate (for comparison) membranes were subjected to the following sterilization techniques.

Autoclaving. The membranes were placed in an autoclaving unit, and subjected to a temperature of 121°C under a pressure of 20 pounds for 10 minutes.

Co^{60} γ-irradiation. The membranes were irradiated under a nitrogen atmosphere to a dosage of 2.5 mrds.

Glow discharge. A glow discharge treatment was given for a total duration of 30 minutes (15 min. each side) using an Edwards Vacuum unit E 306A. This procedure has been described elsewhere.[22]

Glutaraldehyde treatment. The membranes were exposed to a 1% aqueous solution of glutaraldehyde for 1 hour. After exposure, the films were washed with distilled water several times and dried.

The sterilized membranes were studied for their mechanical strength, permeability and platelet adhesion. In these cases, the dialysis was only carried out for 6 hours, for comparison.

RESULTS AND DISCUSSION

The PVA membranes were colorless and transparent having a degree of hydration around 65%. The degree of hydration was unaltered by the polyelectrolyte blending, but the transparency of the membranes was reduced to some extent.

The permeability of urea, creatinine uric acid, inulin and albumin through three selected membranes, PVA alone, polyelectrolyte blended PVA (PVA + PE-10) and surface modified PVA, were evaluated (Table 1) and compared to dialysis grade cellulose acetate [CA] membrane. It seems that the PVA membranes had improved permeability over the standard CA membranes, which may be attributed to the higher water content of the PVA membrane in the swollen state. Also it appears that the polyelectrolyte modification would not cause any significant variation in the permeability characteristics.

The results of the platelet adhesion studies (Table 2) indicated that the polyelectrolyte modification reduced the number of adhered platelets. It seems that the platelet adherence was minimized in the case of (PVA + PE-10) and surface modified PVA, possibly suggesting reduced thrombogenic character of the membrane surface.

The fibrinogen adsorption/desorption on bare PVA surface, 2% polyelectrolyte blended surface and standard cellulose acetate surface are shown in Figure 2. It appears that the fibrinogen adsorption on PVA surfaces at the initial stage was more or less the same as that on the cellulose acetate surface, but as the exposure time increases, the polyelectrolyte modified surface adsorbed relatively less fibrinogen.

The results of fibrinogen adsorption during dialysis and on direct exposure were compared as shown in Table 3. A slight increase of fibrinogen adsorption on PVA membrane was observed in the case of dialysis. This may be due to the fact that during dialysis some fibrinogen molecule passed through the membrane to the other compartment of the dialysis cell (2.4% Table 3) and a part of it possibly became adsorbed on the other side of the membrane.

From the contact angle data (Table 4), it seems that the hydrophilic character of the blended membrane surface was slightly increased. How-

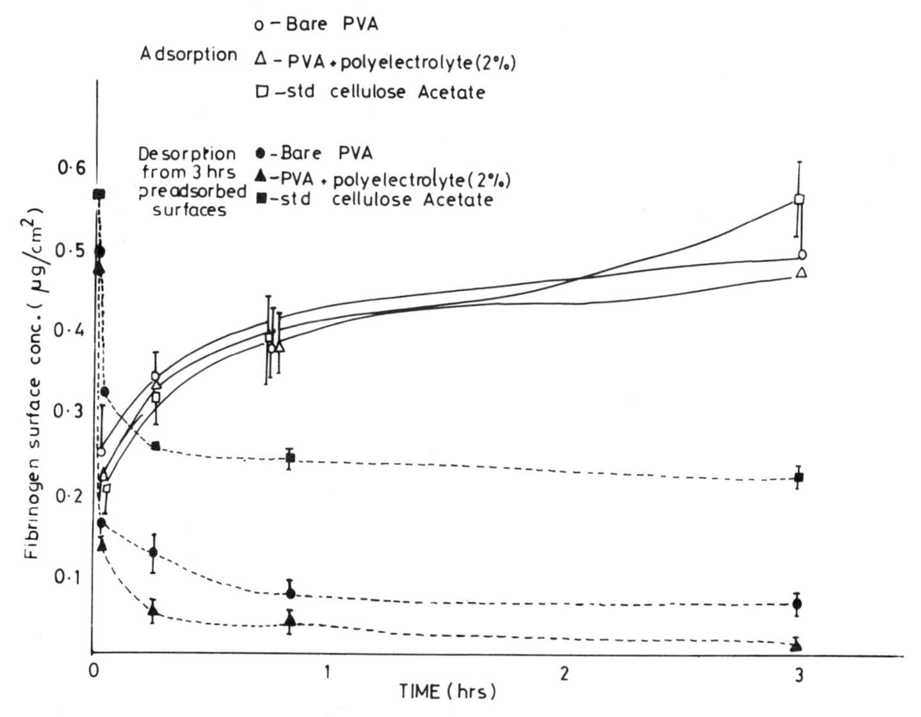

Figure 2. Adsorption/desorption of fibrinogen on PVA surfaces from a mixture of urea, creatinine, dextran, NaCl, KCl and 20 mg% fibrinogen at pH = 7.4

ever, an increase in the contact angle on the surface modified PVA was observed which may be attributed to the effect of γ-irradiation.

Table 5 shows the mechanical properties of bare and modified PVA membranes. Even though the stress value of PVA membranes was less compared to the CA membranes, its strain value was high indicating its elastic nature. It also appears that the polyelectrolyte blending maintained the strain as such with some increase in the stress behavior. The surface modification did not alter the mechanical character significantly.

The permeability properties of the membranes (bare PVA, [PVA + PE-10] and standard cellulose acetate) after different sterilization process were summarized on Table 6. It appears that the permeability properties were not significantly affected by glow discharge, Co^{60} γ-irradiation and glutaraldehyde treatment. But autoclaving seems to reduce the permeability properties which may be attributed to the reduced water content (about 50% reduction of water of hydration in the case of PVA) of the autoclaved membranes in the swollen state. Also a shrinkage of the membranes was observed after autoclaving.

The mechanical properties variations are shown in Table 7. An improvement of mechanical strength for PVA membranes was observed in the case of Co^{60} γ-irradiation and autoclaving which may possibly be due to the further crosslinking of the molecular chains under these conditions. Also, it seems that the mechanical properties were not significantly affected by the glow discharge and glutaraldehyde treatments.

The platelet adhesion data (Table 7) indicates that there was an increase in the number of platelets adhered on the surface, due to the

Table 1. Permeability (%) of various molecules through modified and bare PVA membranes. The values are expressed as the mean ± the standard deviation.

		% passed after			
Components	Sample	1 hour	2 hours	4 hours	6 hours
Urea	1.	32.92±2.1	43.12±3.1	49.01±2.8	58.15±3.5
	2.	34.7 ±1.5	42.8 ±1.1	47.7 ±2.9	56.3 ±2.6
	3.	34.5 ±3.8	44.4 ±1.7	48.1 ±3.8	56.6 ±2.6
	4.	21.7 ±0	38.8 ±0.0	46.2 ±0.9	53.2 ±2.5
Creatinine	1.	27.6 ±1.6	36.16±2.5	44.15±2.2	49.2 ±2.8
	2.	26.7 ±1.8	36.7 ±1.4	44.3 ±1.2	43.15±1.5
	3.	27.0 ±2.2	34.3 ±1.6	41.5 ±2.4	47.45±0.7
	4.	16.9 ±0.6	19.8 ±0.7	31.5 ±1.5	43.47±0.7
Uric acid	1.	14.0 ±1.2	27.4 ±2.0	35.1 ±1.5	42.2 ±2.5
	2.	14.9 ±1.9	28.55±1.7	39.6 ±0.5	40.9 ±1.3
	3.	15.15±1.01	26.9 ±0.8	36.05±3.3	40.5 ±2.3
	4.	5.3 ±1.1	12.3 ±1.2	21.2 ±1.2	33.5 ±0.9
Inulin	1.	4.3 ±1.2	4.5 ±0.7	7.3 ±0.4	6.7 ±0.8
	2.	3.0 ±0.9	4.9 ±0.6	6.4 ±0.5	7.2 ±0.8
	3.	4.2 ±0.8	3.4 ±0.2	7.2 ±0.2	9.0 ±0.8
	4.	4.4 ±0.7	3.2 ±1.0	6.8 ±1.2	6.5 ±0.6
Albumin	1.	0	0	1.4 ±0.4	2.8 ±0.6
	2.	0	2.52±0.22	0	0.65
	3.	0	0	0	0
	4.	0.79±0	.0	1.36±0.5	2.6 ±0.5

1. Bare PVA.
2. Polyelectrolyte blended PVA (PVA + PE-10).
3. Surface modified PVA.
4. Standard cellulose acetate membrane.

glow discharge treatment. Also it appears that Co^{60} γ-irradiation increased the platelet adherence to some extent for the polyelectrolyte blended surface. It should be noted that the platelet adhesion is higher if the glutaraldehyde treated PVA membranes are not washed properly. This

Table 2. Platelet adhesion on PE modified PVA surfaces. The values are expressed as the mean ± the standard deviation.

Surface	No. of platelets/mm²
Bare PVA	27.5 ± 2.5
PVA + PE-2	20.8 ± 1.3
PVA + PE-5	17.1 ± 1.9
PVA + PE-10	12.6 ± 2.4
Surface modified PVA	15.65 ± 1.8
Standard Cellulose Acetate	21.3 ± 1.6

Table 3. Comparison of fibrinogen adsorption by direct exposure and adsorption during a 3 hour dialysis. The values are expressed as the mean ± the standard deviation.

	Surface concentration (μg/cm²)		
Surface	Adsorption by direct exposure	Adsorption during dialysis	Fibrinogen passed (%)
Bare PVA	0.50 ± 0.09	0.608 ± 0.03	3.58
PVA+PE-2	0.48 ± 0.01	0.59 ± 0.07	2.15
CA	0.57 ± 0.07	0.55 ± 0.03	1.56

Table 4. Water contact angle on the PVA surfaces in degrees. The values are expressed as the mean ± the standard deviation.

Surface	Contact angle
Bare PVA	57.6 ± 2.9
PVA + PE -2	50.1 ± 3.4
PVA + PE -5	48.2 ± 3.8
PVA + PE -10	48.8 ± 4.6
Surface Modified PVA	65.23 ± 2.44
Standard CA	28.3 ± 2.1

may possibly be due to traces of glutaraldehyde molecules trapped in the bulk of the membranes. It could be suggested that an initial swelling in distilled water should be given to the membranes. Exposure of this preswollen membranes to glutaraldehyde, followed by repeated washing with distilled water, is a better way to minimize the bulk entrapment of the glutaraldehyde molecules within the membranes.

Table 5. Mechanical properties for the PVA membranes. The values are expressed as the mean ± the standard deviation.

Specimen	Tensile stress kg/cm²	Tensile strain %
Bare PVA	488.7 ± 58.5	288.6 ±59.4
PVA + PE-2	540.4 ± 52.5	266.4 ± 37
PVA + PE-5	555.9 ± 46.4	297.7 ± 28.4
PVA + PE-10	537.3 ± 99.6	268 ± 28.4
Surface modified PVA	485.2 ± 19.2	321.2 ± 16.6
Standard CA	1017.75 ± 53.8	44.42 ± 6.77

Table 6. The effect of different sterilization methods on the permeability of various solutes through the PVA membranes during 6 hours dialysis. The values are expressed as the mean ± the standard deviation.

Type		Percentage passed for				
		Urea	Creatinine	Uric Acid	Inulin	Albumin
(A)	1.	51.9±1.1	40.5±2.1	19.3±1.8	4.4±0.96	0
	2.	50.6±2.8	33.1±3.6	20.8±0.76	4.2±0.83	0
	3.	47.6±1.8	35.8±1.2	22.7±0.51	4.2±1.2	0
(B)	1.	63.6±2.3	47.3±2.3	41.3±1.4	7.5±0.8	0.5±0.02
	2.	57.6±0.87	46.6±2.2	39.2±2.4	7.4±0.6	0
	3.	56.1±2.2	44.6±1.2	31.6±1.2	7.1±0.6	1.0±0.03
(C)	1.	56.3±2.3	41.66	36.5±0.38	6.4±1.3	0
	2.	59.5±2.6	44.7±1.3	37.8±1.4	7.2±0	0
	3.	51.2±1.0	38.7±0.5	32.5±1.7	7.08±0.59	0
(D)	1.	60 ±5.8	52.8±0.3	40.1±0.63	8.1±1.3	0
	2.	63 ±2.9	43.1±2.5	40.2±0.82	7.8±1.9	0
	3.	57.4±0.8	45.2±1.15	31.7±2.3	6.2±1.07	0
(E)	1.	58.2±3.5	49.2±2.8	42.2±2.5	6.7±0.8	2.8±0.6
	2.	56.3±2.6	43.2±2.6	40.9±1.3	7.2±0.8	0.65
	3.	53.2±2.5	43.5±0.7	33.5±0.9	6.5±0.6	2.6±0.5

Sterilization techniques used were:

(A) Autoclaving.
(B) Glow discharge.
(C) Co^{60} γ-irradiation.
(D) 1% Glutaraldehyde.
(E) Without any treatment.

Sample surfaces were:

(1) Bare PVA
(2) PVA + PE-10
(3) Standard Cellulose Acetate

CONCLUSION

PVA hydrogels with 65% water of hydration could be prepared from paraformaldehyde crosslinking of an aqueous PVA solution. It seems that the polyelectrolyte modified PVA has a tendency to adhere a smaller number of platelets than the standard cellulose acetate membranes. This reduction of platelets under static conditions may become more effective when considering the blood flow conditionsd. The 2% polyelectrolyte blended PVA was as blood compatible as the dialysis grade cellulose acetate membrane. The 10% polyelectrolyte blended PVA and surface modified PVA seems superior to the standard cellulose membranes owing to their improved blood compatibility and permeability characteristics. The glutaraldehyde treatment and Co^{60} γ-irradiation seems to be effective methods for sterilization of these membranes.

Table 7. The effect of different sterilization techniques on the platelet adhesion data and mechanical property variations of the PVA membranes. The values are expressed as the mean ± the standard deviation.

Method	Type	Tensile stress kg/cm^2	Elongation %	Platelet count mm^2
(A)	1.	488.72 ± 58.5	288.56 ± 59.36	26.24 ± 2.4
	2.	537.29 ± 99.6	228 ± 28.4	12.5 ± 2.45
	3.	1017.75 ± 53.8	44.4 ± 6.77	21.25 ± 1.58
(B)	1.	651.3 ± 20.34	211.12 ± 9.16	24.01 ± 2.5
	2.	635.5 ± 27.28	236.5 ± 10.6	13.12 ± 1.9
	3.	937.3 ±132.5	45.2 ± 8.4	20.2 ± 2.1
(C)	1.	473.58 ± 30.5	248.4 ± 20.4	29.4 ± 3
	2.	543.28 ± 22.4	273.8 ± 18.2	15.8 ± 2.4
	3.	850.17 ± 43.62	46.3 ± 7.6	29.81 ± 3.2
(D)	1.	589.4 ± 43.62	316.3 ± 36.43	25.01 ± 2.4
	2.	629.12 ± 70.5	246 ± 31.3	15.01 ± 2.5
	3.	880.25 ± 45.5	45.5 ± 30.6	22.8 ± 1.4
(E)	1.	494.6 ± 73.8	256.9 ± 16.2	25.8 ± 2.0
	2.	564.58 ± 35.3	243.2 ± 20.64	14.3 ± 1.2
	3.	1073.3 ± 53.8	44.4 ± 6.7	23.4 ± 1.8

Sterilization techniques used were:

(A) Without any treatment.
(B) Autoclaving.
(C) Glow discharge.
(D) Co^{60} γ-irradiation.
(E) 1% Glutaraldehyde.

Sample surfaces were:

(1) Bare PV.
(2) PVA + PE-10.
(3) Standard Cellulose Acetate.

ACKNOWLEDGMENTS

We appreciate the valuable discussions with Mr. Thomas Chandy and the help received from Dr. A. V. Lal for providing calf blood. This work is funded by DST India.

REFERENCES

1. H. Yasuda, C. E. Lamaze & L. D. Ikenberry, Makromol. Chem., **118**, 19 (1968).
2. S. Peter, N. Hese & R. Stefan, Desalination, **19**, 161 (1976).
3. C. T. Chen, Y. T. Chang, M. C. Chen & A. V. Tobolsky, J. Appl. Polym. Sci., **17**, 789 (1973).

4. E. W. Merril, "*Final report on Contract PH-43-66-491, NTIS PB 224 949/86A,*" National Technical Information Service, Springfield, VA, 1973.
5. A. Charlesby, "*Atomic Radiation and Polymers,*" Pergamon, Oxford 1960.
6. A. Chapiro, "*Radiation Chemistry of Polymeric Systems,*" Wiley-Interscience, New York, 1962.
7. E. W. Merril, E. W. Salzman, P. S. L. Wong, T. P. Ashford, A. H. Brow, & W. G. Austen, J. Appl. Physiology, **29**, 723 (1969).
8. M. Sorm, S. Nespurek, L. Mrkvickova, J. Kalal & Z. Vorlova in: "*Medical Polymers: Chemical Problems,*" B. Sedlack, et al, Eds. John Wiley & Sons, J. Poly Sci., Poly Symp., **66**, 349.
9. L. Van der Does, J. Hoffman, T. E. van Uttern, J. Polym. Sci. Polym. Lett., **11**, 169 (1973).
10. T. Beugeling, L. Van der Does, A. Bantejs & W. L. Serderal, J. Biomed. Mater. Res., **8**, 375 (1974).
11. C. G. Gebelein & D. Murphy, Proc. Polymeric Materials: Science & Engineering, **53**, 415 (1985).
"The Synthesis of Some Polymeric Biomaterials With Potential Blood Compatibility".
12. C. G. Gebelein & D. Murphy in: "*Advances in Biomedical Polymers,*" C. G. Gebelein, Ed., Plenum Publ., New York, 1987, p. 277.
"The Synthesis of Some Potentially Blood Compatible Heparin-Like Polymeric Biomaterials."
13. E. O. Lundell, G. T. Kwiatkowski, J. S. Byck, F. D. Osterholtz, W. S. Creasy, & D. D. Stewart in: "*Hydrogels for Medical and Related Applications,*" Joseph P. Andrade, Ed., ACS Symp. #31, American Chemical Society, Washington, DC, 1976, p. 305.
14. C. P. Sharma & Ajanta K. Nair, Trans. Soc. Biomaterials, **X**, 100 (1988).
15. A. Marmur & S. L. Cooper, J. Colloid Interfacial Sci., **89**, 458 (1982).
16. S. Hirano, K. Tobetto, M. Hasegawa & N. Matsuda, J. Biomed. Mater. Res., **14**, 477 (1980).
17. T. Chandy & C. P. Sharma, Thromb. Res., **14**, 9 (1986).
18. E. S. Lee & S. W. Kim, Trans. Amer. Soc. Artif. Internal Org., **XXV**, 124 (1979).
19. ASTM-D-638-80, Annual Book of ASTM Standards, **35**, 250 (1982).
20. T. Chandy & C. P. Sharma, J. Colloid Interfacial Sci., **108**, 104 (1985).
21. S. Uniyal, & J. L. Brash, Thromb. Haemostasis, **47**, 285 (1982).
22. C. P. Sharma, T. Chandy, & M. C. Sunny, J. Biomat. Appl., **1**, 533 (1987).

CONDENSATION OF BIOACTIVE COMPOUNDS INTO THE MEMBRANE COMPARTMENT BY CONJUGATION WITH SYNTHETIC POLYPEPTIDES

Yukio Imanishi and Shunsaku Kimura

Department of Polymer Chemistry
Kyoto University
Yoshida Honmachi, Sakyo-ku
Kyoto 606, Japan

Peptides mimicing the membrane-affinity sequences of channel-forming antibiotics, and signal peptides of secretory proteins, and membrane fusion peptides were synthesized. Some of the synthetic peptides were combined with an opiate peptide, enkephalin. The interaction of conjugated enkephalins with lipid membrane was investigated by the binding assay with opiate receptors of rat brain homogenates. The connection of the specific peptides to the C-terminal of enkephalin altered the receptor specificity from δ to μ type. However, the receptor specificity of the conjugated enkephalin was not clearly related to the ability to binding with lipid bilayer membrane. On the other hand, enkephalin was conjugated with polypeptides marked with rhodamine and the interaction with platelets, fibroblast cells, and erythrocytes was investigated. By a fluorescence microscopy technique the conjugated enkephalin was found to be bound by platelets and fibroblast cells, but not by erythrocytes. Upon acceptance of the conjugate by the receptor in fibroblast cells, patching and capping of the complex took place and the conjugate was ultimately internalized into cytosol, while unconjugated enkephalin was not internalized. It was therefore concluded that the action of a bioactive compound on the cell can be controlled by the properties of the polypeptide combined with the bioactive compound.

INTRODUCTION

Many chemical signals from the surroundings to living cells are accepted at the cell membrane and induce the cellular responses according to the stimuli. Since the first step of this process, called signal transduction, occurs at the cell membrane, the whole physiological reaction of the cell can be regulated by modifying the interaction of chemicals with the cell membrane. The membrane structure might be considered as being divided into three different compartments: hydrophobic core of lipid bilayer membrane, hydrophilic membrane surface, and the aqueous region, apart from the surface of lipid bilayer which is more than the

Biomimetic Polymers
Edited by C. G. Gebelein
Plenum Press, New York, 1990

Debye-length. It is likely that the reaction will occur at one of these defined regions of the membrane, therefore, it should be useful to fix such chemicals, which will modulate the cell activity, to a certain region of the membrane in order to regulate the signal transduction specifically.

LIGAND-MEMBRANE INTERACTIONS

1. Classification

The action of a bioactive compound to living cells might be classified into four types: (1) peptide hormones and neurotransmitters, which are bound by receptor proteins in the cell membrane; (2) lectins and cholera toxins, which are bound by glycoproteins or glycolipids in the cell membrane; (3) ionophores and cytolytic peptides, which are solubilized into the lipid membrane and change the membrane structure or the ion-permeability across the membrane; (4) steroid hormones, which are bound by receptor proteins in the cytoplasm. Three types of chemical signals among the above four kinds are transmitted to the cytoplasma via the chemical reactions occurring in the cell membrane. The biosignal molecules belonging to the class (3) interact directly especially with lipid bilayer membrane. The biosignal molecules belonging to the class (1) are thought to associate with the lipid bilayer membrane to regulate the subsequent binding to receptor proteins.

The ligand-membrane interactions have been proposed to play an important role in the binding of the ligand to the receptor in the membrane. When the ligand is bound to the lipid membrane and lateral diffusion on the membrane is allowed, the diffusion process of a ligand changes from a three-dimensional space to a two-dimensional surface. According to Adam's calculation, the required time for the chemical to reach the receptor in membrane will be shortened by this reduction of dimension.[1]

Schwyzer has also proposed that the process of ligand-receptor reaction should be composed of multiple sequential steps including surface accumulation of charged ligands, ligand-membrane interactions, and finally binding to the receptor.[2] Therefore, the total free energy of binding can be divided into the moderate values among several steps. This model thereby can provide the explanation for the large apparent association contants and the high association and dissociation rates. Furthermore, it is pointed out that the specific conformation and orientation of the ligand in the membrane, which will facilitate the binding to the receptor, should be induced as the result of the interaction with the lipid membrane.[3]

2. Membrane Compartment Concept

The importance of ligand-lipid membrane interactions is emphasized in the proposed model about the interaction of peptide hormones with the receptor.[4-6] For example, the opiate receptor, which is of particular interest due to its clinical relevance, is shown to be classified into several subtypes, such as δ, μ and κ receptors. Endogeneous opiate peptides were found to bind selectively to each receptor. These opiate peptides contain the common peptide segment, YGGF, at the N-terminal of the peptide sequence.[7] This message segment is followed by the address segment, which determines the receptor selectivity of the opiate peptides.

Schwyzer indicated that the opiate peptide accumulated on the membrane taking a preferred conformation and a specific orientation due to the interaction with the lipid membrane.[4-6] As a result, the message segment of each opiate peptide was condensed at the certain region of membrane. It was suggested that the binding site of each receptor is supposedly exposed to the different membrane compartment, and when the message segment and the binding site coexist at the same compartment, their association should be facilitated (membrane compartment concept).

Thus, it is likely that the activity of bioactive compounds can be regulated by the suitable modification of the interaction of bioactive compounds with the lipid membrane.

3. Introduction of Peptide Hormone to Polymer Carrier

Conjugation of bioactive compounds with a polymer has been reported to improve the activity in various aspects. For example, peptide hormones have been covalently conjugated with a polymer carrier. Though the mobility and flexibility of introduced peptides should be hindered, the activity was potentiated due to the cooperative effects of the bound hormone molecules on the binding to the receptors in membrane. That is, α-MSH was introduced to tobacco mosaic virus (TMV), which is constituted with 2130 capsomers (each capsomer contains 158 amino acid residues) and the RNA chain. The capsomer containing one α-MSH molecule showed the reduced tyrosinase activity on melanoma cells in culture. However, the prepared complex of TMV/α-MSH, which has about 300 α-MSH molecules per one TMV, exhibited higher activity and binding to the cell, and produced a long-lasting response.[8]

On the other hand, antibiotics have been covalently bound to a biocompatible polymer. Such conjugates successfully reduced the toxicity of drugs and improved keeping a high concentration in the circulation.[9] It was also reported that conjugation of enzyme with poly(ethylene glycol) improved the biocompatibility and enhanced the activity (reduction of antigenity).[10]

The development of the polymer carrier for bioactive compounds will be quite interesting when the property of bioactive compounds can be improved as described above. This chapter is divided into three parts. First, we have searched for the peptide molecules which are distributed to the restricted region of lipid membrane while adopting a defined conformation and taking a specific orientation. These peptide molecules will be useful to develop the membrane-associating ligand by carrying a bioactive compound. Secondly, the message segment of the opiate peptide was covalently bound to the synthetic address segment, which showed the different affinity with lipid membrane, and investigated for the interaction with this opiate receptor. Finally, several opiate peptide molecules were introduced to a membrane-associating polypeptide chain, and the interactions with living cells were analyzed.

MEMBRANE-ASSOCIATING PEPTIDES

1. α-Helical Peptide

It is necessary to develop the carrier peptide for the ligand to become accumulated on the membrane by adopting a specific conformation and orientation, and to be condensed to the defined membrane compartment. It is often pointed out that membrane-bound enzymes have the successive hydrophobic amino acid residues in the primary sequence, which supposedly spans the membrane by taking a α-helical conformation. This interpretation is reasonable because the polar groups of the carbonyl oxygen and the amino proton should be shielded by forming hydrogen bondings to become buried deeply into the hydrophobic core of the lipid bilayer membrane. Alamethicin, which is composed of 20 amino acid residues, is known to form a voltage-gated ion pore in lipid bilayer membrane. The study by X-ray crystallography revealed the formation of intramolecular hydrogen bondings throughout the molecule.[11] Such helical rods of alamethicin are thought to aggregate in the membrane and form the ion-channel across the lipid bilayer membrane. Eight out of 20 amino acid residues of alamethicin are α-aminoisobutyric acid (Aib). Aib is of particular interest because it improves the solubility of peptides. This is attributed to the property of Aib to disrupt the formation of the β-sheet structure due to the steric hindrance at the C^{α} carbon, which is attached by two methyl groups.[12] In this study, the sequential polypeptides, Boc-$(Ala\text{-}Aib)_n$-OMe (n=2,4,6,8) and Boc-Ser(anth)-$(Ala\text{-}Aib)_n$-OMe (n=2,4,6,8,10, Anth represents 9-methylanthracene), were synthesized (Figure 1), and investigated for their interaction with the lipid bilayer membrane.

1A. Conformation of Boc-$(Ala\text{-}Aib)_n$-OMe

CD spectra of Boc-$(Ala\text{-}Aib)_n$-OMe were measured in ethanol. As the peptide length increased, the strength of negative Cotton effect having double minima increased, indicating the increase of the content of the α-helical conformation (Figure 2). α-Helical contents of Boc-$(Ala\text{-}Aib)_n$-OMe, n = 8 and 10, were calculated to be about 40% and 60%, respectively. It should be noted, however, that α-helical and 3_{10}-helix can not be distinguished from CD spectra, and oligopeptides containing Aib were frequently shown to form a 3_{10}-helix.

1B. Distribution of Boc-Ser(Anth)-$(Ala\text{-}Aib)_n$-OMe to Lipid Bilayer Membrane

The fluorescent probe, the 9-methylanthracene group, was introduced

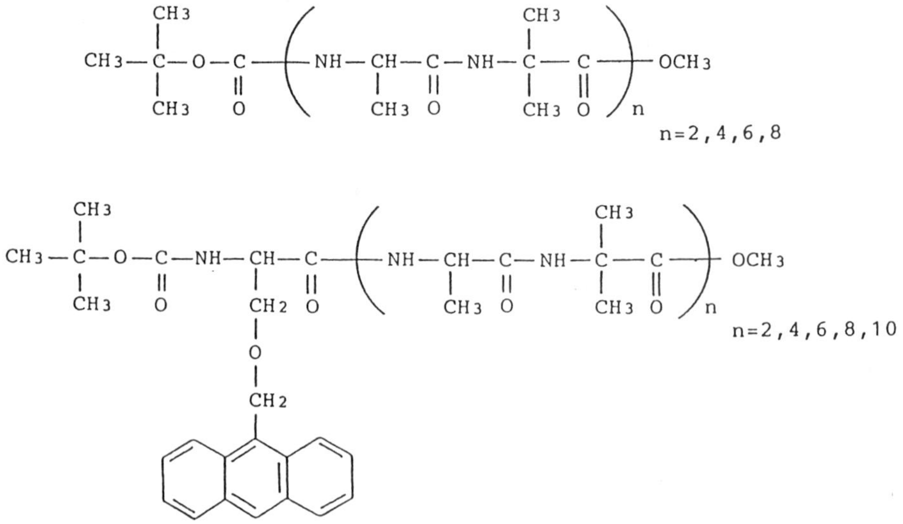

CH3 CH3
CH3—C—O—C—(NH—CH—C—NH—C—C)n—OCH3
CH3 O CH3 O CH3 O
n=2,4,6,8

CH3 CH3
CH3—C—O—C—NH—CH—C—(NH—CH—C—NH—C—C)n—OCH3
CH3 O CH2 O CH3 O CH3 O
O
CH2
n=2,4,6,8,10

Figure 1. Molecular structure of Boc-$(Ala\text{-}Aib)_n$-OCH_3 and Boc-Ser(Anth)-$(Ala\text{-}Aib)_n$-OCH_3.

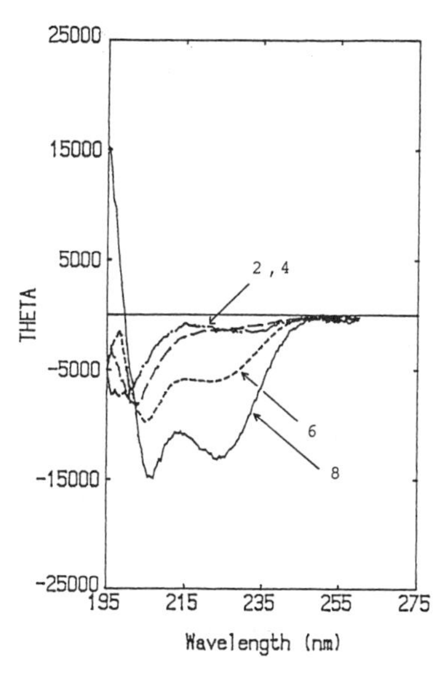

Figure 2. CD spectra of Boc-(Ala-Aib)$_n$-OCH_3 in ethanol.

to the N-terminal residue of these sequential polypeptides. The interaction of polypeptides with the lipid bilayer membrane was analyzed by monitoring the fluorescence from the anthryl group.

The intensive monomer emission was observed for Boc-Ser(Anth)-(Ala-Aib)$_n$-OMe, n = 2,4,6, in a buffer solution, whereas, the excimer emission was predominant in the fluorescence spectra of those when n = 8,10. The latter polypeptides should aggregate in a buffer solution due to the high hydrophobicity. The addition of dimyristoylphosphatidylcholine (DMPC) vesicles decreased the intensity of monomer emission of Boc-Ser(Anth)-(Ala-Aib)$_n$-OMe, n = 2,4,6, at first. However, a further addition of vesicles increased the intensity of monomer emission. This observation can be interpreted as follows. When the amount of the added DMPC vesicles was small, the distributed peptide molecules to membrane should be condensed so that they would interact with each other to reduce the intensity of monomer emission due to the excimer formation or concentration quenching. In the presence of the adequate amount of DMPC vesicles, the peptide molecules could be dispersed in membrane at a monomeric state, and the intensity of monomer emission should be restored. On the other hand, the monomer emission of Boc-Ser(Anth)-Ala-Aib)$_n$-OMe, for n = 8,10, increased with the addition of DMPC vesicles, indicating that the aggregated state in a buffer solution should be dissociated by distributing to the lipid bilayer membrane.

The fluorescence intensity leveled off when the added DMPC vesicles exceeded a certain amount. This amount of lipid decreased as the peptide length increased. It is therefore suggested that the affinity of these sequential polypeptides for lipid bilayer membrane increased with the elongation of peptide length. The longer peptides tend to aggregate both in a buffer solution and membrane.

1C. Orientation of Boc-Ser(Anth)-(Ala-Aib)$_n$-OMe in Membrane

The location of the anthryl group of Boc-Ser(Anth)-(Ala-Aib)$_n$-OMe in the membrane was evaluated from the fluorescence quenching by a series of doxylstearic acid. The doxyl group can be fixed to a restricted region of membrane according to the position introduced on the steric acid. 5-

doxylstearic acid was shown to quench the monomer emission of Boc-Ser-(Anth)-(Ala-Aib)$_n$-OMe, n = 6,8, more effectively than when n = 2,4. It is therefore concluded that the n-terminal of the peptide moved into deeper region of hydrophobic core of the lipid bilayer membrane as the peptide length increased.

However, the Stern-Volmer's quenching coefficient of monomer emission from the Boc-Ser(Anth)-(Ala-Aib)$_{10}$-OMe by 5 doxylstearic acid was smaller than those of the Boc-Ser(Anth)-(Ala-Aib)$_n$-OMe, n = 6,8. Since the most of the added Boc-Ser(Anth)-(Ala-Aib)$_{10}$-OMe molecules were distributed to the lipid bilayer membrane and did not aggregate tightly in the membrane, the result of quenching experiments indicated that the N-terminal of Boc-Ser(Anth)-(Ala-Aib)$_{10}$-OMe should locate close to the inner surface of the lipid bilayer membrane of the vesicle. This topological arrangement can be realized when Boc-Ser(Anth)-(Ala-Aib)$_{10}$-OMe adopts an α-helical conformation oriented perpendicularly to the membrane surface.

The microviscosity at the certain depth from the membrane surface has been evaluated on the basis of the NMR measurement of the ^{13}C and ^{2}H spin-lattice relaxation times of dipalmitoly-phosphatidylcholine (DPPC) and selectively deuterated DPPC vesicles, respectively.[13] It was clearly shown that the glycerol backbone had the shortest relaxation time, while the longest relaxation time was observed for the methyl groups at the end of the phospholipid molecule. It was assumed that the mobility of the anthryl group of sequential polypeptide should be affected by the microviscosity of the domain where the N-terminal of peptide is located. The fluorescence depolarization of the anthryl group of a series of sequential polypeptides was examined in the presence of DMPC vesicles (Figure 3). The fluorescence depolarization values of anthryl group decreased as the chain length increased from Boc-Ser(Anth)-(Ala-Aib)$_2$-OMe to Boc-Ser-(Anth)-(Ala-Aib)$_8$-OMe. This result can not be explained by the assumption that these sequential polypeptides are distributed to lipid membrane by adopting an α-helical conformation with a fixed orientation to membrane. Supposedly, one fraction of polypeptides would stay at membrane surface without taking the defined conformation, and another fraction would adopt a partial α-helical conformation oriented perpendicularly to the membrane surface. As the chain length of the sequential polypeptide increases, the latter fraction will increase, and the mobility of the polypeptide in the membrane should be diminished, on the average, due to the intercalation within the tight packing of lipid molecules.

It is noteworthy that the fluorescence depolarization was found to increase from Boc-Ser(Anth)-(Ala-Aib)$_8$-OMe to Boc-Ser(Anth)-(Ala-Aib)$_{10}$-OMe. Since most of the latter peptide molecules in the membrane are thought to adopt an α-helix oriented perpendicularly to the membrane surface, the most plausible explanation for the large value of fluorescence depolarization should be ascribed to the energy transfer between the neighboring polypeptides. This is supported by the observation that the depolarization values were dependent on the peptide concentration in the membrane. It is therefore suggested that Boc-Ser(Anth)-(Ala-Aib)$_{10}$-OMe is distributed to the membrane while forming a loose aggregate.

Considering these experimental results together, Boc-Ser(Anth)(Ala-Aib)$_{10}$-OMe should span the lipid bilayer membrane adopting an α-helical conformation and should tend to aggregate loosely (Figure 4).

2. Signal Peptide Analog

It is known that secretory and membrane-bound proteins are synthe-

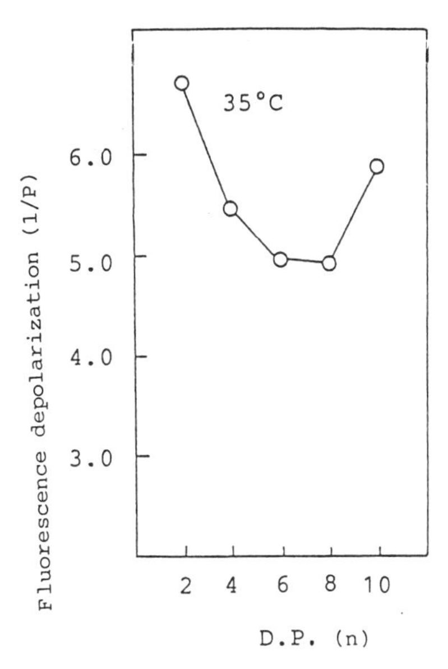

Figure 3. The dependence of the fluorescence depolarization value of anthracene moiety in Boc-Ser(Anth)-(Ala-Aib)$_n$-OCH_3 on the chain length of the peptide in the presence of DMPC vesicles.

sized by membrane-bound ribosomes. According to the signal peptide hypothesis, such proteins are synthesized with a peptide extension at the N-terminal of proteins called signal peptides, which are thought to play the role of translocating the protein through the cytoplasmic membrane.[14]

Inouye et al., have examined the primary sequence of various precursors of the E. coli envelope proteins, and pointed out several common features of the sequence.[15] These common features in the primary sequence are thought to be responsible for the translocation of the protein through membrane. (i) Containing successively two basic amino acids at N-terminal section. (ii) Continuous hydrophobic amino acids spanning 50 A in length. (iii) One proline or glycine residue lies between positions 11 and 13, and the other lies between positions 5 and 7. Taking these points into consideration, the model peptide, Lys-Lys-(Leu-Leu-Aib-Ala-Pro)$_3$-OMe, was synthesized and investigated for the interactions with the lipid bilayer membrane.

Aib residues were introduced into this model peptide to suppress the formation of the aggregation of peptide molecules. Boc-(Leu-Leu-Aib-Ala-Pro)$_n$-OMe, n = 1,2 were also synthesized as reference compounds.

Inouye et al., have also proposed the mechanism, called "loop model", to describe how the signal peptide interacts with membrane.[15] According to the loop model, the positively charged amino acids at the N-terminal section should attach to the negatively charged inner surface of the cytoplasmic membrane. The hydrophobic segment next to these hydrophilic amino acids will be inserted deeply into the hydrophobic core of lipid bilayer membrane while forming the loop as shown in Figure 5.

2A. Conformation of Lys-Lys-(Leu-Leu-Aib-Ala-Pro)$_3$-Ind

CD spectra of Lys-Lys-(Leu-Leu-Aib-Ala-Pro)$_3$-Ind (Ind represents indolylamide) and Boc-Leu-Leu-Aib-Ala-Pro-Ind were shown in Figure 6. A

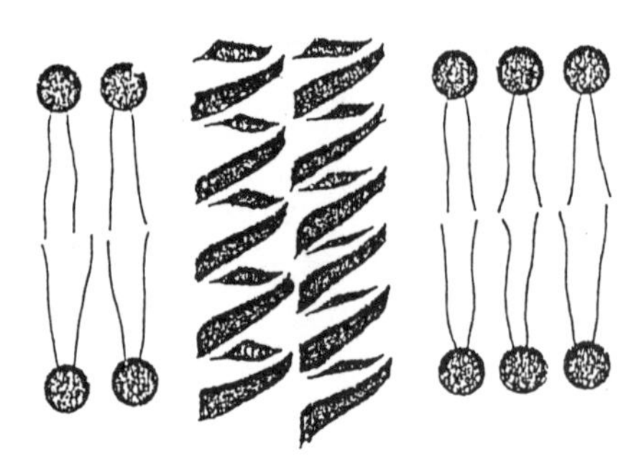

Figure 4. The proposed structure of Boc-Ser(Anth)-(Ala-Aib)$_n$-OCH_3 spanning the lipid bilayer membrane.

negative Cotton effect around 220 nm was observed both in methanol and sodium dodecyl sulfate (SDS) micelles. The strength of the Cotton effect of Lys-Lys-(Leu-Leu-Aib-Ala-Pro)$_3$-Ind in methanol was not remarkably different from that of Boc-Leu-Leu-Aib-Ala-Pro-Ind. Thus, the elongation of peptide chain did not induce the regular structure such as an α-helix or β-sheet structure. The change of solvent from methanol to SDS micelles slightly increased the strength of the negative Cotton effect. The hydrophobic environment is suggested to promote the formation of a specific conformation.

The Nuclear Overhauser Effect-correlated two dimensional ^1H-nuclear magnetic resonance spectroscopy of Boc-(Leu-Leu-Aib-Ala-Pro)$_2$-OMe in $CDCl_3$ was measured to obtain precise information about the conformation. NOE-correlated signals were observed for proton pairs of (Leu^1-NH, Aib^8-NH) and (Leu^2-NH, Aib^8-NH), indicating that the protons in these pairs might be close to each other. It is therefore suggested that this model peptide should take the folded structure, but not the extended one. The Pro residue is well known to be involved in the β-turn, which might be the reason for the model peptide to adopt the folded conformation. Lys-Lys-(Leu-Leu-Aib-Ala-Pro)$_3$-Ind is also supposed to take the folded structure.

2B. Distribution of Lys-Lys-(Leu-Leu-Aib-Ala-Pro)$_3$-Ind to Membrane

A fluorescence probe, indolylamide group, was introduced to the C-terminal of each model peptide. When the indolyl group was transferred from an aqueous phase into a hydrophobic region, the fluorescence intensity increases and the wavelength of the maximum fluorescence shifts to a shorter wavelength. The change of the fluorescence spectra of Boc-(Leu-Leu-Aib-Ala-Pro)$_2$-Ind with the addition of the DPPC liposome is shown in Figure 7, indicating the distribution coefficient of the peptide to membrane can be evaluated by analyzing the dependence of the fluorescence intensity on the lipid concentration. Since the fluorescence intensity of Lys-Lys-(Leu-Leu-Aib-Ala-Pro)$_3$-Ind leveled off with a smaller amount of lipid than that required for Boc-Leu-Leu-Aib-Ala-Pro-Ind, prolongation of the peptide chain from the former to the latter one was shown to increase the affinity of the peptide for lipid bilayer membrane.

The location of Lys-Lys-(Leu-Leu-Aib-Ala-Pro)$_3$-Ind in the membrane was determined by examining the position of the indolyl group in the membrane. The quenching of the indolyl fluorophore, in Lys-Lys-(Leu-Leu-Aib-Ala-Pro)$_3$-Ind, by the water-soluble quencher acrylamide was studied. Quenching data are normally presented as a Stern-Volmer plot of quenching efficiency. The quenching efficiency by acrylamide of Lys-Lys-(Leu-Leu-Aib-Ala-Pro)$_3$-Ind distributed to membrane was found to be similar to

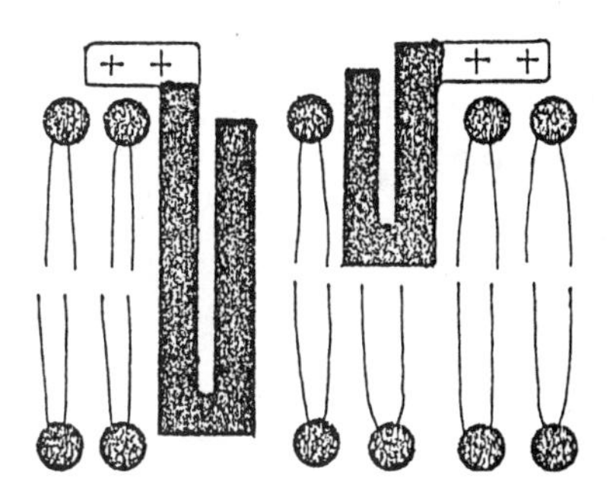

Figure 5. The incorporation of signal peptide into lipid bilayer membrane according to the "loop model" proposed by Inouye and Halegona, 1980.

those of glucagon and α-MSH in a buffer solution. Thus, C-terminal of Lys-Lys-(Leu-Leu-Aib-Ala-Pro)$_3$-Ind distributed to the membrane was suggested to locate close to the aqueous phase. Since the N-terminal segment contains positive charges, it will also stay favorablely at the membrane surface. It is therefore indicated that the whole molecule of Lys-Lys-(Leu-Leu-Aib-Ala-Pro)$_3$-Ind should locate near the hydrophilic membrane surface.

This result was further confirmed by measuring the energy transfer efficiency from the indolyl group in Lys-Lys-(Leu-Leu-Aib-Ala-Pro)$_3$-Ind to the anthryl group of 3- or 10-anthroyloxystearic acid. When the indolyl group locates near the membrane surface, the energy transfer to 3-anthroyloxystearic acid occurs favorably. On the other hand, the indolyl group, being buried deeply in the hydrophobic core of the membrane, will show the high efficiency of energy transfer to 10-anthroyloxystearic acid. As shown in Table 1, the energy transfer efficiency from Lys-Lys-(Leu-Leu-Aib-Ala-Pro)$_3$-Ind to 3-anthroyloxystearic acid, in the presence of DPPC vesicles, was similar to that of from Boc-Leu-Leu-Aib-Ala-Pro-Ind. Furthermore, the energy transfer efficiency from Lys-Lys-(Leu-Leu-Aib-Ala-Pro)$_3$-Ind to 10-anthroyloxystearic acid was not very high. Taken these experimental results together, Lys-Lys-(Leu-Leu-Aib-Ala-Pro)$_3$-Ind should locate near the membrane surface, though the molecule is highly hydrophobic (Figure 8).

Table 1. The energy transfer efficiency (%) from Lys-Lys-(Leu-Leu-Aib-Ala-Pro)$_3$-Ind or Boc-Leu-Leu-Aib-Ala-Pro-Ind to 3- or 10-anthroyloxystearic acid in the presence of DPPC vesicles at 30°C.

Donor	Acceptor 3-anthroyloxy-stearic acid	10-anthroyloxy-stearic acid
Boc-Leu-Leu-Aib-Ala-Pro-Ind	8.9	6.4
Lys-Lys-(Leu-Leu-Aib-Ala-Pro)$_3$-Ind	8.9	7.9

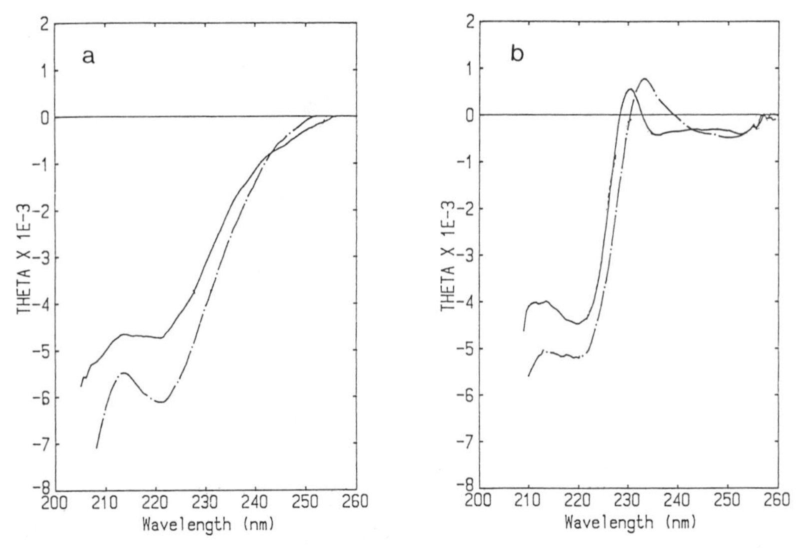

Figure 6. CD spectra of (a) Lys-Lys-(Leu-Leu-Aib-Ala-Pro)$_3$-Ind and (b) Boc-Leu-Leu-Aib-Ala-Pro-Ind. ——, in methanol; -·-, in SDS micelles.

It is assumed that this model peptide can not take a regular conformation with the hydrophilic amide protons and carbonyl oxygens being shielded effectively by forming hydrogen bondings. This should prevent Lys-Lys-(Leu-Leu-Aib-Ala-Pro)$_3$-Ind from becoming buried deeply in the hydrophobic core of the lipid bilayer membrane.

3. Amphiphilic Peptide

An amphiphilic peptide is also admitted to show a membrane-directed property. One face of an amphiphilic peptide is occupied preferentially by lipophilic chains. Since the interface between the aqueous phase and lipid bilayer membrane represents the amphiphilic environment, the amphiphilic peptide will tend to accumulate favorably on the membrane surface.

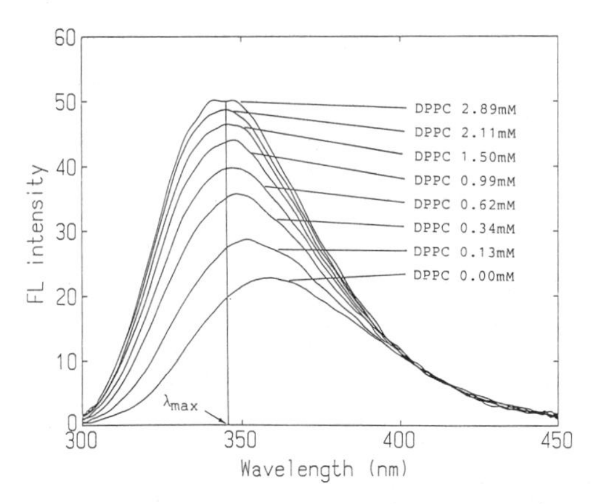

Figure 7. Fluorescence spectra of Boc-(Leu-Leu-Aib-Ala-Pro)$_2$-Ind with the addition of DPPC vesicles at 49°C.

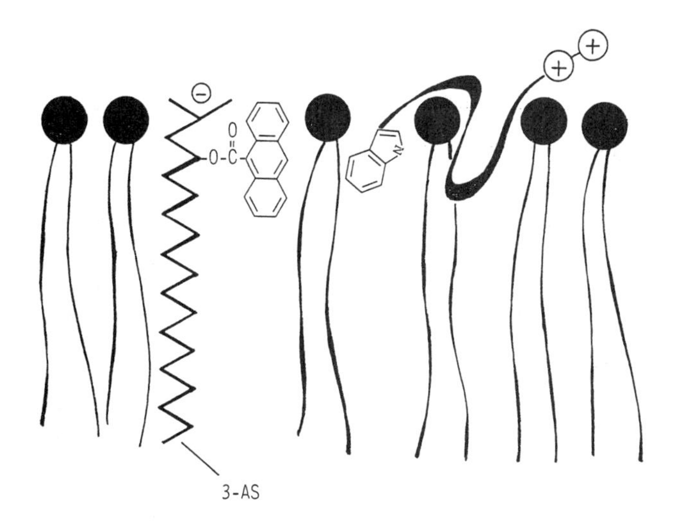

Figure 8. The proposed location of Lys-Lys-(Leu-Leu-Aib-Ala-Pro)$_3$-Ind in the lipid bilayer membrane.

Two ways have been reported how to define the degree of the amphiphilic character of a helix. One is called the helical hydrophobic moment,[16] and the other is the amphiphilic moment.[4-6] The former moment represents the amphiphilicity perpendicular to its axis. On the other hand, the latter one comprises the amphiphilicity along the helix axis as well as perpendicular to the axis. Kaiser and Kezdy have reported the relation of the biological activity of peptides with the helical hydrophobic moment, and successfully synthesized an artificial peptide hormone with high potency due to the ideal amphiphilic structure.[17] On the other hand, Schwyzer et al. have succeeded in explaining the orientation of dynorphin in membrane by using the amphiphilic moment.[4-6]

Membrane proteins of enveloped viruses are another instance which demonstrates the strong interaction of an amphiphilic peptide with the lipid bilayer membrane. For example, an amphiphilic helix ending in a short hydrophobic sequence was found in hemagglutinin, which is thought to be essential for influenza virus to induce membrane fusions.[18]

In this study, a sequential copolypeptide was designed to interact with the lipid bilayer membrane in a pH-dependent manner. The model peptide synthesized here is poly(Lys-Aib-Leu-Aib), which will adopt an α-helix at a high pH region while forming an amphiphilic structure (Figure 9). However, in a low pH region, a charge repulsion between aligned Lys residues along the helix axis should destabilize the helical structure. Therefore, the interaction of poly(Lys-Aib-Leu-Aib) with the lipid bilayer membrane is expected to be controlled by a coil-helix transition induced by a pH change.

3A. Conformation of Poly(Lys-Aib-Leu-Aib)

The major portion of the copolymer eluted through the GPC column showed the molecular weight of 4,600 daltons. The CD spectra of poly(Lys-Aib-Leu-Aib) in trifluoroethanol (TFE)/water (3/1) were measured at varying pH values. The negative Cotton effect with a double minima was found, indicating the presence of a α-helix. The value of Θ_{222} changed drastically from a -10,000 at pH 6-7 to -16,000 (deg cm^2 dmol^{-1}) at pH 8-9, indicating that the content of the α-helix increased as the pH in-

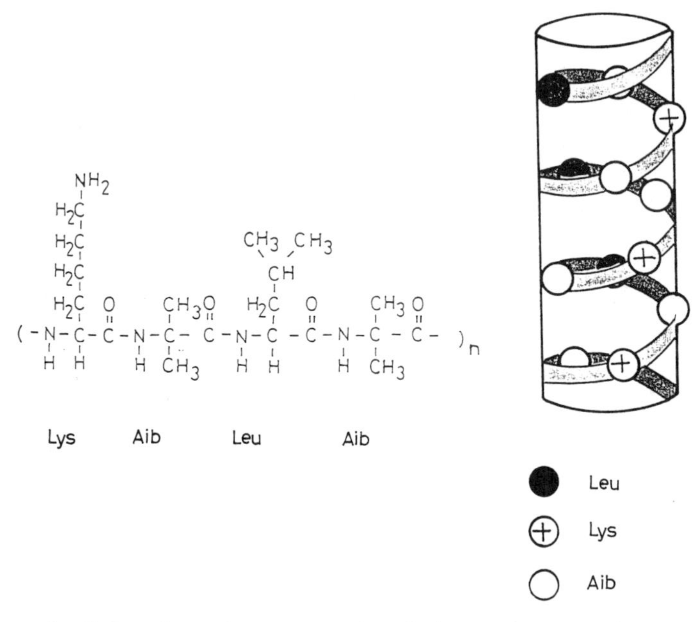

Figure 9. Molecular structure of poly(Lys-Aib-Leu-Aib).

creased. The stabilization of the α-helix at a high pH region should be ascribed to the reduction of a charge repulsion between the side chains of Lys residues by deprotonation.

The effect of ionic strength on the conformation of poly(Lys-Aib-Leu-Aib) was examined. The helical content was expected to increase with increasing ionic strength because the charged groups should be effectively shielded. However, the ellipticities at 208 and 222 nm were found to increase as the ionic strength was increased. The reason for this unexpected result might be due to the formation of an aggregate as the result of the reduced intermolecular repulsion or the distribution of intramolecular hydrogen bondings by the added salt.

3B. Interaction of Poly(Lys-Aib-Leu-Aib) With Lipid Bilayer Membrane

The interaction of poly(Lys-Aib-Leu-Aib) with CF-trapped DPPC vesicles was examined. The leakage of CF from the vesicles was not affected very much by the addition of poly(Lys-Aib-Leu-Aib) at pH 4.3. However, the CF leakage was enhanced by the addition of poly(Lys-Aib-Leu-Aib) as the pH increased. The rate of CF leakage at a half minute after the addition of polymer was found to be 11-fold larger at pH 7.5 and 21-fold larger at pH 8.6 than that at pH 4.3.

It is likely that poly(Lys-Aib-Leu-Aib) forms an amphiphilic α-helical structure as the pH increases, which should interact strongly with lipid bilayer membrane and disturb the membrane structure.

The effect of the ionic strength on the CF leakage induced by the addition of poly(Lys-Aib-Leu-Aib) was studied. As the concentration of NaCl increased, the CF leakage induced by the addition of poly(Lys-Aib-Leu-Aib) was found to be suppressed. This result can be explained by a similar reason to that described above. That is, the fraction of an

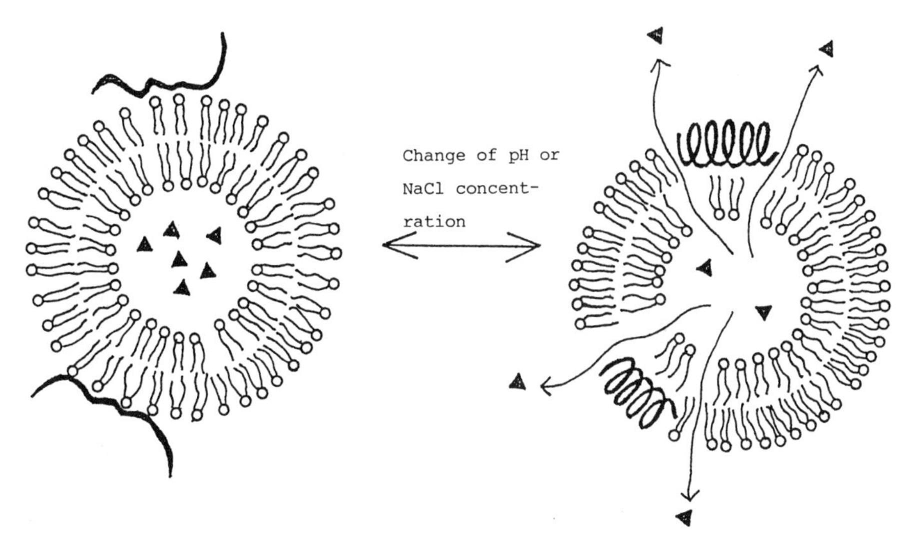

Figure 10. Schematic representation of CF-leakage from DPPC vesicles regulated by pH or NaCl concentration in the presence of poly(Lys-Aib-Leu-Aib).

amphiphilic structure was decreased by increasing the NaCl concentration in solution, which should weaken the interaction with the lipid bilayer membrane.

Therefore, the possibility that the release of the entrapped chemicals inside of the DPPC vesicles can be controlled by pH or NaCl concentration in the presence of poly(Lys-Aib-Leu-Aib) (Figure 10) is indicated.

When poly(Lys-Aib-Leu-Aib) interacts with more than two vesicles simultaneously, the vesicles should aggregate with each other. Poly(Lys-Aib-Leu-Aib) was added to DPPC vesicles at pH 6.0, and the pH of solution was increased (Figure 11). The absorption at 300 nm did not increase much at pH 6.8, but an abrupt increase of the absorption was observed at pH 7.5, indicating that the vesicles aggregated with each other. When the pH was decreased to 6.5 again, the absorption was diminished, however, not to the original level. Thus, the aggregation and dissociation of vesicles were shown to be controlled by pH in the presence of poly(Lys-Aib-Leu-Aib). These processes were not fully reversible, probably because once poly(Lys-Aib-Leu-Aib) was incorporated into lipid bilayer membranes, the formed complex may become so stable that the dissociation of polypeptide from membrane will become difficult.

PEPTIDE-CONJUGATED ENKEPHALIN

Peptide hormones are intercellular messengers regulating and modulating the nervous and endocrine system. They bind to the receptor in membrane and exert their effects. Recently, the significance of the lipid membrane has been recognized in this peptide-receptor interaction. For example, the opiate peptides were discovered to interact with the lipid bilayer membrane, regulating the conformation and orientation of the peptide in the membrane. As a result, the opiate message domains of opiate peptides are accumulated to a specific region of the lipid mem-

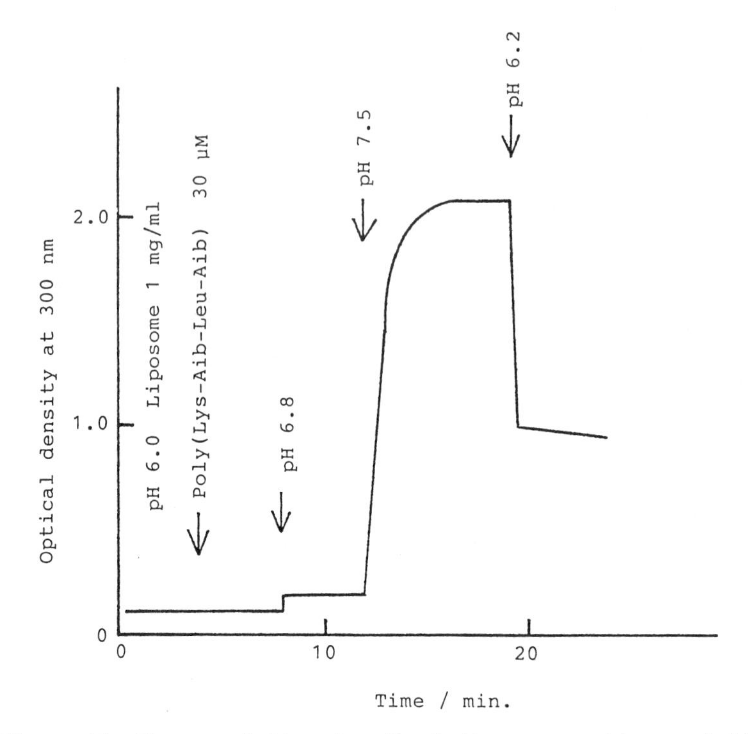

Figure 11. The regulation by pH of the aggregation and dissociation of DPPC vessicles induced by poly(Lys-Aib-Leu-Aib).

brane (membrane compartment), which supposedly constitutes a part of receptor selectivity.[4-6]

1. Enkephalin Derivative Carrying the Artificial Address Peptide

The artificial address segments, which tend to stay at the aqueous phase or to absorb at the membrane surface, were conjugated with the opiate message segments. Conformation, interaction with the lipid bilayer membrane, and binding assay to the opiate receptors of conjugates were investigated.

The molecular structure of the synthesized conjugates are Tyr-Gly-Gly-Phe-Leu-Gly-Pro-(Lys-Sar-Sar-Sar)$_2$-OMe (YGGFLGP-(KS'S'S')$_2$-OMe), Tyr-Gly-Gly-Phe-Leu-Gly-Pro-(Lys-Aib-Leu-Aib)$_2$-OMe (YGGFLGP(KA'LA')$_2$-OMe), and Tyr-Gly-Gly-Phe-Leu-Gly-Pro-(Lys-Pro-Pro-Pro)$_2$-OMe (YGGFLGP(KPPP)$_2$-OMe). The address part of YGGFLGP(KS'S'S')$_2$-OMe is very hydrophilic and will not take a regular conformation, while YGGFLGP(KA'LA')$_2$-OMe should adopt an amphiphilic helical structure and show a high affinity for the lipid bilayer membrane. On the other hand, YGGFLGP(KPPP)$_2$-OMe is expected to take a rod-shaped structure without exhibiting the amphiphilicity.

1A. Conformation of Enkephalin/Address Peptide Conjugates

The CD spectra of enkephalin/address peptide conjugates were measured in ethanol and are shown in Figure 12. It can be seen that these conjugates take distinctly different conformations. While YGGFLGP(KS'S'S')$_2$-OMe is shown to take a random coil conformation, YGGFLGP(KA'LA')-OMe

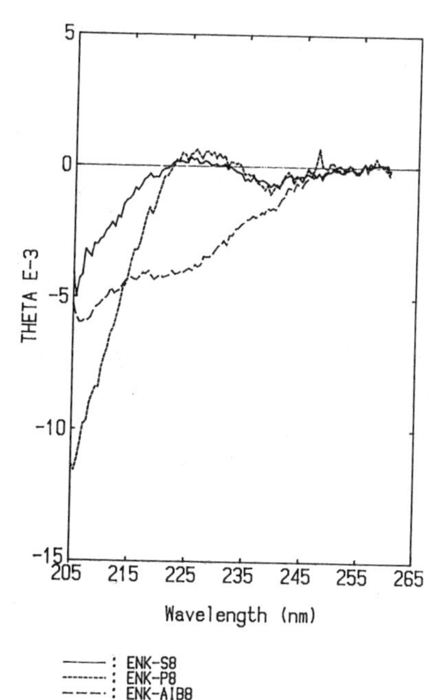

Figure 12. CD spectra of ——, YGGFLGP(KS'S'S')$_2$-OCH$_3$; ···, YGGFLGP(KPPP)$_2$-OCH$_3$; and ---, YGGFLGP(KA'LA')$_2$-OCH$_3$ in ethanol.

showed the negative Cotton effect, indicating the formation of an α-helical conformation, partially. On the other hand, YGGFLGP(KPPP)$_2$-OMe showed the large negative Cotton effect at lower wavelength than 220 nm, suggesting the 3_1-helix as found in the case of poly(Pro).

1B. Interaction with Lipid Bilayer Membrane

The distribution of these synthesized conjugates to the lipid bilayer membrane was studied by using the negatively charged hydrophobic fluorescent probe 1-anilinonaphthalene-8-sulfonate (ANS). Since the conjugates have positive charges, the surface potential of the membrane will be changed by the distribution of these peptides to membrane. This change can be detected by measuring the fluorescence intensity from ANS incubated with DPPC vesicles.

The fluorescence intensity of ANS in the presence of the DPPC vesicles was increased by the addition of YGGFLGP(KA'LA')$_2$-OMe. In the absence of the DPPC vesicles, such a change in fluorescence intensity was not observed. The increase of fluorescence intensity indicates the accumulation of YGGFLGP(KA'LA')$_2$-OMe on membrane surface. On the other hand, the addition of YGGFLGP(KS'S'S')$_2$-OMe or YGGFLGP(KPPP)$_2$-OMe did not induce such increase in fluorescence intensity, indicating that they stay at the aqueous compartment.

These synthesized conjugates were added to CF-trapped DPPC vesicles, and the leakage of CF from the vesicles was measured. Neither addition of YGGFLGP(KS'S'S')$_2$-OMe nor YGGFLGP(KPPP)-OMe to CF-trapped DPPC vesicles increased the CF leakage. On the other hand, the addition of YGGFLGP-(KA'LA')-OMe enhanced the CF leakage, indicating that YGGFLGP(KA'LA')$_2$-OMe disturbed the membrane structure by adsorbing on the membrane surface.

It is therefore concluded that YGGFLGP(KA'LA')$_2$-OMe tends to locate at membrane surface by adopting an amphiphilic helical structure, and the message segment should be condensed at the membrane surface compartment.

1C. Binding Assay to Opiate Receptors

The binding of these conjugates to opiate receptors was assayed by using rat brain homogenates. It is well known that opiate receptors can be classified into three different subtypes, the δ, μ, and κ types. In order to obtain the IC_{50} value of conjugates for each opiate receptor, we have used ^{3}H-labeled [D-Pen^2, D-Pen^2] enkephalin (DPDPE) and [D-Ala^2, Me-Phe^4, Gly-ol^5] enkephalin (DAGO) respectively for δ- and μ-receptors. Portions of the bovine brain homogenate were incubated for 1 hr at 20°C with the conjugate in the presence of the ^{3}H-labeled agonist. The reaction was stopped by adding the cold buffer, and filtered through Whatman glass fiber circles (GF/C). After washing the filter by buffer solutions, the filters were transferred to counting vials, and radioactivity was determined by liquid scintillation spectrometry after the addition of counting fluid. The ratio of specific to nonspecific binding in bovine brain homogenates was about four.

Relative potencies of these conjugates in reducing stereospecific ^{3}H-labeled agonist binding in the bovine brain homogenate are summarized in Table 2. YGGFLGP(KS'S'S')$_2$-OMe and YGGFLGP(KPPP)$_2$-OMe showed weak δ-specificity, which is similar to enkephalin amide, indicating that the hydrophilic address segment does not influence on the binding of the message segment to receptor. However, it should be notable that the affinities of YGGFLGP(KA'LA')$_2$-OMe for μ- and δ-receptors were almost identical with each other, and were lower than those of YGGFLGP(KS'S'S')$_2$OMe and YGGFLGP(KPPP)$_2$-OMe. Taken together, it is shown that the binding of peptide hormone to receptor is affected by the interactions of the peptide hormone with lipid bilayer membrane. The orientation of message segment of YGGFLGP(KA'LA')$_2$-OMe in membrane should be fixed to the relatively unfavorable position for binding to the receptor, resulted in the lower affinity.

2. Interaction of Enkephalin/Polypeptide Conjugates With Living Cells

As described before, an enkephalin/polypeptide conjugate, which has several enkephalin molecules in one peptide chain, is expected to possess the potentiated activity. Here the polypeptides poly(Lys) and poly(Lys-Ala-Leu-Ala) were chosen for the carrier molecule of enkephalin. The molecular structure is shown in Figure 13. These polypeptides are expected to exhibit a high affinity for cell membrane due to a favorable elec-

Table 2. Inhibition of binding of [^{3}H]DAGO (2.0 nM) and [^{3}H]-DPDPE (2.5 nM) to bovine brain homogenates at 20°C.

Peptide	[^{3}H]DAGO (μ) IC_{50}[1]	k_I[1]	[^{3}H]DPDPE (δ) IC_{50}[1]	k_I[1]	DAGO/DPDPE
YGGFLCP(KS'S'S')$_2$-OMe	5.4	5.1	2.2	2.1	2.4
YGGFLGP(KPPP)$_2$-OMe	3.5	3.3	1.1	1.1	3.0
YGGFLGP(KA'LA')$_2$-OMe	15.1	14.2	12.5	12.1	1.2

(1) nM

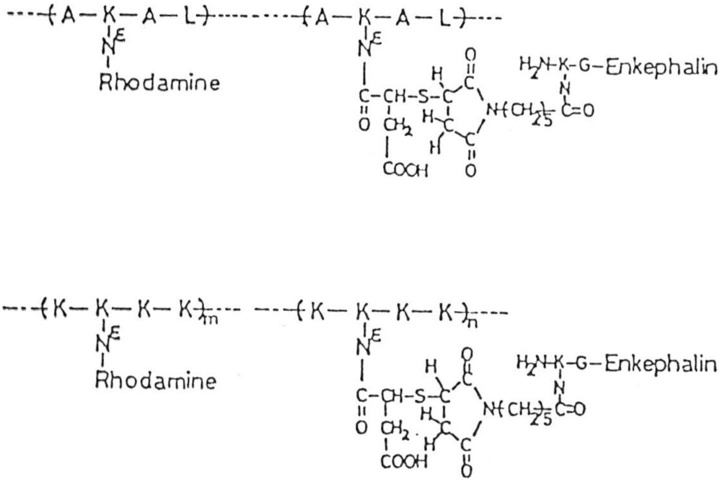

----(A—K—A—L)---------(A—K—A—L)----

Figure 13. Molecular structure of poly(KA'LA')/rhodamine/ enkephalin and poly(K)/rhodamine/enkephalin.

trostatic interaction between them. The latter one, in particular, will adopt an amphiphilic helical structure. Rhodamine was introduced to the polypeptide moiety as a fluorescent probe, and the interaction with platelets, fibroblast cells, and erythrocytes was investigated by the observation with a fluorescence microscopy technique.

2A. Characterization of Carrier Polypeptides

Poly(Lys) and poly(Lys-Ala-Leu-Ala) were eluted through Sephadex G-50 column using water as eluant. The molecular weight of the fraction, which showed the maximum in an elution curve, was estimated to be 25,000 dalton for both polypeptides.

The CD spectrum of poly(Lys-Ala-Leu-Ala) in water showed two negative bands at 203 nm (Θ = -35,000) and 225 nm (Θ = -13,000), indicating that poly(Lys-Ala-Leu-Ala) should take a regular conformation, though the CD spectrum was distorted from that of the typical α-helix.

Poly(Lys-Ala-Leu-Ala) was shown to provide the hydrophobic environment in a buffer solution, because the fluorescence intensity of ANS in a buffer solution was enhanced and the maximum wavelength of emission shifted to the shorter wavelength with the addition of poly(Lys-Ala-Leu-Ala). At the same time, the fluorescence polarization value of ANS was increased from 0.101 in a buffer solution to 0.258 in the presence of poly(Lys-Ala-Leu-Ala). Furthermore, induced CD of ANS was observed at 350 nm. All these results support the idea that poly(Lys-Ala-Leu-Ala) forms a hydrophobic region, and an ANS molecule is tightly bound to poly(Lys-Ala-Leu-Ala) due to the favorable electrostatic and hydrophobic interactions.

2B. Interaction With Platelets and Erythrocytes

Platelets are known to aggregate with the addition of adenosine diphosphate (ADP). However, this aggregation can be suppressed by the presence of prostaglandin E_1 (PGE_1), which is ascribed to the increase of cAMP level within the platelets induced by PGE_1. Opiate peptides were shown to decrease the cAMP level of neuroblastoma x glioma hybrid cells.[19] Morphine was also shown to inhibit the suppressive effect of PGE_1 on the aggregation of platelets, indicating that platelets possess

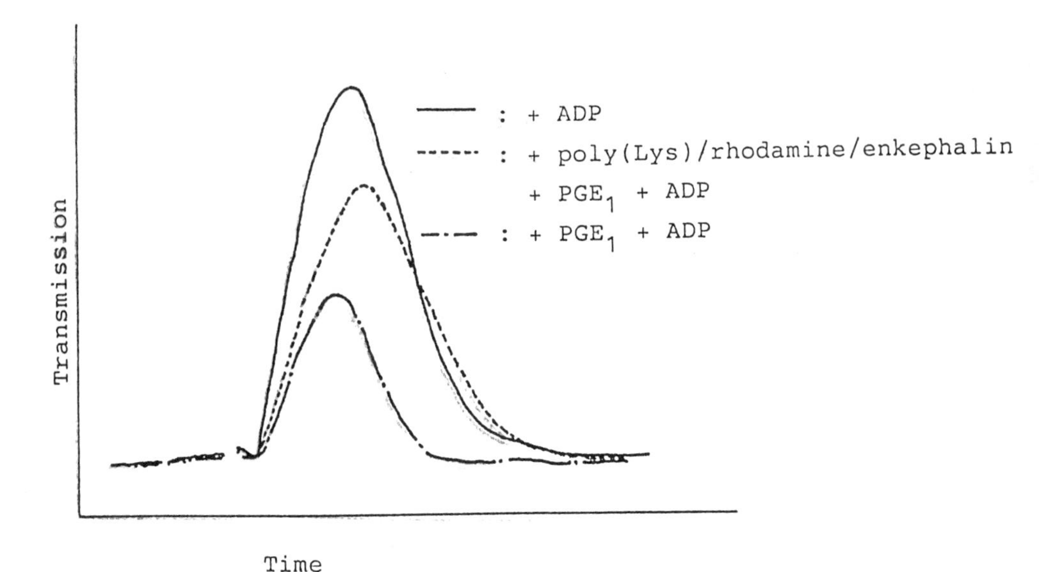

Figure 14. Prevention by poly(Lys)/rhodamine/enkephalin of the anti-aggregating effects of PGE_1 in ADP-induced aggregation of human platelets.

the opiate receptor.[20] Whereas, no evidence was reported for erythrocytes to possess the opiate receptor.

Poly(Lys)/rhodamine/enkephalin conjugate was incubated with the mixture of platelets and erythrocytes, and after washing the cells by a buffer solution, they were observed by the fluorescence microscope technique. Platelets were clearly shown to be labeled by the conjugate, while erythrocytes were not, supporting the idea that platelets contain the opiate receptor, but that the erythrocytes are deficient.

In fact, the addition of a poly(Lys)/rhodamine/enkephalin conjugate to platelets pretreated with PGE_1 partly inhibited the effect of PGE_1 on the aggregation induced by ADP (Figure 14). It is therefore concluded that the physiological activity of the enkephalin moiety bound to poly-(Lys) was not impaired. However, when a poly(Lys)/rhodamine/ enkephalin conjugate was added excessively to platelets, the effect of enkephalin was found to be canceled out by the effect of the poly(Lys) to suppress the aggregation of platelets.

2C. Interaction With Fibroblast Cells

A rhodamine-conjugated enkephalin, Tyr-D-Ala-Gly-Phe-Leu-Gly-Lys-(rhodamine), was incubated with fibroblast cells. The observation by a fluorescence microscopy technique showed fluorescent patches on the cell surface, indicating that the receptors formed clusters. However, no internalization of receptors was found.

In contrast, poly(Lys)/rhodamine/enkephalin bound to the cell surface showed fluorescent patches, and then formed a cap, and ultimately was internalized into the cell upon incubation at 37°C.

Poly(Lys-Ala-Leu-Ala)/rhodamine/enkephalin was also found to bind to the cell surface and seemed to become internalized, but the observation was not as clearly shown as in the case of poly(Lys)/rhodamine/ enkephalin. This is partly ascribed to the non-specific binding of poly(Lys-Ala-Leu-Ala)/rhodamine/enkephalin conjugate due to adopting the amphiphilic structure at the cell membrane.

It is usually observed that opiate receptors are coupled with adenylate cyclase through a G-protein.[21] Thus, the activity of opiate receptors will change the cAMP level within the cell. It has been reported that the concentration of cAMP in fibroblast cells was increased by the addition of PGE_1. Here, we have examined the possibility whether the increased concentration of cAMP in fibroblast cells by PGE_1 can be reduced by the addition of opiate peptides. However, neither enkephalin, [Ala^2, $MePhe^4$, Gly-ol^5]enkephalin, nor dynorphin showed any effect on the cAMP level. It is therefore suggested that the receptor found in fibroblast cells might be silent or coupled with another effector molecule such as a K^+ channel, which is a problem remaining to be resolved.

ACKNOWLEDGMENT

We thank Prof. Sato and Dr. Ueda of Kyoto University for their kind instructions about binding assay.

REFERENCES

1. G. Adam and M. Delbruck, in: "*Structural Chemistry and Molecular Biology*", A. Rich and N. Davidson, Eds., Freeman & Co., New York, 1986, p. 198.
2. D. F. Sargent & R. Schwyzer, Proc. Natl. Acad. Sci., USA, **83**, 5774 (1986).
3. D. Erne, D. F. Sargent & R. Schwyzer, Biochemistry **24**, 4261 (1985).
4. R. Schwyzer, Helv. Chim. Acta., **69**, 1985 (1986).
5. R. Schwyzer, Biochemistry, **25**, 4281 (1986).
6. R. Schwyzer, Biochemistry, **25**, 6335 (1986).
7. B. M. Cox, Life Sci., **31**, 1645 (1982).
8. A. N. Eberle, V. M. Kriwaczek & D. Schulster, in "*Perspectives in Peptide Chemistry*", A. Eberle, R. Geiger & T. Wieland, Eds., Karger, Basel, pp. 407, (1981).
9. H. Maeda, Y. Matsumura, K. Sassamoto, T. Oda & K. Sassamoto, in: "*Protein Tailoring for Food and Medical Uses*", R. E. Feeney & J. R. Whitaker, Eds., Marcel Dekker, New York, 1984, p. 353.
10. A. L. Melton, R. Schief, C. Hatem, J. Kurtzberg, M. L. Markert, R. H. Kobayashi, A. L. Kobayashi & A. Abuchowski, New Engl. J. Med., **316**, 589 (1987).
11. R. O. Fox & F. M. Richards, Nature, **300**, 325, (1982).
12. M. Narita, M. Doi, H. Sugasawa & K. Ishikawa, Bull. Chem. Soc. Japan, **58**, 1473 (1985).
13. J. Seelig & A. Seelig, Quart. Rev. Biophys., **13**, 19 (1980).
14. P. Walter & G. Blobel, J. Cell Biol., **91**, 557 (1981).
15. M. Inouye & S. Halegoua, CRC Critical Reviews in Biochem., **7**, 339, (1980).
16. D. Eisenberg, R. M. Weiss & T. C. Terwilliger, Nature, **299**, 371, (1982).
17. E. T. Kaiser & F. J. Kezdy, Science, **223**, 249 (1984).
18. M. J. Gething, R. W. Doms, D. York & J. White, J. Cell Biol., **120**, 11 (1986).
19. M. Brandt, C. Buchen & B. Hamprecht, J. Neurochem., **34**, 643 (1980).

20. R. J. Gryglewski, A. Szczeklik & K. Bieron, Nature, **256**, 56, (1975).
21. A. K. Sharma, M. Nirenberg & W. A. Klee, Proc. Natl. Acad. Sci., USA, **72**, 590 (1975).

POLYPHOSPHATES MIMICING STRUCTURES AND FUNCTIONS OF TEICHOIC ACIDS

Stanislaw Penczek and Pawel Klosinski

Center of Molecular and Macromolecular Studies
Polish Academy of Sciences
90-363 Lodz, Sienkiewicza 112, Poland

The structures of several types of teichoic acids (TAs), the components of the cell walls and membranes of many Gram-positive bacteria are presented and their physiological functions are discussed. It seems, that due to the high affinity to Mg^{2+} ions, TAs act as scavengers of these ions from the surrounding medium and create a buffer system for the maintenance of Mg^{2+} concentration on the proper level in the membrane region. The simple model of TA backbones as well as their close analogs have been prepared in the three different ways: polycondensation of diols with phosphorylating reagents, ring-opening polymerization of cyclic monomers and polyaddition of phosphorus acids to diepoxides. These methods provide polyphosphates used in active ion transport in the synthetic membranes. Studies on cation transportation through these membranes show high affinity to Mg^{2+} ions for the simplest polymer of this type, namely poly(propylene-1,3 phosphate). The change of the polyphosphate structure used in the membrane allows separation of the different ions from their mixtures.

INTRODUCTION

The teichoic acids are the major components of the cell wall of a number of Gram-positive bacteria.[1] The second main component is the peptidoglycan, forming the three-dimensional network.[1,2] The linear polyphosphate chains of TA are linked covalently to the molecules of muramic acid, the constituent of the peptidoglycan.[3] The lipid membrane is located under the cell wall, which in turn often contains lipoteichoic acids (LTAs).[4] The LTA consists of polyphosphate chains connected with lipid fragments. Both TAs and LTAs have rather short chains with the length up to 40 repeating units.[1,4]

Teichoic and lipoteichoic acids exhibit interesting biological properties, including their immunological behavior.[5,6] TAs and LTAs are also related to some diseases (e.g. rheumatic fever).[6] Both TAs and LTAs were isolated from the bacterial sources, using different methods and could not be standardized sufficiently. Thus, properties of polymers extracted

Biomimetic Polymers
Edited by C. G. Gebelein
Plenum Press, New York, 1990

in different ways could be varied, and therefore a method is needed for the preparation of polymers with well defined composition and structures and mimicking the TA and/or LTA functions. At least three methods, described in the following text, provide the required products.

The occurrence and structures of TAs, the synthesis of their analogs and some biological properties of TAs have previously been reviewed.[7] This paper is encompassing briefly the previously reviewed information and new work, mostly from our own and cooperating laboratories.

STRUCTURE OF TEICHOIC AND LIPOTEICHOIC ACIDS

The name teichoic acids was used for the first time by Baddiley et al.,[8] in 1958 for the polyphosphates of 1,5-ribitol [1] or 1,3-glycerol [2], extracted from bacterial sources. The polyol parts of these polymers chains are substituted with different sugar moieties, like D-glucose, D-galactose, D-glucosamine, D-galactosamine and their N-acetyl derivatives[9-11] and/or with D-alanyl esters, as shown below.[1,9]

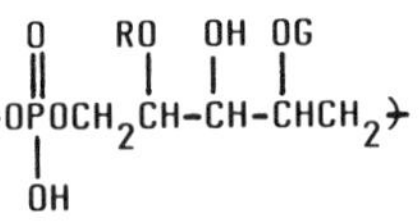

$$\left(\!-OP(=O)(OH)OCH_2CH(OR)\text{-}CH(OH)\text{-}CH(OG)CH_2\!-\right) \quad [1]$$

$$\left(\!-OP(=O)(OH)OCH_2CH(OR)CH_2\!-\right) \quad [2]$$

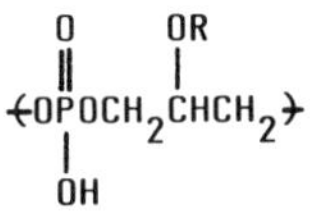

R = D-alanyl or H

G = sugar or H

R = sugar or D-alanyl or H

Many polyphosphates with more or less complicated structures were extracted from the bacterial cell walls and membranes since the first report on teichoic acids. For example, polyphosphates with sugar moieties in the polymer chain were isolated from *Staphylococcus lactis* I3 [3],[12] and from *Bacillus stearothermophilus* B65 [4].[13] Teichoic acids from *Diplococcus pneumonia* (Pneumococcus) [14-18] and from different types of *Haemophilus influenza*[19] have more complicated structures, with oligosaccharide fragments between phosphate groups.

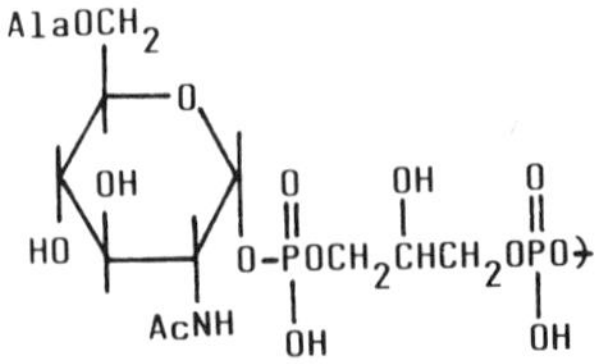

[3]: AlaOCH$_2$ (pyranose ring with OH, HO, AcNH) O-P(=O)(OH)OCH$_2$CH(OH)CH$_2$OP(=O)(OH)O→

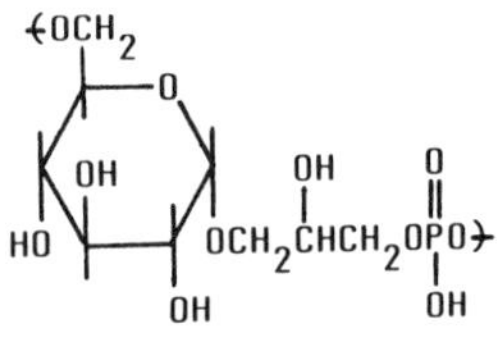

[4]: ←OCH$_2$ (pyranose ring with OH, HO, OH) OCH$_2$CH(OH)CH$_2$OP(=O)(OH)O→

Ala = D-alanyl

Ac = acetyl

In some TAs, ribitol or glycerol are joined with phosphate units at positions differing from 1,5 or 1,3, respectively. TA extracted from cell walls of *Bacillus subtilis* var. *niger* WM has a structure of poly(glycerol-2,3 phosphate) substituted with glucose or D-alanyl at the position 1 [5], as shown below.[20] The monomeric unit of the polymer found in a capsule of bacteria *Haemophilus influenza* type *a* consists of 4-O-β-D-glucopyranosyl-D-ribitol-5-phosphate linked at the 4' position in the sugar residue [6].[21]

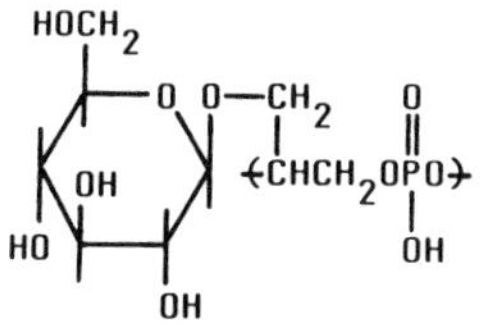

[5]

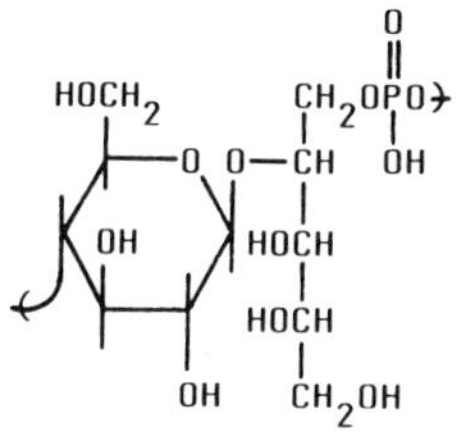

[6]

In some cases, the poly(saccharide-1 phosphate)s were extracted from a bacterial cell wall. The chains of the TA isolated from *Staphylococcus lactis* 2102 consist of approximately 23 repeating units, each built from α-D-N-acetyloglucosamine-1-phosphate. The monomeric units are linked through the phosphate ester bond at the 6 position in the sugar residue as shown in [7].[22]

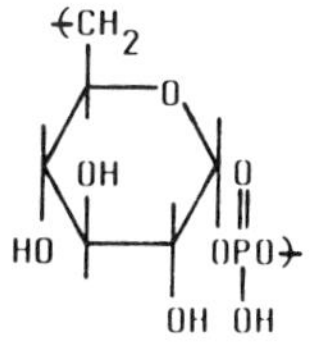

[7]

Lipoteichoic acids (LTAs) were found under the cell wall in bacterial membranes. These polymers consist of hydrophilic polyphosphates, mostly of glycerol covalently bound to the hydrophobic lipid moieties. In one case, the poly(ribitol phosphate) was found to be a hydrophilic part of the LTA.[23] The lipid parts of LTA consist of glycolipids[24,25] or phosphatidylglycerols[26] esterified with long chain fatty acids.

PHYSIOLOGICAL FUNCTIONS

Both the teichoic and lipoteichoic acids seem to act with peptidoglycan as a buffer system maintaining the concentration of Mg^{2+} cations near the membrane region in the range of 10-15 mM. This concentration level of magnesium ions is an optimal value for the activity of the enzymes associated with the membrane.[27]

Baddiley et al.,[28] found that Mg^{2+} ions are associated with the bacterial cell walls in two ways. One fraction of the magnesium ions is bound stronger, being associated with two adjacent phosphate groups in the TA chain. The second fraction of Mg^{2+} cations is associated with only one phosphate group and thus their binding with polymer chain is weaker. The second neighbor -OP(O)(O⁻)O- anion is associated, in this case, with the protonated amino group of the D-alanyl substituent. The second positive charge of Mg^{2+} is neutralized by the anion from the medium. The masking of the phosphate anions by the D-alanyl substituents causes the observed deviation from the simple stoichiometric ratio P:Mg = 2:1.

When extracted from bacteria glycerol, TA binds Mg^{2+} ions in solution stronger than ribitol teichoic acid. The corresponding apparent association constants (K_a) at pH = 5 were found to be equal to 2.7 x 10^3 L·mol^{-1} and 0.61 x 10^3 L·mol^{-1} for glycerol[29] and ribitol teichoic acids,[30] respectively. The lower value of K_a for ribitol TA was accounted for by the impossibility of two adjacent phosphate groups associating with one Mg^{2+} ion, because of steric restrictions.[30] Ribitol TA in cell walls binds magnesium ions with the strength similar to that estimated for glycerol TA. The corresponding K_a was found to be 2.7 x 10^3 L·mol^{-1}.[29] A similar value of K_a was measured for the synthetic analog of glycerol TA, namely poly(propylene-1,3 phosphate), prepared in these Author's laboratories. It was established that the strength of Mg^{2+} association depends on the polymer chain length.[31] The K_a determined from the corresponding Scatchard's plots were equal to 1.6 x 10^3 L·mol^{-1} and 2.8 x 10^3 L·mol^{-1} for poly(propylene-1,3 phosphate) with $\overline{DP}_n$ = 3.0 and with $\overline{DP}_n$ = 8.5, respectively. For calcium cations the K_a values were slightly smaller: 1.5 x 10^3 L·mol^{-1} and 2.2 x 10^3 L·mol^{-1}, respectively. Monovalent cations were weaker associated with poly(propylene-1,3 phosphate). The K_a for K^+ ions were equal to 0.5 x 10^3 L·mol^{-1} for the oligomer with $\overline{DP}_n$ = 3.0 and 1.0 x 10^3 L·mol^{-1} for the studied polyphosphate with the average chain length of 10.5 units.[31]

The pH of the solution has no influence on the Mg^{2+} binding strength, but the amount of ions associated with TA decreased with decreasing pH, because of protonation of the phosphate anions.[29] Teichoic acids exhibit elevated affinity for Mg^{2+} ions.[29] They bind magnesium ions in the presence of an excess of potassium or sodium cations and even divalent calcium ions. However, the selectivity of the cell wall in Mg^{2+} binding is caused not only by the polyphosphates and it is somehow connected with the presence of carboxylic groups of the peptidoglycan.[32,33]

The other physiological function of TA and LTA is their influence on enzyme activity. Lipoteichoic acids play a role of the carrier in the biosynthesis of the polyphosphate chains of wall TA.[34] Lipoteichoic acids inhibit activity of the bacterial autolytic enzymes.[35]

SYNTHESIS OF THE TEICHOIC ACIDS ANALOGS

1. Polycondesation

In 1959 Michelson[36] described the preparation of poly(glycerol phosphate) based on polycondensation of the β- or α-glycerol phosphate **[8]**, activated with diphenyl chlorophosphate or tetraphenyl pyrophosphate, as shown in Equation 1. Polymer **[10]**, obtained as a result of the polycondensation (Equation 2) of the corresponding cyclic pyrophosphate **[9]**, contained a five-membered phosphate ring in the structural unit. The cyclic parts in **[10]** were hydrolyzed (Equation 3) with acidic water giving the linear poly(glycerol phosphate) [11]. The product was isolated as a calcium salt and dialyzed several times against water and 2N NaCl. The chain length of the purified salt of [11] was calculated to be 12 structural units. Calculation was done on the basis of the end-group analysis. The terminal monophosphate groups were liberated from the polymer [11] as an inorganic phosphate using prostatic monoesterase. Polyphosphate [11] exhibited some amount of incorrect 1,2-phosphodiester bonds in the glycerol fragments. Polyphosphate [11] behaved as the naturally occurring glycerol TA in the reaction with anti-glycerol TA antibodies.[37]

$$\underset{[8]}{(CH_2OH)_2CH{-}O{-}P(O)(OH)_2} + \begin{matrix} 2\ (PhO)_2P(O)Cl \\ \text{or} \\ 2\ [(PhO)_2P(O)]_2 \end{matrix} \longrightarrow$$

$$\longrightarrow \underset{[9]}{HOCH_2{-}CH{<}(O)(OCH_2){>}P(=O){-}OP(O)(OPh)_2} + \begin{matrix} 2\ HCl \\ \text{or} \\ 2\ (PhO)_2P(O)OH \end{matrix} \qquad \text{(Equation 1)}$$

$$n\,[9] \longrightarrow \underset{[10]}{\left(CH_2{-}CH{<}(O)(OCH_2){>}P(=O){-}O \right)_n} + n\ (PhO)_2P(O)OH \qquad \text{(Equation 2)}$$

$$[10] + n\ H_2O \longrightarrow \underset{[11]}{HO\left(CH_2CH(OH)CH_2OP(=O)(OH)O\right)_nH} \qquad \text{(Equation 3)}$$

Poly(ribitol phosphate) [15] was obtained by Baddiley et al.,[38] in a similar way, as shown in Equations 4-6.

$$\underset{[12]}{HOCH_2{-}\overset{|}{HC}{-}O{-}C(CH_3)_2{-}O{-}\overset{|}{HC}{-}HC(OP(O)(OH)_2){-}H_2COH} + 2\ (PhO)_2P(O)Cl \longrightarrow \underset{[13]}{HOCH_2{-}HC{-}O{-}C(CH_3)_2{-}O{-}HC{-}HC{<}(O)(OCH_2){>}P(=O){-}OP(O)(OPh)_2} \qquad \text{(Equation 4)}$$

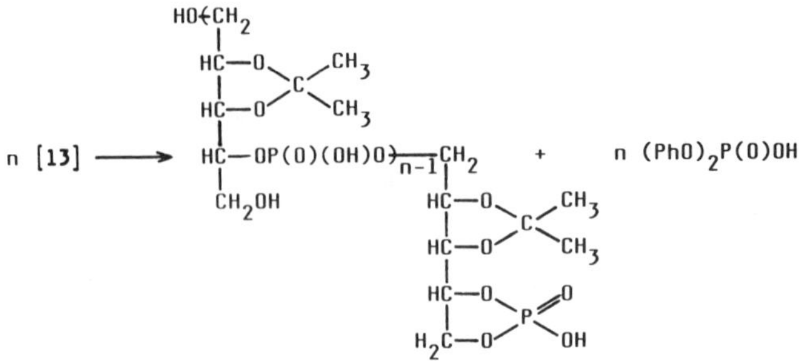

$$n\,[13] \longrightarrow HO\!\left(\begin{array}{l} CH_2 \\ | \\ HC{-}O \\ | \quad\;\; \rangle C(CH_3)_2 \\ HC{-}O \\ | \\ HC{-}OP(O)(OH)O \\ | \\ CH_2OH \end{array}\right)_{n-1}\!\!\begin{array}{l} CH_2 \\ | \\ HC{-}O \\ | \quad\;\; \rangle C(CH_3)_2 \\ HC{-}O \\ | \\ HC{-}O \\ | \quad\;\; \rangle P(=O)OH \\ H_2C{-}O \end{array} \quad + \quad n\,(PhO)_2P(O)OH$$

[14]

(Equation 5)

$$[14] + (n+1)\,H_2O \longrightarrow HO\!\left(\begin{array}{l} CH_2 \\ | \\ HCOH \\ | \\ HCOH \\ | \\ HCOP(O)(OH)O \end{array}\right)_{n}\!H \quad \begin{array}{l} \\ | \\ CH_2OH \end{array}$$

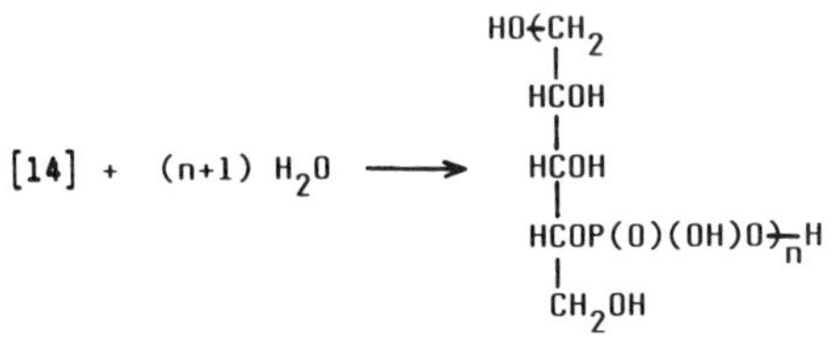

(plus 1,5 isomer)

[15]

(Equation 6)

First, the phosphate of the properly blocked ribitol **[12]** was converted into the corresponding cyclic pyrophosphate **[13]** (Equation 4). Molecules of this active compound underwent self condensation (Equation 5) to a linear polyphosphate with the cyclic end-group **[14]**. Finally, the blocking izopropylidene groups were removed and the terminal cyclic phosphates were opened in diluted HCl (Equation 6). Product **[15]** was isolated as a lithium salt and was purified by dialysis. The average chain length was equal to 6 units. The microstructure of the oligomer **[15]** was not uniform. **[15]** had both 1,4- and 1,5-phosphodiester bonds.[38]

Recently, polycondensation methods involving reaction between diol **[16]** and dialkyl H-phosphates (H-phosphites) **[17]**[39,40] or tetraalkyldiamides of phosphonous acid **[18]**[41] were developed (Equation 7).

$$HOR'OH \quad + \quad XP(=O)(H)X \longrightarrow \left(OP(=O)(H)OR'\right) \quad + \quad 2\,XH$$

[16] [17], X = OR

[18], X = NR_2

(Equation 7)

High molecular weight polyphosphites **[19]** were prepared in the Author's laboratory for the first time, starting from dimethyl H-phosphonate **[19]**.[39,40] This process was studied earlier by other authors[42-45] and comprehensively reviewed by Borisov.[46] Preparation of high molecular weight polymers was claimed in one paper,[47] but other authors could not repeat this result.[43-45] Actually Vogt and Balasubramanian[43] permitted some side reaction to occur, leading to formation of ethers and monoalkyl phosphorous acid and thus disturbing the high molecular weight polymer formation. These side reactions were eliminated in our work by performing polycondensation in two stages.[39] In the first stage, diols were reacted with an excess of [19], at lower temperature and with a catalytic amount of sodium metal (Equation 8).

$$(n+1)\ CH_3O\overset{O}{\underset{H}{P}}OCH_3 + n\ HOROH \longrightarrow CH_3(O\overset{O}{\underset{H}{P}}OR)_n O\overset{O}{\underset{H}{P}}OCH_3 + 2n\ CH_3OH \qquad \text{(Equation 8)}$$

[19] [20]

The excess of dimethyl H-phosphonate blocked all of the hydroxyl groups in the diol moiety and prevented them, in this way, from yielding nonreactive chain ends; e. g., by dehydration. Besides, water, when formed in dehydration, would hydrolyze [19] yielding an acid, which in turn would further catalyze the dehydration. Elimination of these processes allowed using [20] in the next stage of polycondensation, when the temperature of the reaction mixture is increased to 200° and the pressure is decreased. Under these conditions [19] is formed from the two end-groups of [20], yielding molecules with the length being a sum of the lengths of both starting molecules (Equation 9).

$$p\ [20] \longrightarrow CH_3(O\overset{O}{\underset{H}{P}}OR)_{pn} O\overset{O}{\underset{H}{P}}OCH_3 + (p-1)\ [19] \qquad \text{(Equation 9)}$$

[21]

The resulting dimethyl H-phosphonate is distilled off. Poly(H-phosphonate)s of oligo(ethylene glycol)s $\bar{M}_n$ = 1000 and of hexanediol-1,6 were prepared this way. The $\bar{M}_n$ of [21] reached 3×10^4.[39,40] Poly(H-phosphonate)s [21] were oxidized into the corresponding polyphosphates [22], using gaseous N_2O_4 without noticeable degree of the polymer chain rupture (Equation 10).[39]

$$(O\overset{O}{\underset{H}{P}}OR) \longrightarrow (O\overset{O}{\underset{OH}{P}}OR) \qquad \text{(Equation 10)}$$

[22]

The temperature of the polycondensation process can be decreased to 130° by using another class of phosphorylating agents, namely the tetraalkyldiamides of phosphonous acid [18].[41] This is shown in Equation 11.

Unfortunately, there are some side reactions occurring in the polycondensation of Equation 11 which lower the polymerization degree of the products [23]. The major side reaction was established to be the dealky-

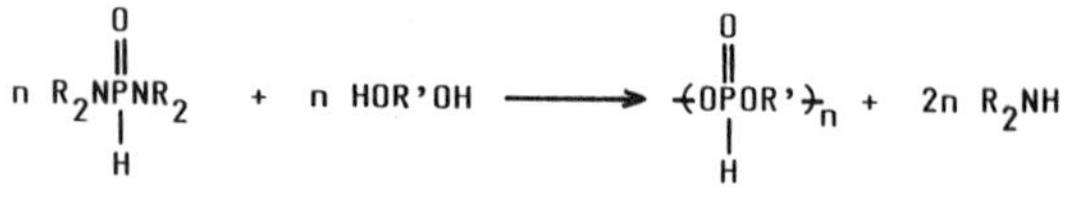

$$n\ R_2N\overset{O}{\overset{\|}{P}}NR_2 \ (\text{P–H}) + n\ HOR'OH \longrightarrow -(O\overset{O}{\overset{\|}{P}}OR')_n- \ (\text{P–H}) + 2n\ R_2NH \qquad \text{(Equation 11)}$$

[18] [23] [24]

lation of [23] by the dialkylamine [24], the second product of condensation. This side reaction was eliminated at least partially, by rapidly removing the formed amine from the reaction mixture and by conducting the condensation in a polymer nonsolvent. For example, polycondensations of cyclohexanediol-1,4 and *trans*, *cis*-1,4-dihydroxymethylenecyclohexane were carried out in o-dichlorobenzene; a nonsolvent for the poly(H-phosphonate)s of both diols. The resulting polymers [23] exhibited $\overline{M}_n$ equal up to 3.8×10^4.[41] In a manner similar to the poly(H-phosphonate)s previously described, polymers [23] were oxidized to the corresponding polyphosphates.

2. Ring-Opening Polymerization

This method is based on ionic or coordinate-ionic polymerization of cyclic monomers, corresponding structurally to the linear unit of the polymer chain. Polymerization of five- or six-membered cyclic monomers provides linear polymers with a sequence of atoms in the chain similar to the backbone of the glycerol teichoic acids with 1,2- or 1,3-phosphodiester bonds. Three types of phosphorus containing compounds can be polymerized to the high molecular weight products, convertible into poly(alkylene phosphate) form, namely: cyclic triesters of phosphoric acid, cyclic diesters of phosphonous acid, and N,N-dialkylamides of alkylene phosphorous acid.

2-Alkoxy-2-oxo-1,3,2-dioxaphosphorinanes [25] only oligomerized in the presence of both anionic and cationic initiators.[48,49] The reason for this phenomenon was established to be the material chain transfer to monomer leading to the formation of the cyclic-end group [26] and to creation of the exocyclic group ion [27] as shown in Equation 12. [27] is able to reinitiate the polymerization of [25]. However, the transfer reaction according to Equation 12 is eliminated when the *t*-butoxy exocyclic groups is used. Yasuda et al.,[50] polymerized 2-*t*-butoxy-4-methyl-2-oxo-1,3,2-dioxaphosphorinane [28] using a 1:0.9 $(C_2H_5)_3Al:H_2O$ mixture (Equation 13). Polyphosphate [29] was heated at 130° in order to remove the *t*-butyl group as gaseous izobutylene. The resulting corresponding polyacid [30] had a molecular weight equal to 2.5×10^4 (measured by GPC).

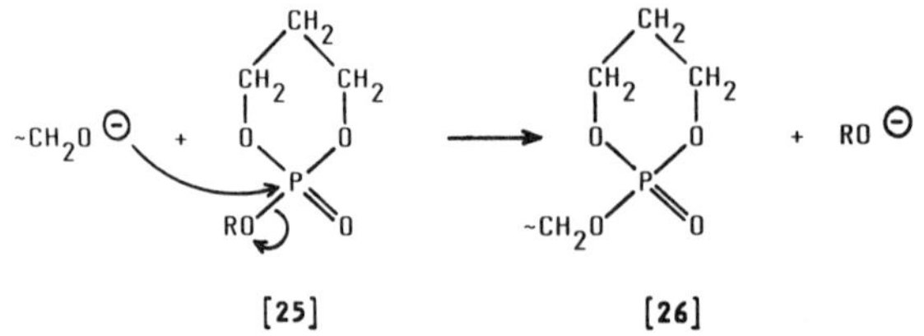

(Equation 12)

$$\sim CH_2O^{\ominus} + [25] \longrightarrow [26] + RO^{\ominus}$$

[25] [26]

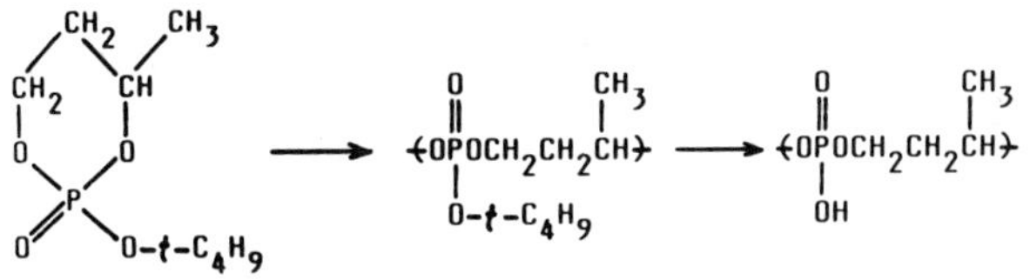

$$\text{[28]} \longrightarrow \ \text{+OP(=O)(O-}t\text{-}C_4H_9\text{)OCH}_2\text{CH}_2\text{CH(CH}_3\text{)+ [29]} \longrightarrow \ \text{+OP(=O)(OH)OCH}_2\text{CH}_2\text{CH(CH}_3\text{)+ [30]}$$

(Equation 13)

[28] [29] [30]

The use of smaller, more strained, five-membered ring cyclic monomers allows preparing polymers with molecular weights up to 1.0 x 10^5.[51] Moreover, the transfer reaction is not all eliminated, leading, especially at higher degree of monomer conversion, to the product with branched units [32] and other structural nonhomogeneities, as is shown in Equations 14 and 15 for 2-methoxy-2-oxo-1,3,2-dioxaphospholane [33].[52]

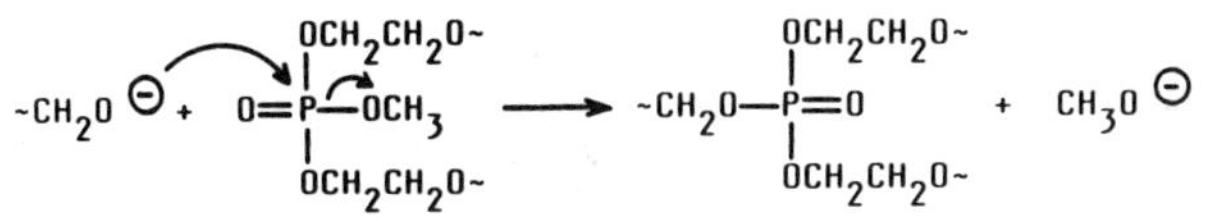

$$\text{~CH}_2\text{O}^{\ominus} + \text{O=P(OCH}_3\text{)(OCH}_2\text{CH}_2\text{O~)}_2 \ \text{[31]} \longrightarrow \text{~CH}_2\text{O—P(=O)(OCH}_2\text{CH}_2\text{O~)}_2 \ \text{[32]} + \text{CH}_3\text{O}^{\ominus}$$

(Equation 14)

[31] [32]

$$\text{CH}_3\text{O}^{\ominus} + \text{[33]} \longrightarrow \text{CH}_3\text{OP(=O)(OCH}_3\text{)OCH}_2\text{CH}_2\text{O}^{\ominus}$$

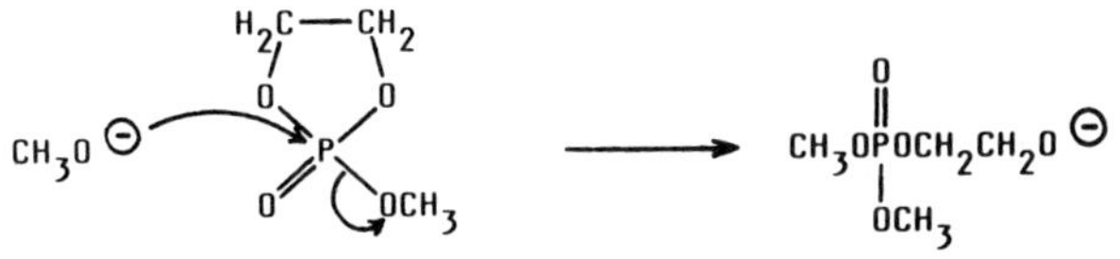

(Equation 15)

[33]

The linear polyphosphate [31] was obtained by quenching the polymerization at monomer [33] conversion lower then 40%.[52] [31] was then converted into the corresponding acid form [35] by first dealkylating with trimethylamine and then by passing the ammonium salt [34] through the cation exchange resin (Equation 16).[53]

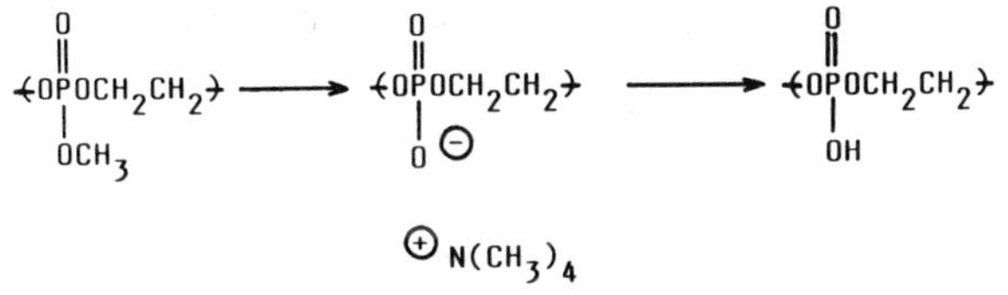

$$\text{+OP(=O)(OCH}_3\text{)OCH}_2\text{CH}_2\text{+} \longrightarrow \text{+OP(=O)(O}^{\ominus}\text{)OCH}_2\text{CH}_2\text{+} \ {}^{\oplus}\text{N(CH}_3\text{)}_4 \longrightarrow \text{+OP(=O)(OH)OCH}_2\text{CH}_2\text{+}$$

(Equation 16)

[31] [34] [35]

Similar polymers were subsequently obtained by Yasuda et al.,[50] in the polymerization of the five-membered cyclic monomers [36] and [37], both with *t*-butoxy exocyclic groups. [36] and [37] were polymerized in benzene with $(C_2H_5)_2Mg$ as initiator. Conversion of the resultant polymers [38] and [39] into the polyacid forms [40] and [41] was achieved by heating them under Ar atmosphere above 80° (Equation 17). Removal of the *t*-butyl group was almost quantitative (95%). The molecular weights of **[40]** and **[41]**, measured by GPC, reached 2.8×10^4 and 1.8×10^4, respectively. The NMR spectra of **[40]** agreed well with those reported earlier for [35].

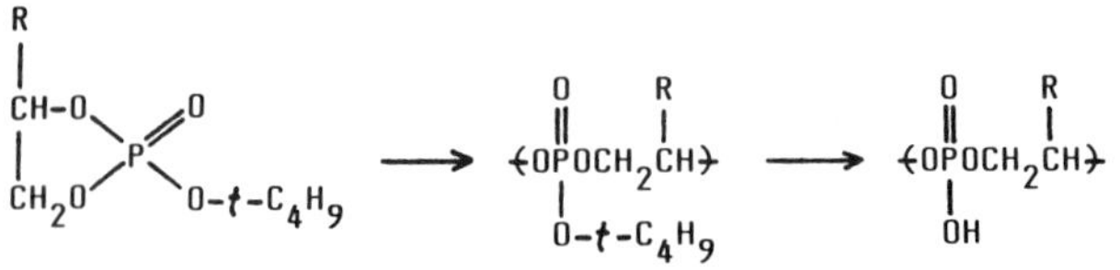

$$\text{[36]/[37]: cyclic } \mathrm{CH(R)-O,\ CH_2O} \text{ ring } \mathrm{P(=O)O\text{-}\mathit{t}\text{-}C_4H_9} \longrightarrow \mathrm{{+}OP(=O)(O\text{-}\mathit{t}\text{-}C_4H_9)OCH_2CH(R){+}} \longrightarrow \mathrm{{+}OP(=O)(OH)OCH_2CH(R){+}} \quad \text{(Equation 17)}$$

[36], R = H [38], R = H [40], R = H

[37], R = CH_3 [39], R = CH_3 [41], R = CH_3

The cyclic methyl phosphate [42], derived from α-glycerol acetate (4-acetoxymethylene-2-methoxy-2-oxo-1,3,2-dioxaphospholane), was polymerized using $Al(i\text{-}C_4H_9)_3$ as an initiator.[54] High molecular weight polyester [43] ($\overline{M}_n$ up to 2.5×10^4) was converted into polysalt **[44]** form using NaI in acetone (Equation 18). The degree of dealkylation of [43] attained 75%.[54] Nonquantitative dealkylation is probably due to the irregularities of the polymer **[43]** structure. It was established that the structure of polyester **[43]** is not regular. Two different openings of the unsymmetrically substituted cyclic monomer produce three kinds of the polymer units: head-to-head ($-CH_2CH(R)OPOCH(R)CH_2-$), head-to-tail ($-CH_2CH(R)OPOCH_2CHR-$), and tail-to-tail ($-CH(R)CH_2OPOCH_2CH(R)-$).

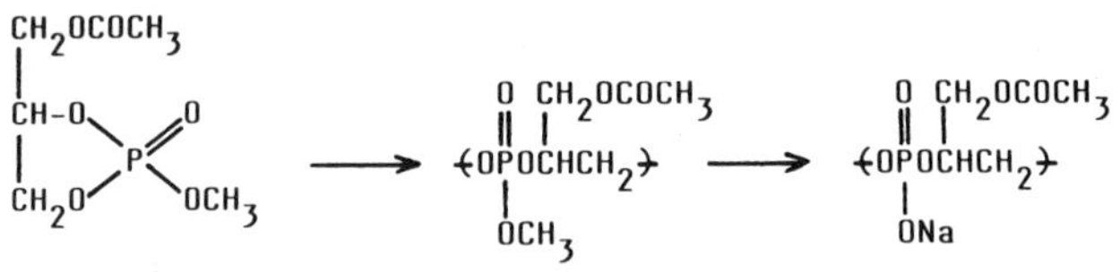

$$\text{[42]: cyclic } \mathrm{CH(CH_2OCOCH_3)-O,\ CH_2O} \text{ ring } \mathrm{P(=O)OCH_3} \longrightarrow \mathrm{{+}OP(=O)(OCH_3)OCH(CH_2OCOCH_3)CH_2{+}} \longrightarrow \mathrm{{+}OP(=O)(ONa)OCH(CH_2OCOCH_3)CH_2{+}} \quad \text{(Equation 18)}$$

[42] [43] [44]

Gehrmann and Vogt cationically polymerized the bicyclic phosphate of glycerol (1-oxo-2,6,7-trioxa-1-phosphabicyclo[2.2.1]heptane) **[45]** (traces of water served as an initiator).[55] The monomer is reactive enough to be polymerized at -78° to the crosslinked product, containing the five-membered ring **[46a]**, linear **[46b]** and six-membered ring **[46c]** units (Equation 19).

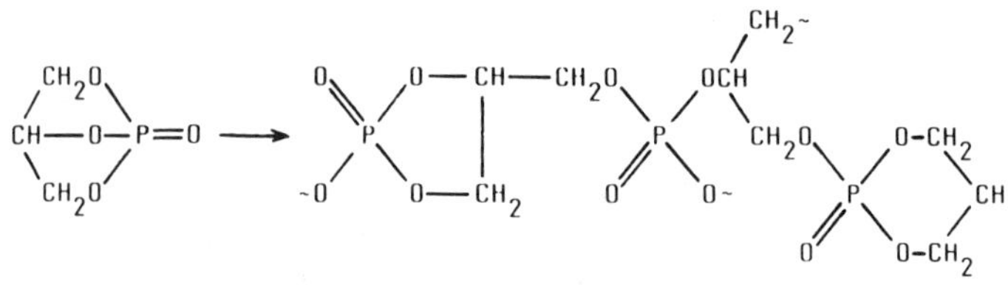

$$\text{[45]} \longrightarrow \text{[46a]} + \text{[46b]} + \text{[46c]} \quad \text{(Equation 19)}$$

[45] [46a] [46b] [46c]

The hydrolysis of **[46]** allowed, under controlled conditions, obtaining the linear poly(glycerol phosphate) **[47]** with $\bar{M}_n$ up to 9 x 10^3. However, the structure of the final product **[47]** was not uniform and consisted of both possible glycerol phosphates with 1,2- and 1,3-phosphodiester bonds (structure **[47a]** and **[47b]**, respectively), as shown in Equation 20.

$$[46] \xrightarrow{H_2O} \left(\!\!-O\overset{O}{\overset{\|}{P}}OCH(CH_2OH)CH_2-\!\!\right)\left(\!\!-O\overset{O}{\overset{\|}{P}}OCH_2CH(OH)CH_2-\!\!\right) \quad \text{(P–OH on each unit)}$$

[47a] [47b]

(Equation 20)

Anionic or coordinate-anionic polymerization of 2-hydro-2-oxo-1,3,2-dioxaphosphorinane **[48]** leads to the high molecular weight ($\bar{M}_n$ up to 1.05 x 10^5) poly(H-phosphonate) **[49]** (Equation 21).[56]

$$\text{cyclo-}[CH_2(CH_2O)_2P(O)H] \longrightarrow \left(\!\!-O\overset{O}{\overset{\|}{\underset{H}{\underset{|}{P}}}}OCH_2CH_2CH_2-\!\!\right)$$

[48] [49]

(Equation 21)

The use of the H-phosphonate monomer **[48]** in place of the cyclic triesters eliminates the transfer reaction, and leads to high molecular weight polymers with nondetectable structural defects (^{13}C and ^{31}P NMR was used). Polymer **[49]** was easily converted into the polyacid form **[50]** by simple oxidation with N_2O_4 in CH_2Cl_2 solution, as shown in Equation 22).[53] Conversion of **[49]** into **[50]** by oxidation was accompanied by a slight decrease of $\bar{M}_n$.

$$[49] \xrightarrow{N_2O_4} \left(\!\!-O\overset{O}{\overset{\|}{\underset{OH}{\underset{|}{P}}}}OCH_2CH_2CH_2-\!\!\right)$$

[50]

(Equation 22)

Poly(H-phosphonate) **[49]** was also prepared by acetolysis of polymer **[52]**, Equation 23, being a polymerization product of 2-N,N-diethylamino-1,3,2-dioxaphosphorinane **[51]**.[57] **[51]** was polymerized using potassium or cesium alcoholates. Alcoholates of lithium and sodium as well as organometallic compounds like $Al(i\text{-}C_4H_9)_3$ were ineffective.[57]

Poly(glycerol-1,2 phosphate) **[56]** was obtained in the reaction sequence starting with the polymerization of 4-acetoxymethyl-2-hydro-2-oxo-1,3,2-dioxaphospholane **[53]** initiated with $Al(i\text{-}C_4H_9)_3$ (Equation 24).[54] Poly(H-phosphonate) **[54]** gave, after oxidation with N_2O_4, linear poly(1-acetylglycerol-2,3 phosphate) **[55]**, Equation 25. The $\bar{M}_n$ of **[55]**, measured

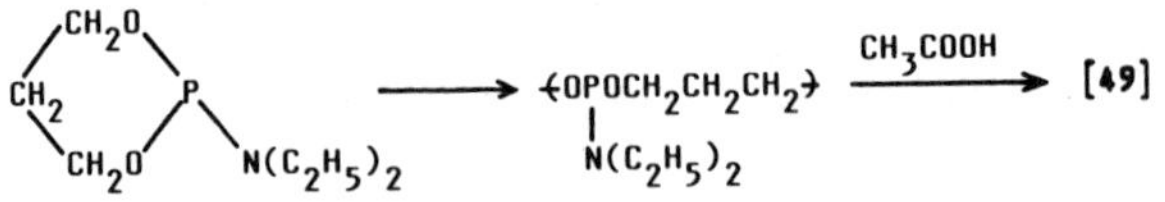

$$\text{[51]}\ \xrightarrow{} \ \text{+OPOCH}_2\text{CH}_2\text{CH}_2\text{+ (P–N(C}_2\text{H}_5)_2)\ \text{[52]} \xrightarrow{CH_3COOH} \text{[49]} \quad \text{(Equation 23)}$$

[51] [52]

osmometrically in DMF solution, reached 1.5 x 10^4. Acetic ester groups in [55] can easily be removed using ammonia in dry methanol (Equation 26).[58]

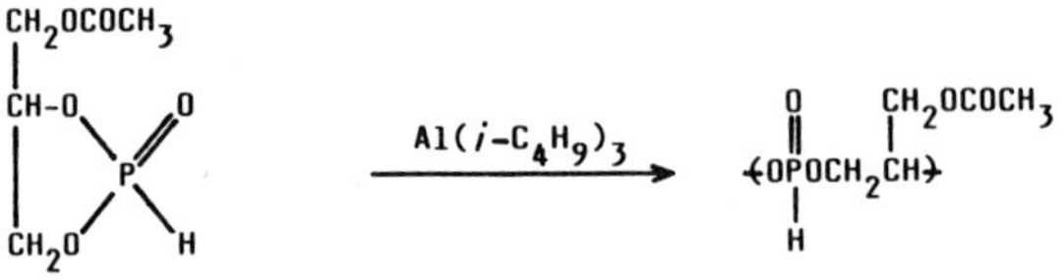

$$\text{[53]} \xrightarrow{Al(i\text{-}C_4H_9)_3} \text{+OP(O)(H)OCH}_2\text{CH(CH}_2\text{OCOCH}_3\text{)+} \quad \text{(Equation 24)}$$

[53] [54]

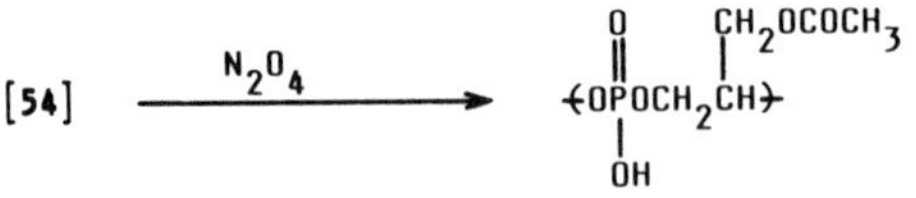

$$\text{[54]} \xrightarrow{N_2O_4} \text{+OP(O)(OH)OCH}_2\text{CH(CH}_2\text{OCOCH}_3\text{)+} \quad \text{(Equation 25)}$$

[55]

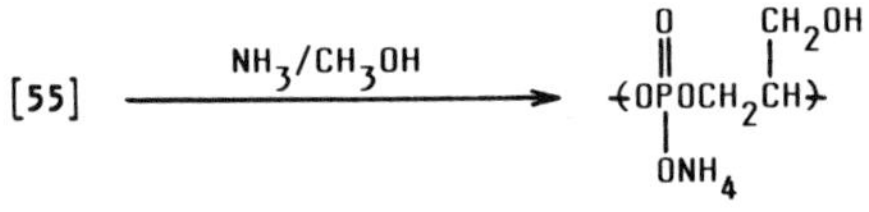

$$\text{[55]} \xrightarrow{NH_3/CH_3OH} \text{+OP(O)(ONH}_4\text{)OCH}_2\text{CH(CH}_2\text{OH)+} \quad \text{(Equation 26)}$$

The results of the studies of the microstructure of polyacid [55] and derivatives thereof have been published recently.[58] It was found, in agreement with previous findings, that unsymmetrically substituted five-membered cyclic monomers can open in two different ways with approximately the same probability. This leads to the already mentioned structures for polyphosphate **[43]**, consisting of three types of diads (shown below for polymer [57], taken as an example): head-to-head **[57a]**, head-to-tail **[57b]**, and tail-to-tail **[57c]**.

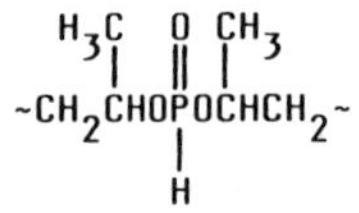

$\sim CH_2CH(CH_3)OP(O)(H)OCH(CH_3)CH_2\sim$ [57a]

$\sim CH(CH_3)CH_2OP(O)(H)OCH(CH_3)CH_2\sim$ [57b]

$\sim CH(CH_3)CH_2OP(O)(H)OCH_2CH(CH_3)\sim$ [57c]

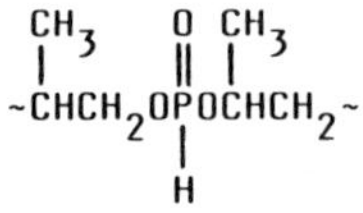

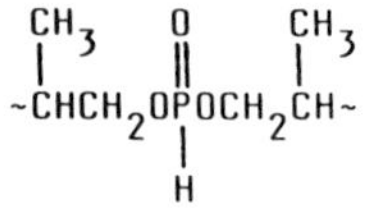

3. Polyaddition of Phosphorus Acids to Diepoxides

The poly(H-phosphonate)s or polyphosphates of tetrols can be obtained by polyaddition of phosphonous [58] or phosphoric acid [59] to diepoxides [60]. The idealized process proceeds as shown in Equation 27.

$$n\ \mathrm{HOP(O)(X)OH} + n\ \underset{\backslash O/}{CH_2-CH}-R-\underset{\backslash O/}{CH-CH_2} \longrightarrow \left(\mathrm{OP(O)(X)OCH_2CH(OH)-R-CH(OH)CH_2}\right)_n \qquad \text{(Equation 27)}$$

[58], X = H [60]

[59], X = OH

For both [58] and [59], polyaddition is accompanied by side-reactions: addition of epoxy ring to hydroxyl group, formed in the reaction between acids and the epoxy rings, and intramolecular cyclization of the products. These disturb the stoichiometry of the reaction mixture and lead to the products with branched structures or even to gels when phosphoric acid is used [59].[59] Better results were obtained for polyaddition of phosphonous acid [58] to [60] carried out in the presence of the stoichiometric amount of the reagent blocking the hydroxyl groups. Acetic anhydride or dihydropyran were found to be suitable for this purpose, Equation 28.[60] However, even under these conditions, products [61] (R = $CH_2O(CH_2CH_2O)_nCH_2$, n = 1,2,3 or $CH_2OC_6H_4C(CH_3)_2C_6H_4OCH_2$) with $\bar{M}_n$ up to 1.1 x 10^4) have branched structures with cyclic end-groups.[60]

$$\mathrm{HOP(O)(H)OH} + \underset{\backslash O/}{CH_2-CH}R\underset{\backslash O/}{CH-CH_2} + 2\ (CH_3CO)_2O \longrightarrow$$

[58] [60]

(Equation 28)

$$\longrightarrow \left(\mathrm{OP(O)(H)OCH_2CHRCH(OCOCH_3)CH_2}\right) \text{ with } \mathrm{OCOCH_3} \text{ on CH} + 2\ CH_3COOH$$

[61]

The polyaddition of bis(trimethylsilyl) methyl phosphate [62] to diepoxides avoids the above mentioned branching because the addition of the phosphoryl fragment to the epoxide ring is connected with the simultaneous blocking of the hydroxyl groups by the trimethylsilyl moieties (Equation 29).[61]

This polyaddition proceeds at 80° in the presence of $SnCl_2$ or methylimidazole, used as catalysts. The structures of thus obtained polymers [63] (R = $CH_2O(CH_2CH_2O)_2CH_2$ or a simple chemical bond - derivative of erythritol) with $\bar{M}_n$ ranging from 3 up to 9 x 10^3, were studied by ^1H, $^{31}P\{^1H\}$, $^{13}C\{^1H\}$ and ^{29}Si NMR. It was established that epoxy ring is

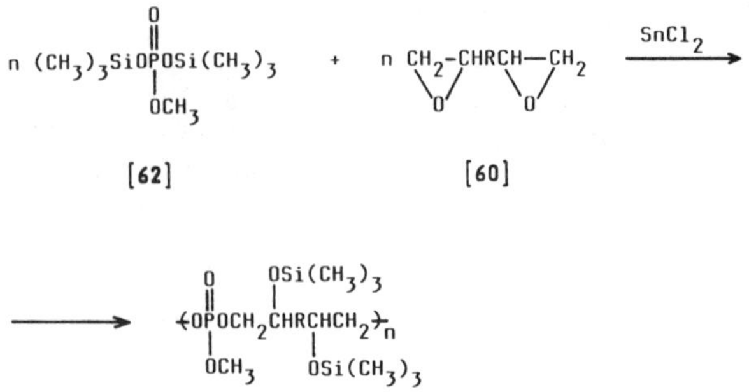

$$n\ (CH_3)_3SiO\overset{O}{\overset{\|}{P}}(OCH_3)OSi(CH_3)_3 \quad + \quad n\ CH_2\text{-}CHRCH\text{—}CH_2\ (\text{diepoxide}) \xrightarrow{SnCl_2}$$

[62] [60]

(Equation 29)

$$\longrightarrow\ \text{-(}OP(O)(OCH_3)OCH_2CHRCH(OSi(CH_3)_3)CH(OSi(CH_3)_3)CH_2\text{)-}_n$$

[63]

opening in two ways, leading to different environments of the phosphate groups in three kinds of diads: head-to-head, head-to-tail and tail-to-tail. The ratio of two ring openings depends on the structure of the starting epoxy compound, as it was established in the model studies with the low molecular weight compounds.[61]

Polymers with MW higher than a few thousand have not yet been obtained. It seems that cyclization producing unreactive, in this process, cyclic end-group **[64]**, is responsible for stopping the chain growth, as shown in Equation 30. The presence of cyclic groups **[64]** in the macromolecules and the formation of disiloxane [65] were detected by $^{31}P\{^{1}H\}$ and ^{29}Si NMR spectroscopies.

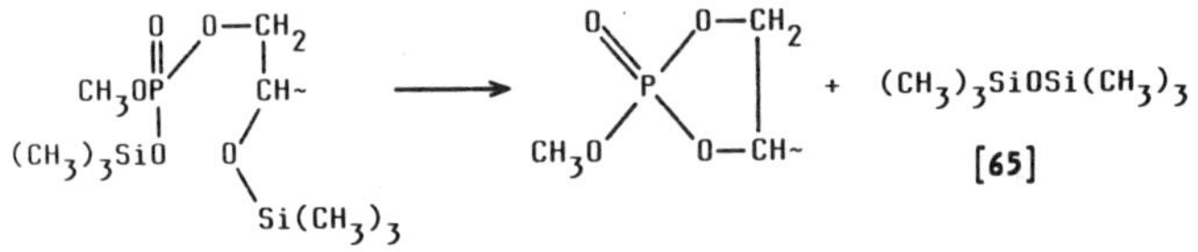

$$CH_3OP(O)(OSi(CH_3)_3)\text{—}O\text{—}CH_2\text{—}CH(OSi(CH_3)_3)\sim \longrightarrow \text{cyclic } CH_3OP(O)(O\text{—}CH_2\text{—}CH\sim\text{—}O) + (CH_3)_3SiOSi(CH_3)_3$$

[64] [65]

(Equation 30)

Silylated polyphosphates [63] (with R as before) were transformed into the corresponding ionic form by desilylation with water, followed by further dealkylation of the methyl group by trimethylamine. When diepoxybutane is used in the polyaddition, poly(erythritol phosphate), a very close TA analog, can be obtained in this way (Equation 31).[61]

(Equation 31)

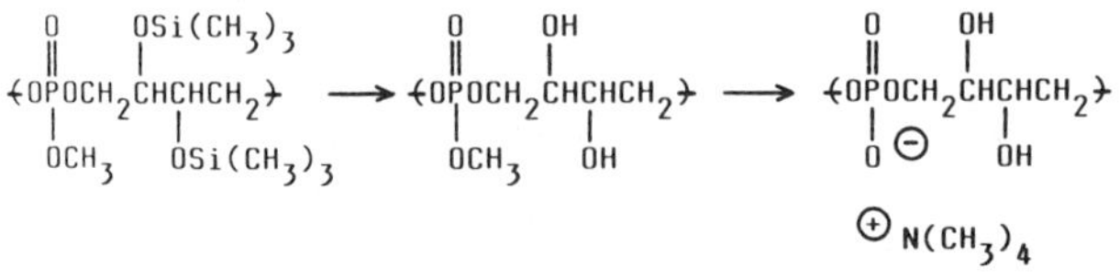

$$\text{-(}OP(O)(OCH_3)OCH_2CH(OSi(CH_3)_3)CH(OSi(CH_3)_3)CH_2\text{)-} \longrightarrow \text{-(}OP(O)(OCH_3)OCH_2CH(OH)CH(OH)CH_2\text{)-} \longrightarrow \text{-(}OP(O)(O^{\ominus})OCH_2CH(OH)CH(OH)CH_2\text{)-}\ \ ^{\oplus}N(CH_3)_4$$

MEMBRANES, BASED ON SYNTHETIC POLYPHOSPHATES, USED TO MIMIC ION TRANSPORT IN BACTERIAL CELL WALL*

(*) This section is prepared mostly on the basis of the work of Prof. A. Narebska's laboratory (Copernicus University, Torun, Poland), cooperating on this subject with the laboratory of the present authors.

The preparation of simple analogs of teichoic acids, like poly(propylene-1,3 phosphate) **[50]** or poly(glycerol-1,2 phosphate) **[56]**, allows study of their affinity to different cations and the ion transportation properties of the membranes, based on these polymers. Thus, several types of membranes were prepared as follows:

(1) monocomponent liquid membranes, containing a solution of the studied polyphosphate, located between two Nafion^R membranes.[62,63]
(2) bicomponent membranes of the "snake in cage" type with polyphosphate captured in the matrix of the crosslinked poly(acrylate amide).[64]
(3) bicomponent membranes, built from polyphosphate, suspended in the solid PVC.[65]
(4) bicomponent membranes, containing polyphosphate captured in the crosslinked, partially hydrolyzed poly(acrylate amide).[66]

The last type of membrane is closely mimicing the cell walls of the Gram-positive bacteria. The cell wall of these microorganisms is a matrix of polyglycan, crosslinked with peptide bridges, with teichoic acid chains connected to the molecules of muramic acid, the component of the polyglycan chain.[3]

It was established, that the affinity of the poly(propylene-1,3 phosphate) **[50]** to the different cations depends on the polyphosphate chain length. **[50]**, with the chains longer than 8 units, exhibits greater affinity to magnesium than to calcium cations.[64] The shorter polyphosphate behave in the opposite way. The affinities to the monovalent cations, like sodium or potassium ions, are smaller in both cases. This is in agreement with results of *in vitro* studies of teichoic acids, extracted from the bacteria cell wall.[29] The following sequence of affinity was found for glycerol-1,3 teichoic acid with chain length equal to 8 units: $Mg^{2+} > Ca^{2+} \gg K^+, Na^+$. Synthetic poly(glycerol phosphate)s, both substituted with acetate groups **[55]** and with unsubstituted hydroxyl groups **[56]**, studied in the liquid membrane system shows only slight or practically no preferential transportation of Mg^{2+} ions in the presence of Ca^{2+} ions.[63] This probably stems from the shorter distance between the two adjacent phosphate groups in the chain. In **[55]** and **[56]**, only two carbon atoms separate negatively charged phosphate moieties, whereas in poly(propylene-1,3 phosphate) **[50]** there are three methylene groups between two -OP(O)(O$^-$)O- residues.

The bicomponent membranes of the third type were prepared by suspending solid, powdered poly(propylene-1,3 phosphate) **[50]** in PVC solution in THF.[65] Then the solvent was evaporated slowly, giving films of 0.01 - 0.02 cm thickness. The conductivity of this type of membranes can be described on the basis of the conductivity percolation theory. The specific conductivity of the membrane k_m in this theory is given by Equation 32, where, k_p is a specific conductivity of poly(propylene-1,3 phosphate) **[50]**, V and V_c (critical volume) are the volume fractions of the polyphosphate **[50]** above and at the percolation threshold, respectively, and t is a critical exponent.

$$k_m = k_p (V-V_c)^t \qquad \text{(Equation 32)}$$

For [**50**]-PVC membranes, V_c is about 6% for low molecular weight polyphosphate [**50**]. Above this content, the dispersed polyphosphate starts to form conducting clusters. The critical exponent t was found to be equal to 1.64 for Na^+ salt of [**50**]. For the acid form of [**50**], t was found to be 1.04. This value is below the theoretical t = 1.6.[65] This difference is probably due to the reduced conductivity of the hydrogen form of [**50**]. It was found by potentiometric titration,[67] that poly(propylene-1,3 phosphate) undergoes conformational transition from the hypercoiled form to the extended coil (more open structure) when pH of solution is increased and, in turn, the degree of ionization of the phosphate groups increases to α = 0.3-0.4.

The role of the poly(alkylene phosphate) chains and peptidoglycan matrix, also negatively charged, was established in the studies of the fourth type of membrane. This membrane was built from the poly(propylene-1,3 phosphate) [**50**] (a model of TA) with the chain length equal to 7 units, trapped in the gel of the partially hydrolyzed, crosslinked polyacrylamide [**66**] (a model of the peptidoglycan network). The gel of [**66**] contained 6 mol.% of the hydrolyzed amide groups.[66] The concentration of the carboxyl groups was lower then critical, allowing these anionic groups to participate in the cation transport by themselves, without any cooperation with the polyphosphate fraction. Thus, the ion transport in the studied system was possible only with the participation of the poly(propylene-1,3 phosphate). This mimiced bacterial cell wall system was placed between two ion-exchange Nafion^R membranes, separating the system from the feed (solution of Mg^{2+} and/or Ca^{2+}) and stripping phase (solution of H^+ or Na^+). The presence of carboxyl groups in the membrane enhanced the preference for the magnesium ions in comparison with the system without these groups.[66] It seems that, in the natural cell walls, the carboxyl groups of the peptidoglycan support the teichoic acid action in scavenging Mg^{2+} ions from the outer medium and in transporting them through the cell wall. Indeed, a similar effect was observed for the cell wall of the *Bacillus subtilis*. The ratio of bound Mg^{2+} ions to Ca^{2+} ions decreased from 20.7 for the intact cell wall to 1.4 for the cell wall with blocked carboxyl groups.[32]

The transport of magnesium ions coupled with the transport of the protons in the opposite direction was much faster then the transport of Ca^{2+} ions. When the acid solution in the stripping phase was replaced by the sodium salt solution the calcium ions were transported through the membrane faster then Mg^{2+} ions.[66] This difference accounts for the operation of the proton pump in the bacteria cells but not the sodium one. The proton countertransport guarantees the higher selectivity of the teichoic acid in the Mg^{2+} ions transport.

These results were used to elaborate a number of systems structurally related to TA, but differing in the units linking the phosphoryl groups, and used in separation of some cations, like Ni^{2+} and Co^{2+}, usually very difficult to separate.[67] However, this topic is beyond the scope of the present review and is mentioned here to show that one can borrow from nature a structure and then use the synthetic analogs to mimic functions in different, technically relevant areas.

CONCLUSIONS

Three types of methods provide synthetic polyphosphates related to teichoic acids, biopolymers extracted from cell walls and membranes of the Gram-positive bacteria. The availability of the synthetic analogs of TA allows checking the hypothesis of the role played by TA in the ion transport in the cell wall. Poly(propylene-1,3 phosphate) behaves in the similar way as TA, both in solution and in synthetic membranes. The ion association constants as well as the ion affinity of this synthetic poly-phosphate are very close to those of the naturally occurring glycerol teichoic acid.

ACKNOWLEDGMENT

This work was supported by the Polish Academy of Sciences, Grant 01.13.

REFERENCES

1. J. Baddiley, Eassay Biochem., **8**, 35 (1972).
2. R. S. Munson & L. Glaser, in: "*Biology of Carbohydrates,*" Ed. V. Ginsburg, J. Wiley & Sons Inc., 1981.
3. J. Coley, E. Tarelli, A. R. Archibald & J. Baddiley, FEBS Lett., **88**, 1 (1978).
4. P. A. Lambert, I. C. Hancock & J. Baddiley, Biochim. Biophys. Acta, **472**, 1 (1977).
5. A. J. Wicken & K. W. Knox, Science, **187**, 1161 (1976).
6. R. Dziarski, Curr. Top. Microbiol. Immunol., **76**, 113 (1976).
7. P. Klosinski & S. Penczek, Adv. Polym. Sci., **79**, 139 (1986).
8. J. J. Armstrong, J. Baddiley, J. G. Buchanan, B. Carss & G. R. Greenberg, J. Chem. Soc., **1958**, 4344.
9. J. Baddiley, J. G. Buchanan, R. O. Martin & H. L. RajBhandary, Biochem. J., **85**, 49 (1962).
10. A. L, Davidson & J. Baddiley, J. Gen. Microbiol., **32**, 271 (1963).
11. M. E. Sharpe, A. L. Davidson & J. Baddiley, J. Gen. Microbiol., **34**, 333 (1964).
12. A. R. Archibald, J. Baddiley, J. E. Heckels & S. Heptinstall, Biochem. J., **125**, 353 (1971).
13. W. R. De Boer, J. T. M. Wouters, A. J. Anderson & A. R. Archibald, Eur. J. Biochem., **85**, 433 (1978).
14. F. Fiedler, M. J. Schaffler & E. Stackenbrandt, Arch. Microbiol., **129**, 85 (1981).
15. J. H. Voerkamp, G. E. J. M. Hoelen & H. J. M. Op Den Camp, Biochim. Biophys. Acta, **755**, 439 (1983).
16. A. R. Archibald, J. Baddiley, J. E. Heckels & S. Heptinstall, Biochem. J., **125**, 353 (1971).
17. M. J. Watson & J. Baddiley, Biochem. J., **137**, 399 (1974).
18. M. J. Watson, J. M. Tyler, J. G. Buchanan & J. Baddiley, Biochem. J., **130**, 14 (1972).
19. W. Egan, R. Schneerson, E. K. Werner & G. Zon, J. Am. Chem. Soc., **104**, 2898 (1982).
20. W. R. De Boer, F. J. Kruyssen, J. T. M. Wouters & C. Kruk, Eur. J. Biochem., **62**, 1 (1976).
21. P. Branefors-Helander, C. Erbing, L. Kenne & B. Lindberg, Carbohydr. Res., **56**, 117 (1977).
22. A. R. Archibald & G. H. Stafford, Biochem. J., **130**, 681 (1972).
23. H. Arakawa, A. Shimada, N. Ishimoto & E. Ito, J. Biochem. (Tokyo), **89**, 1555 (1981).
24. M. Nakano & W. Fischer, Hoppe-Seyler's Z. Physiol. Chem., **359**, 1 (1978).

25. W. Fischer, R. A. Lame & M. Nakano, Biochim. Biophys. Acta, **528**, 298 (1978).
26. D. Button & N. L. Hennings, J. Bacteriol., **128**, 149 (1976).
27. A. H. Hughes, J. C. Hancock & J. Baddiley, Biochem. J., **132**, 83 (1973).
28. J. Baddiley, J. C. Hancock & P. M. A. Sherwood, Nature, **243**, 43 (1973).
29. P. A. Lambert, T. C. Hancock & J. Baddiley, Biochem. J., **149**, 519 (1975).
30. J. E. Heckels, P. A. Lambert & J. Baddiley, Biochem. J., **162**, 359 (1977).
31. R. Wodzki & J. Ostrowska-Czubenko & A. Narebska, in preparation.
32. T. J. Beveridge & R. G. E. Murray, J. Bacteriol., **141**, 876 (1980).
33. L. Ou, N. Chaterjie, F. E. Young & R. E. Marquis, Can. J. Microbiol., **22**, 975 (1976).
34. H. U. Koch & W. Fisher, J. Biol. Chem., **257**, 9473 (1982).
35. W. Fischer, P. Roesel & H. U. Koch, J. Bacteriol., **146**, 467 (1981).
36. A. M. Michelson, J. Chem. Soc., **1959**, 1371.
37. M. McCarty, J. Exp. Med., **109**, 361 (1959).
38. D. A. Applegarth, J. G. Buchanan & J. Baddiley, J. Chem. Soc., **1965**, 1213.
39. J. Pretula & S. Penczek, Makromol. Chem., Rapid Commun., **9**, 731 (1988).
40. J. Pretula & S. Penczek, Makromol. Chem., in press.
41. J. Baran, P. Klosinski & S. Penczek, Makromol. Chem., **190**, 1903 (1989).
42. K. A. Petrov, E. Y. Nifant'ev & R. G. Goltsova, Vysokomol. Soedin., **6**, 1545 (1964).
43. W. Vogt & S. Balasubramanian, Makromol. Chem., **163**, 111 (1973).
44. G. Borisov, Vysokomol. Soedin., Ser. A, **15**, 275 (1973).
45. F. K. Samigulin, I. M. Kafengauz & A. P. Kafengauz, Kinet. Katal., **9**, 898 (1968).
46. G. Borisov, Vysokomol. Soedin., Ser. A, **15**, 275 (1973).
47. K. A. Petrov, E. Y. Nifant'ev, R. G. Goltsova & S. M. Korneev, Vysokomol. Soedin., **6**, 68 (1964).
48. G. Lapienis & S. Penczek, Macromolecules, **10**, 1301 (1977).
49. G. Lapienis & S. Penczek, J. Polym. Sci., Polym. Chem. Ed., **15**, 371 (1977).
50. H. Yasuda, M. Sumitani & A. Nakamura, Macromolecules, **14**, 458 (1981).
51. J. Libiszowski, K. Kaluzynski & S. Penczek, J. Polym. Sci., Polym. Chem. Ed., **16**, 1275 (1978).
52. S. Penczek & J. Libiszowski, Makromol. Chem., **189**, 1765 (1988).
53. K. Kaluzynski, J. Libiszowski & S. Penczek, Macromolecules, **9**, 365 (1976).
54. P. Klosinski & S. Penczek, Macromolecules, **16**, 316 (1983).
55. T. Gehrmann & W. Vogt, Makromol. Chem., **182**, 3069 (1981).
56. K. Kaluzynski, J. Libiszowski & S. Penczek, Makromol. Chem., **178**, 2943 (1977).
57. J. Pretula, K. Kaluzynski & S. Penczek, J.Polym. Sci., Polym. Chem. Ed., **22**, 1251 (1984).
58. T. Biela, P. Klosinski & S. Penczek, J. Polym. Sci., Part A, Polym. Chem., **27**, 763 (1989).
59. A. Nyk, P. Klosinski & S. Penczek; in preparation.
60. P. Klosinski & S. Penczek, Makromol. Chem., Rapid Commun., **9**, 159 (1988).
61. A. Nyk, P. Klosinski & S. Penczek; in preparation.
62. R. Wodzki & K. Kaluzynski, Makromol. Chem., **190**, 107 (1989).
63. R. Wodzki & P. Klosinski, Makromol. Chem., in press.
64. R. Wodzki & A. Narebska, in preparation.
65. R. Wodzki, Eur. Polym. J., **22**, 845 (1986).

66. R. Wodzki, in preparation.
67. R. Wodzki, Eur. Polym. J., 22, 841 (1986).
68. A. Narebska, R. Wodzki, A. Wyszynska, S. Penczek, J. Pretula & K. Kaluzynski; Patent pending.

POLYPHOSPHATES MODELING ELEMENTS OF NUCLEIC ACIDS STRUCTURE

Stanislaw Penczek and Pawel Klosinski

Center of Molecular and Macromolecular Studies
Polish Academy of Sciences
Lodz 90-363, Sienkiewicza 112, Poland

The synthesis of models of nucleic acids (NA) is reviewed. The earlier prepared models, already discussed in review papers by Jones, Takemoto, and Overberger, are either oligonucleotides with some changes in sugar and/or bases (Jones) or are based on non-polyphosphate chains. The latter models contain elements of nucleosides bound to the vinyl chains (e.g. polyvinyladenine), polymethacrylates, as well as to the more hydrophilic chains like polyethyleneimine (Overberger). More recently, synthetic methods have been elaborated, allowing chains to be prepared with the same sequence of atoms as in the NA main chains. These poly(alkylene phosphates), as bare chains, and as chains with either sugars or NA bases, have been synthetized recently in the Author's laboratory and are described in more detail. The elaborated methods allow preparation of poly(alkylene phosphates) by ring-opening polymerization, polycondensation and reactions in the polymer chain. In contrast to the known oligonucleotide syntheses, these methods allow large scale syntheses.

INTRODUCTION

In 1979 Jones reviewed the preparations of NA analogs, their structures and biological properties.[1] The structures, discussed in his review, comprised four types of polymers or oligomers being nucleic acid analogs, namely:

(1) Oligo- and polynucleotides containing unnatural bases.
(2) Oligo- and polynucleotides containing unnatural sugars.
(3) Oligomers and polymers consisting of nucleosides joined by linkages other than phosphodiester linkages.
(4) Oligomers and polymers containing pyrimidines and/or purines linked to backbones other than sugar phosphates.

The fourth class mainly described polyvinyl derivatives. There are 259 references cited in the Jones review. The most interesting result of this extensive work, reviewed by Jones, is that some of the analogs exhibit an antiviral activity and some of them are active against viral

Biomimetic Polymers
Edited by C. G. Gebelein
Plenum Press, New York, 1990

enzymes. These findings give hope that it will be possible to synthetize polynucleotide analogs inhibiting viral infections.

Takemoto studied mostly vinyl polymers containing nucleic acid bases, like polyvinyladenine and polyvinyluracil. These synthetized polymers were studied as biologically active polymers and as models of nucleic acids. In the later area, the base pairing, as well as base stacking, was studied. On the other hand, at least one of the polymers (polyvinylcytosine), complexed with polyinosine, is an interferon inducer. Another important result of this series of work is the elaboration of polystyrene resins containing nucleic acid bases and used in the affinity chromatography to separate NA bases.[2]

More recently Takemoto and Inaki have reviewed the structures and properties of polypeptides, polyacrylamides and polymetacrylamides substituted with NA bases.[3] The major reason to shift to these polymers was an attempt to have more hydrophilic backbones, remembering that NA contain highly hydrophilic alkylene phosphate units. Overberger et al.,[4] used the polyimine chain as a carrier for NA bases, for the same reason.

Physiological properties of synthetic analogs of polynucleotides (mostly vinyl type) were reviewed by Pitha in 1983 (there are 44 references in Pitha's review).[5] Results have not been very encouraging. After evaluating the effects of polynucleotide analogs on enzymatic systems, their interaction with cells in culture, the effects on viral replication and interferon induction, the major conclusion has been that the previously synthesized analogs are unable to penetrate through lipid membranes. According to Pitha, it is important to introduce structural elements that would make these bioanalogous polymers both water and lipid soluble.

It has become apparent that methods are needed to prepare chains with alkylene phosphate repeating units as one of the possible ways to overcome this problem.

Recently, several poly(alkylene phosphate)s with structures related to nucleic acids, [Structure 1], have been prepared in the Author's laboratory and the synthesis of these polymers are described in the following text.

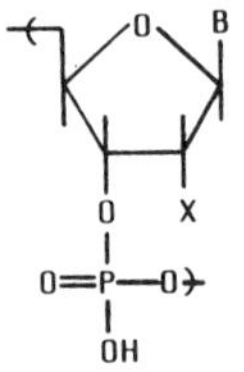

O B

O X

O=P—O

OH

[1a], X = H

[1b], X = OH

B = nucleic base

Structure [1]

POLYMERS MODELING NUCLEIC ACID BACKBONES

Ring-opening polymerization of cyclic phosphorus containing monomers was shown to provide high molecular weight polyphosphates with a sequence of atoms in the structural unit identical with that in NA, namely: phosphate groups joined together through a three carbon atom unit. Thus, the anionic or coordinate-anionic polymerization of 2-hydro-2-oxo-1,3,2-dioxaphosphorinane [2],[6] followed by oxidation of the formed poly(propylene-1,3 H-phosphonate) [3],[7] yielded the corresponding polyphosphate [4] with the required sequence of atoms in the polymer chain, as shown in Equation 1. The molecular weights of [3] and [4] were equal up to 1.05 x 10^5 and 8 x 10^4, respectively.[6,8]

$$\text{[2]: cyclic } CH_2(CH_2O)_2P(=O)H \longrightarrow -\!(OP(=O)(H)OCH_2CH_2CH_2)\!- \;[3] \longrightarrow -\!(OP(=O)(OH)OCH_2CH_2CH_2)\!- \;[4]$$ (Equation 1)

Another way of preparing of poly(H-phosphonate) [3] is the acidolysis of the polyamide [6], formed in the anionic polymerization of 2-N,N-diethylamino-1,3,2-dioxaphosphorinane [5], as shown in Equation 2.[9] The molecular weight of polyamide [6] attained 4 x 10^4.[9]

$$\text{[5]: cyclic } CH_2(CH_2O)_2P{-}N(C_2H_5)_2 \longrightarrow -\!(OP(N(C_2H_5)_2)OCH_2CH_2CH_2)\!- \;[6] \xrightarrow{CH_3COOH} [3]$$ (Equation 2)

Both preparations methods described in Equations 1 and 2 are described in more detail in the Chapter on teichoic acid mimicing polymers (in this Book).

POLYPHOSPHATES WITH NUCLEIC ACID BASES IN THE SIDE CHAIN

Poly(propylene-1,3 H-phosphonate) [3] was converted into several different types of derivatives, taking advantage of the presence of the reactive P-H bonds. The hydrogen atom can be easily substituted by the chlorine atom just by passing gaseous dry chlorine through the CH_2Cl_2 solution of [3] at room temperature, as shown in Equation 3.[10]

$$-\!(OP(=O)(H)OCH_2CH_2CH_2)\!- \;[3] + Cl_2 \longrightarrow -\!(OP(=O)(Cl)OCH_2CH_2CH_2)\!- \;[7] + HCl$$ (Equation 3)

The yield in Equation 3 is quantitative and no noticeable destruction of the polymer chain takes place during this conversion. The chlorinated polymer [7] serves as a substrate for other derivative preparations. The reaction of [7] with alcohols or amines gives, under mild conditions, the corresponding esters or amides, respectively.[11] In both cases, the stoichiometric amount of tertiary amine is added in order to trap the evolved HCl. Similarly, reaction of [7] with a double amount of imidazole (Equation 4) leads to the highly reactive imidazole derivative [8].[10] Polymer [8], in turn, undergoes easy and quantitative substitution with alcohols. The reaction proceeds under mild conditions at room temperature.

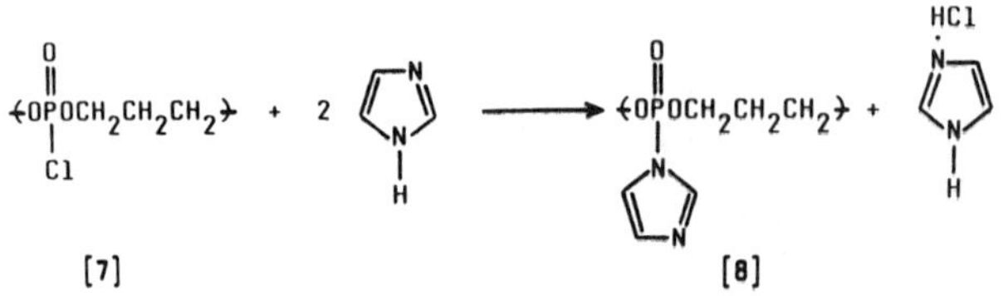

$$\left(\mathrm{OP(O)(Cl)OCH_2CH_2CH_2}\right) + 2\ \text{imidazole} \longrightarrow \left(\mathrm{OP(O)(Im)OCH_2CH_2CH_2}\right) + \text{imidazole}\cdot\mathrm{HCl}$$

[7] [8]

(Equation 4)

Polyphosphate [10], substituted with adenine through the oxyethylene spacer, was prepared in this way, using N9-β-hydroxyethyladenine [9], as shown in Equation 5.[12] The reagents and solvent used in reaction 5 should be carefully dried. Otherwise, the product contains unsubstituted units [4] or two polymer chains can be joined through the pyrophosphate bridges [11] as shown in Equations 6 and 7.

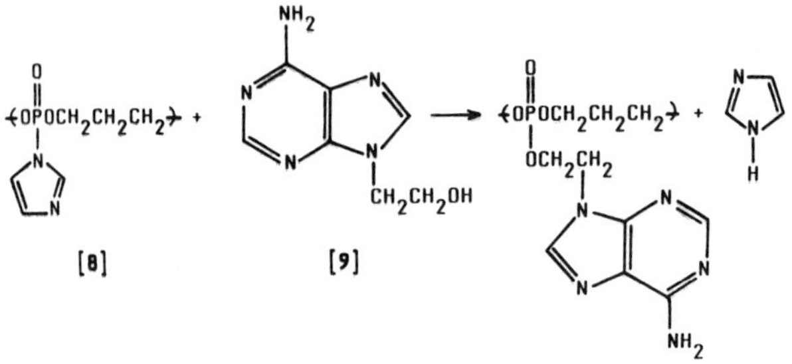

$$\left(\mathrm{OP(O)(Im)OCH_2CH_2CH_2}\right) + \text{Ade-}\mathrm{CH_2CH_2OH} \longrightarrow \left(\mathrm{OP(O)(OCH_2CH_2\text{-Ade})OCH_2CH_2CH_2}\right) + \text{imidazole}$$

[8] [9] [10]

(Equation 5)

$$\left(\mathrm{OP(O)(Im)OCH_2CH_2CH_2}\right) + H_2O \longrightarrow \left(\mathrm{OP(O)(OH)OCH_2CH_2CH_2}\right) + \text{imidazole}$$

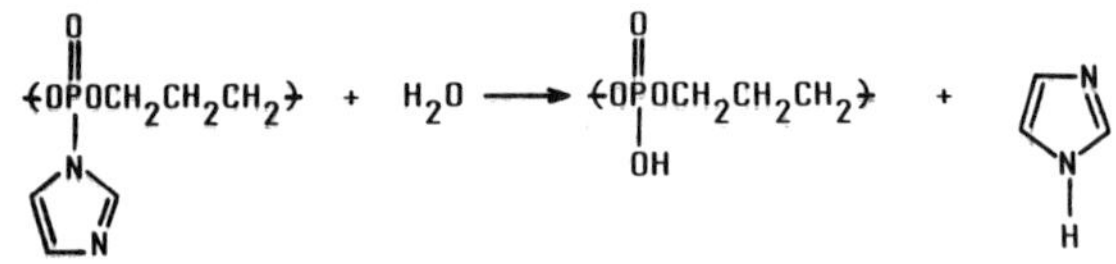

[8] [4]

(Equation 6)

But even if carefully dried and purified substrates were used in the process, the formation of an undesired product was observed in $^{31}P\{^{1}H\}$ NMR. This side product absorbed at $\delta = -10.4$ ppm, whereas the polyester [10] gives a peak at $\delta = -2.0$ ppm in $^{31}P\{^{1}H\}$ NMR spectroscopy. It was established that the side product is the isomer [13], formed in the reaction between [8] and the tautomeric form [12] of N9-β-hydroxyethyladenine [9] (Equation 8). To avoid the formation of [12] and further its

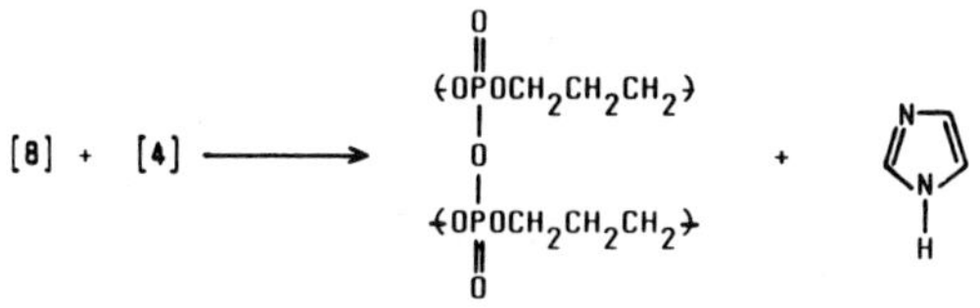

[8] + [4] ⟶ $\leftarrow OP(=O)OCH_2CH_2CH_2 \rightarrow$ / O / $\leftarrow OP(=O)OCH_2CH_2CH_2 \rightarrow$ + imidazole (Equation 7)

[11]

addition, leading to the isomeric unit [13] (Equation 9), the $-NH_2$ function in N^9-β-hydroxyethyladenine [9] was blocked by a method described by Holý, which is shown in Equation 10.[13]

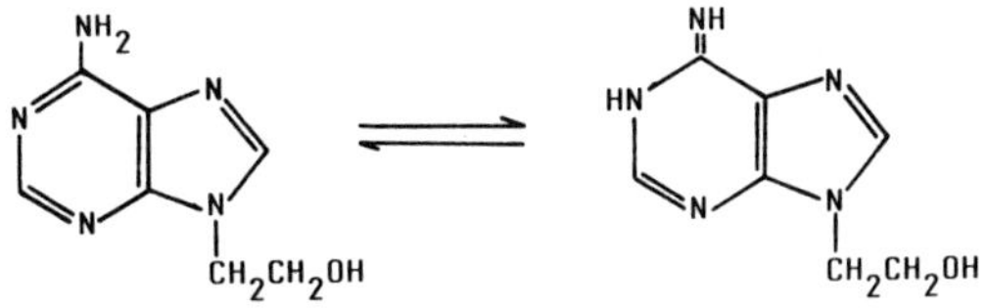

NH_2 ... CH_2CH_2OH ⇌ NH, HN ... CH_2CH_2OH (Equation 8)

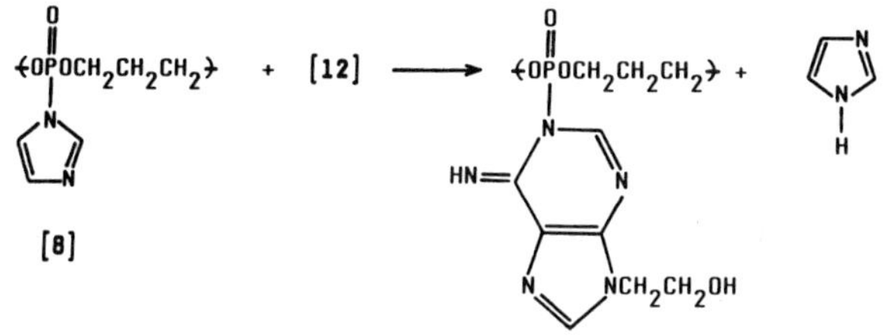

[9] [12]

$\leftarrow OP(=O)OCH_2CH_2CH_2 \rightarrow$ + [12] ⟶ $\leftarrow OP(=O)OCH_2CH_2CH_2 \rightarrow$ + imidazole (Equation 9)

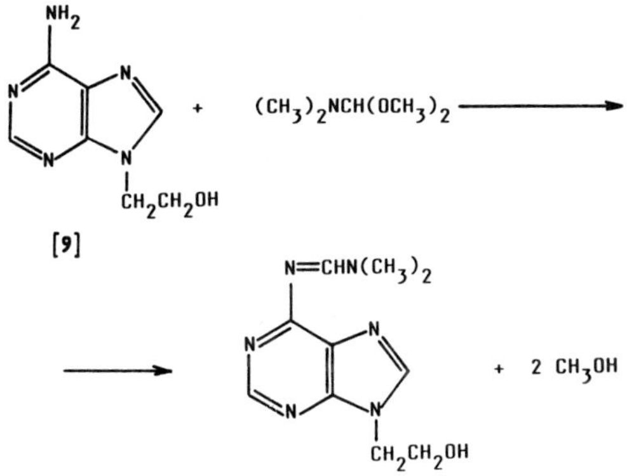

HN= ... NCH_2CH_2OH

[8] [13]

NH_2 ... CH_2CH_2OH + $(CH_3)_2NCH(OCH_3)_2$ ⟶

[9] (Equation 10)

⟶ $N{=}CHN(CH_3)_2$ ... CH_2CH_2OH + 2 CH_3OH

[14]

Complete polymer substitution was achieved using blocked adenine derivative [14] in the reaction with [8], as shown in Equation 11. The removal of blocking groups in [15] was performed in aqueous ammonia, Equation 12.[12]

$$\leftarrow OP(=O)(Im)OCH_2CH_2CH_2 \rightarrow \text{ [8] } + \text{ [14] } \longrightarrow \leftarrow OP(=O)(OCH_2CH_2\text{-}N^9\text{-adenine-}N^6{=}CHN(CH_3)_2)OCH_2CH_2CH_2 \rightarrow \text{ [15] } + \text{ imidazole (N–H)} \qquad \text{(Equation 11)}$$

$$\text{[15]} \xrightarrow{NH_3/H_2O} \leftarrow OP(=O)(OCH_2CH_2\text{-}N^9\text{-adenine-}NH_2)OCH_2CH_2CH_2 \rightarrow \text{ [10]} \qquad \text{(Equation 12)}$$

The polyphosphate substituted with N^1-β-ethyleneuracil, Equation 13, was obtained in a similar way.[14]

$$\leftarrow OP(=O)(Im)OCH_2CH_2CH_2 \rightarrow \text{ [8] } + \text{ uracil-}N^1\text{-}CH_2CH_2OH \text{ [16] } \longrightarrow \leftarrow OP(=O)(OCH_2CH_2\text{-}N^1\text{-uracil})OCH_2CH_2CH_2 \rightarrow \text{ [17] } + \text{ imidazole (N–H)} \qquad \text{(Equation 13)}$$

The use of the poly(propylene-1,3 chlorophosphate) [7] directly in the reaction with the appropriate derivatives of NA bases gave a lower degree of substitution. Several derivatives [18] were synthesized in this way with the yield of substitution less than 20%, Equation 14.[15]

POLYPHOSPHATES WITH SUGAR IN THE CHAIN

Polymers with the tetrahydrofuran ring (modeling the furanoside residue) in the chain were obtained by anionic ring-opening polymerization of cyclic H-phosphonate [19] or cyclic methyl phosphate [20], as in Equation 15.[16] The best results were obtained with n-C_4H_9Li and t-C_4H_9OK used as initiators.

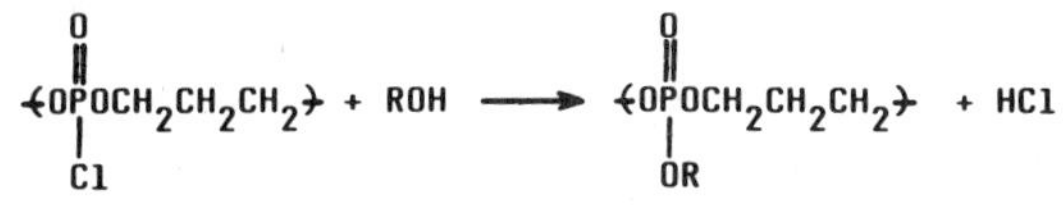

$$\text{-(}\underset{\text{Cl}}{\overset{\text{O}}{\text{OPOCH}_2\text{CH}_2\text{CH}_2}}\text{)-} + \text{ROH} \longrightarrow \text{-(}\underset{\text{OR}}{\overset{\text{O}}{\text{OPOCH}_2\text{CH}_2\text{CH}_2}}\text{)-} + \text{HCl}$$

[7] [18]

where R = (imidazolyl)CH_2CH_2-, (uracil-1-yl)CH_2CH_2-, (cytosin-1-yl)CH_2CH_2-, (adenin-9-yl)CH_2CH_2-, $-CH_2$-(2',3'-O-isopropylidene inosine)

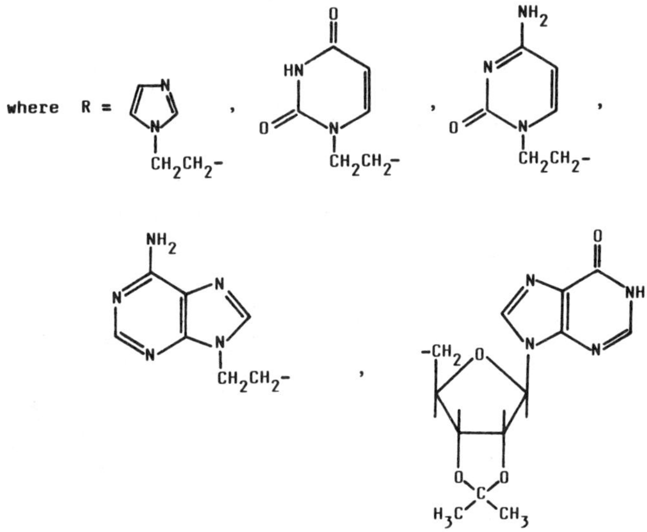

(Equation 14)

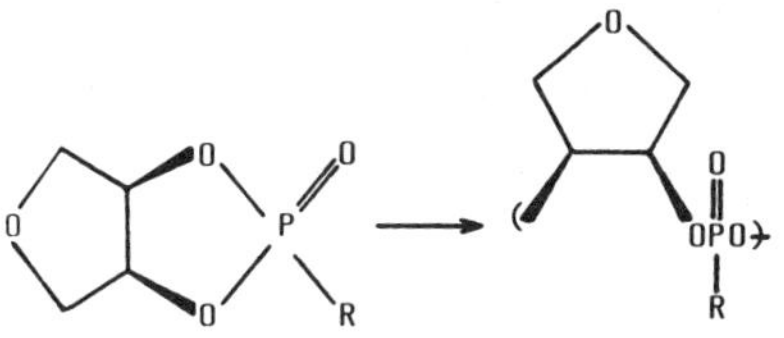

(Equation 15)

[19], R = H [21], R = H

[20], R = OCH_3 [22], R = OCH_3

The yields of [21] and [22] (*cis*-isomers) were rather low, and the $\overline{M}_n$ of these polymers reached 3.5 x 10^3. Similar polyphosphates, but with *trans*-phosphodiester linkages, were prepared from *trans*-3,4-dihydroxy-2,5-dihydrofuran [24] in the polycondensation process with bis(N,N-dimethylamino) methyl phosphite [23] (Equation 16).[17] Polyphosphite [25], with $\overline{M}_n$ up to 1.2 x 10^4, was transformed into the corresponding polyacid [27] by oxidation with N_2O_4, followed by dealkylation with *t*-$C_4H_9NH_2$, as shown in Equation 17.[17]

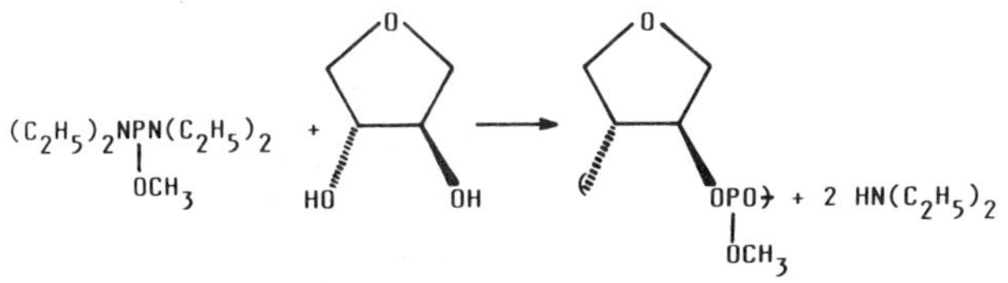

$(C_2H_5)_2NPN(C_2H_5)_2$ (OCH_3) + [24] $\longrightarrow$ [25] (OPO, OCH_3) + 2 $HN(C_2H_5)_2$ (Equation 16)

[23] [24] [25]

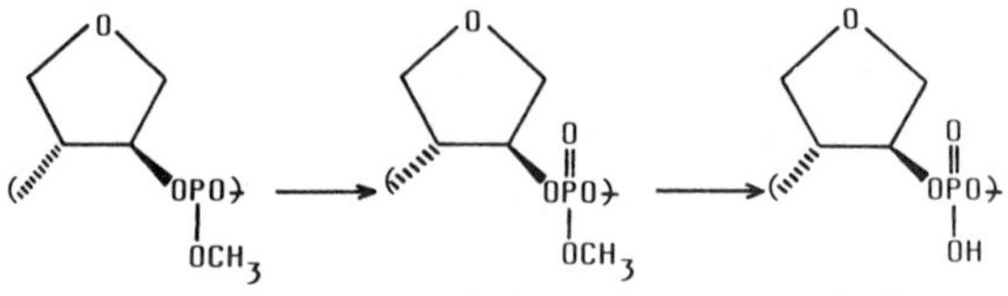

OPO ⟶ OPO ⟶ OPO (Equation 17)

OCH3 OCH3 OH

[25] [26] [27]

Another polymer, bearing only sugar in the chain, namely oligo(H-phosphonate) **[29]** with glucose moieties, and $\overline{DP}_n = 9$, was obtained in the coordinate-anionic polymerization of the corresponding cyclic H-phosphonate **[28]**, according to Equation 18.[18]

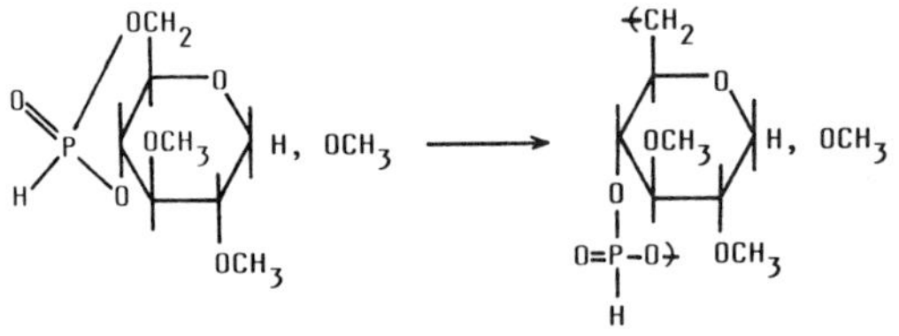

OCH_2, OCH_3, H, OCH_3 ⟶ CH_2, OCH_3, H, OCH_3, O=P–O, H, OCH_3

(Equation 18)

[28] [29]

Poly(deoxyribose phosphate) **[34]**, a model of DNA backbone devoid of bases, was obtained in the sequence of reactions according to Equations 19-22.[16] The monomer, *trans*-3-diethylamino-8-methoxy-2,4,7-trioxa-3-phosphabicyclo[4.3.0]nonane(α,β-methyl-2-deoxy-D-ribofuranoside cyclic diethylphosphoramidite), **[30]**, was polymerized in bulk or in benzene solution using potassium *t*-butanolate. The nearly quantitative conversion of **[30]** was achieved at room temperature after several days, as it was established by $^{31}P\{^1H\}$ NMR spectroscopy:

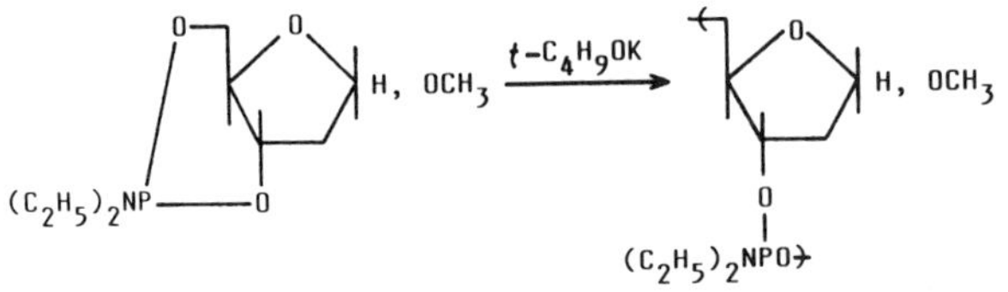

$(C_2H_5)_2NP$, H, OCH_3 $\xrightarrow{t\text{-}C_4H_9OK}$ H, OCH_3, $(C_2H_5)_2NPO$

(Equation 19)

[30] [31]

Polyamide [31] was converted, via Equation 20, into the corresponding poly(H-phosphonate) **[32]** without isolation and purification since **[31]** was very sensitive to traces of water or oxygen.

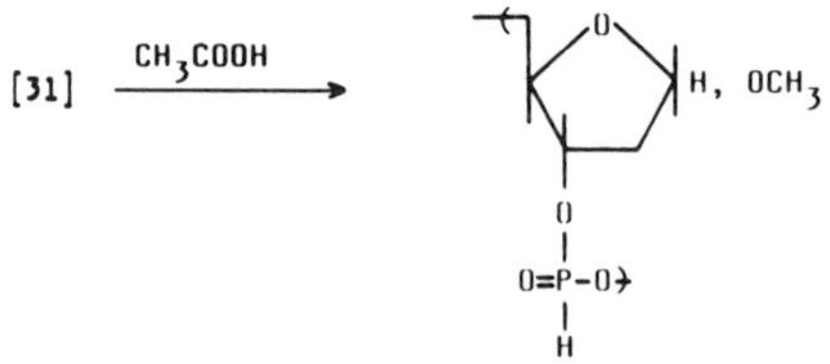

$[31] \xrightarrow{CH_3COOH}$ H, OCH_3, O=P–O, H

(Equation 20)

[32]

Polymer [32] was isolated and, after purification, its molecular weight was measured osmometrically. The $\overline{M}_n$ of [32] attained 6 x 10^3 and $[\alpha]^{25}_D$ was equal to 2.94° in DMF solution. The polyacid form [33] was obtained by oxidation of poly(H-phosphonate) [32] with ozone, as shown in Equation 21.

[32] —O_3→ [33] (H, OCH_3; O=P–O; OH) (Equation 21)

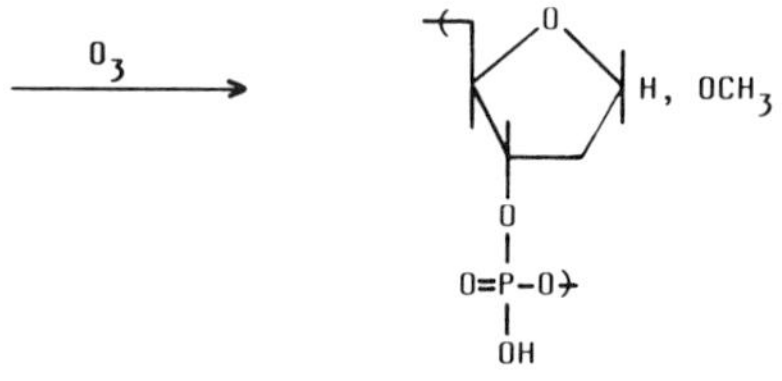

Other oxidizing reagents, like peracetic acid or nitrogen tetraoxide, were less effective in this case. The removal of methyl groups from 1 position of deoxyribose was achieved by simple seasoning of polymer [33] in water solution, Equation 22.

[33] —H_2O→ [34] (H, OH; O=P–O; OH) (Equation 22)

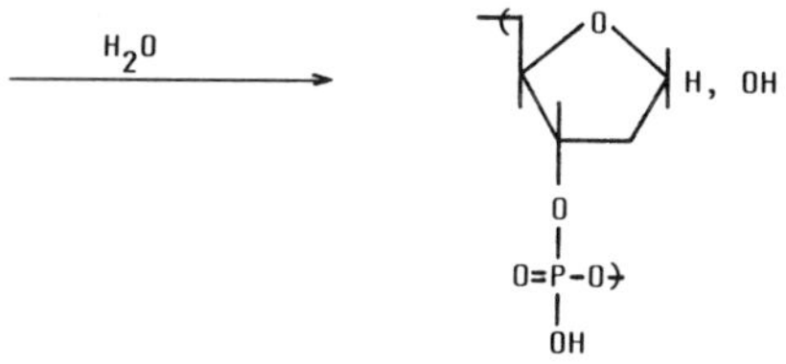

The acidic group in the polymer [34] was methylated with diazomethane (Equation 23) to prepare the nonionic derivative (polytriester [35]), in order to measure the molecular weight osmometrically (in H_2O), at the conditions avoiding stronger H-bonding. The $\overline{M}_n$ of [35] was found to be equal to 3.4 x 10^3 (DP_n = 19).[16]

[34] —CH_2N_2→ [35] (H, OH; O=P–O; OCH_3) (Equation 23)

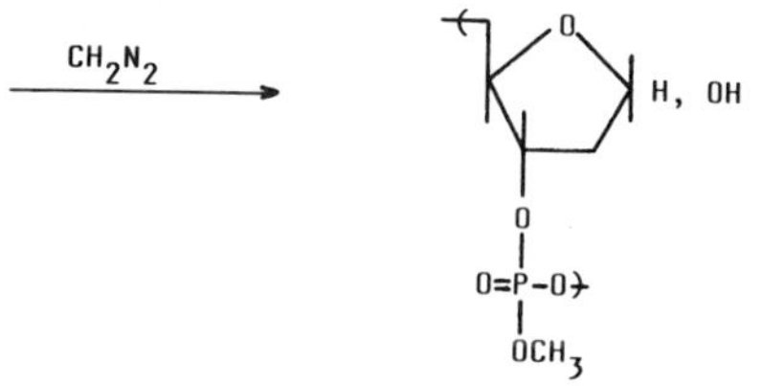

CONCLUSIONS

Developed during the last decade, ring-opening polymerization or polycondensation methods, followed by derivatization of the respective precursor polymers, provide models of NA with one or two elements of their structure; e.g., polymer backbone alone, polyphosphate chain with joined NA bases or polyphosphate chain with sugar residue. The elaborated methods allow to prepare these polymers in a quantity. Studies of the properties and some applications have been made possible this way.

ACKNOWLEDGMENT

The support of the Polish Academy of Sciences is acknowledged, Grant No 01.13.

REFERENCES

1. A. S. Jones, Int. J. Biolog. Macromolecules,**1**, 194 (1979).
2. K. Kondo, T. Horiike & K. Takemoto, J. Macromol. Sci. -Chem., **A16**, 793 (1981).
3. K. Takemoto & Y. Inaki, Acta Polymerica, **39**, 33 (1988).
4. C. G. Overberger, K. A. Brandt, S. Kikyotani, A. G. Ludwick & Y. Morishima; IUPAC 29th International Symposium on Macromolecules, Plenary and Invited Lectures, C. Ion, Ed., Bucharest, 1983, Part 1, page 215.
5. J. Pitha, Adv. Polym. Sci., **50**, 1 (1983).
6. K. Kaluzynski, J. Libiszowski & S. Penczek, Macromol. Chem., **178**, 2943 (1977).
7. K. Kaluzynski, J. Libiszowski & S. Penczek, Macromolecules, **9**, 365 (1976).
8. K. Kaluzynski & S. Penczek; unpublished result.
9. J. Pretula, K. Kaluzynski & S. Penczek, J.Polym. Sci., Polym. Chem. Ed., **22**, 1251 (1984).
10. J. Pretula, K. Kaluzynski & S. Penczek, Macromolecules, **19**, 1797 (1986).
11. K. Kaluzynski; Ph. D. Thesis, Centre of Molecular & Macromolecular Studies, Lodz, 1988.
12. G. Lapienis, S. Penczek, G. P. Aleksiuk & V. A. Kropachev, J. Polym. Sci., Part A: Polym. Chem., **25**, 1729 (1987).
13. J. Zemlicka & A. Holý, Coll. Czech. Chem. Commun., **32**, 3159 (1967).
14. G. Lapienis & S. Penczek, J. Polym. Sci., Part A: Polym. Chem.; in press.
15. V. A. Kropachev, G. P. Aleksiuk, V. L. Zaviriukha & G. I. Kovtun, Makromol. Chem., Suppl., **9**, 47 (1985).
16. G. Lapienis, J. Pretula & S. Penczek, Macromolecules, **16**, 153 (1983).
17. G. Lapienis & S. Penczek, J. Polym. Sci., Part A: Polym. Chem., in press,
18. J. Baran & S. Penczek; unpublished result.

NUCLEIC ACID ANALOGS: THEIR SPECIFIC INTERACTION AND APPLICABILITY

Kiichi Takemoto, Eiko Mochizuki, Takehiko Wada and Yoshiaki Inaki

Faculty of Engineering
Osaka University
Suita, Osaka 565, Japan

A variety of poly(amino acid) and poly(ethyleneimine) derivatives, having pendant nucleic acid bases and nucleosides, were prepared using polymer modification reactions. The conformation of the polymers was studied by CD and ORD measurements. Further study about the polymer complex formation between complementary polymers revealed that the complexes were formed by specific base pairing between pendant bases, retaining their helical conformation in the case of poly(amino acid) type polymers.

INTRODUCTION

The chemistry of synthetic nucleic acid analogs has received much attention in recent years, and there are now a variety of possible applications in the biochemical field, such as polymeric drugs; that is, antiviral, anti-tumor, anti-cancer and interferon inducing agents.[1-6] With this in mind, we have prepared biomedically active polymers having nucleic acid bases and nucleosides, and their properties and functionalities were studied in detail. In order to improve the water solubility of the analogs of the poly(amino acids) or poly(ethyleneimine) type, polymers were prepared having both nucleic acid bases and hydroxyl groups, and their specific base-base interactions were also studied in connection with their bioactivities. In this chapter, the emphasis is focused on our systematic work on the synthesis of new nucleic acid analogs, specific base-base interactions between complementary polymers, and other properties of these polymers.

EXPERIMENTAL

1. Synthesis of Poly(amino acid) Derivatives

1A. Poly(L-lysine) Containing Nucleic Acid Bases.[7-10]

Biomimetic Polymers
Edited by C. G. Gebelein
Plenum Press, New York, 1990

1Aa. p-Nitrophenyl-3-(adenin-9-yl)propionate, (Ade-PNP), [1] and p-Nitrophenyl-3-(thymin-1-yl)propionate, (Thy-PNP), [2]. These p-nitrophenyl esters were prepared by the reaction of p-nitrophenyl trifluoroacetate, (Tfa), with the corresponding carboxyethyl derivatives of adenine and thymine, in pyridine solution, according to the method of Overberger and Inaki.[11]

1Ab. p-Nitrophenyl-3-(uracil-1-yl-propionate), (Ura-PNP), [3]. The p-nitrophenyl ester was prepared by the reaction of p-nitrophenyl trifluoroacetate with the carboxyethyl derivative of uracil using the same procedure described for Thy-PNP, [2], and gave a 97% yield with a m.p. of 197-199°C.

1Ac. Poly(L-lysine-trifluoroacetate), [4]. Poly(ε-N-trifluoroacetate-L-lysine), [poly(ε-N-Tfa-L-lysine)] was prepared from the polymerization of ε-N-trifluoroacetate-α-N-carboxy-L-lysine anhydride in dry dioxane with the use of triethylamine as an initiator, according to the method of Sela.[19] The intrinsic viscosity, determined in dichloroacetic acid at 25°C, was 0.36, which corresponded approximately to a molecular weight of 40,000.

The trifluoroacetate protecting group was removed using piperidine in methanol, and the resultant clear solution was evaporated to dryness under reduced pressure to afford poly(L-lysine trifluoroacetate), [4]. The polymer obtained was dried thoroughly, and was used for further polymer reactions.

1Ad. Poly(ε-N-3-(adenin-9-yl)propionyl-L-lysine), (poly(ε-N-Ade-L-lysine), [5]. To the solution of poly(L-lysine trifluoroacetate), [4], (400 mg, 2.0 mmol), in 10 mL of dimethyl sulfoxide (DMSO), (Ade-PNP), [1], (850 mg, 2.6 mmol) and triethylamine (0.3 mL, 2.0 mmol) were added, and the solution was stirred at 25°C for 3 days. The resulting mixture was poured into excess diethyl ether to give an oily polymer. Acetone was added to the system to precipitate this polymer, which was purified by the repeated reprecipitation from dimethylformamide (DMF)-diethyl ether or DMSO-acetone. The precipitated polymer was washed with diethyl ether and dried in reduced pressure to give a colorless powder, poly(ε-N-Ade-L-lysine), [5], (425 mg, 79% yield).

1Ae. Poly(ε-N-3-(thymin-1-yl)propionyl-L-lysine), (poly(ε-Thy-L-lysine), [6]. This was prepared from poly(L-lysine trifluoroacetate), [4] and (Thy-PNP), [2], using the same procedure described for poly(ε-N-Ade-L-lysine), [5]. Polymer [6] was obtained in 90% yield.

1Af. Poly(ε-N-3-(uracil-1-yl)propionyl-L-lysine), (poly(ε-N-Ura-L-lysine), [7]. Poly(ε-N-Ura-L-lysine), [7], was also prepared by a similar procedure from poly(L-lysine trifluoroacetate), [4], and (Ura-PNP), [3], and gave a 97% yield.

The contents of the nucleic acid bases in the poly(lysine) derivatives were determined by the ultraviolet spectra of the polymers after hydrolysis. The polymers were hydrolyzed, in 6N hydrochloric acid at 105°C for 24 hr., into lysine dihydrochloride and the carboxyethyl derivatives of the nucleic acid bases.

1B. Poly(L-lysine) Containing Uridine, [11].[8] The synthesis of compound [11] was as follows: starting from isopropylidene uridine, [8], the compound [9] was obtained by the reaction with glutaric anhydride at 80°C with a yield of 89%. Further reaction of 1.2 g (3.0 mmol) of compound [9] with 1.3 g (4.5 mmol) of disuccinimide carbonate (DSC), in 30 mL of acetonitrile solution, at room temperature in the presence of catalytic amount of triethylamine, afforded the corresponding activated ether [10] in 2.1 mmol (70%) yield. The compound [10] thus prepared (500 mg; 1.0 mmol) was then allowed to react with 210 mg (1.0 mmol) of HBr salt of poly(L-lysine) in the presence of 0.3mL (2.0 mmol) of triethylamine at 80°C for 3 days, which gave the final product [11] after reprecipitation in acetone in 65% (390mg) yield. In a similar manner, the poly(L-lysine) derivatives having adenosine, cytidine, and inosine were prepared.

1C. Poly(L-glutamic acid) Derivatives Containing Uridine.[9] Starting from isopropylidene uridine, [8], and protected β-alanine, the poly(L-glutamic acid) derivative having uridine was successfully prepared: 560 mg (2.0 mmol) of [9] was allowed to react with 380 mg (2.0 mmol) of β-alanine, having a tert-butoxycarbonyl protecting group, in the presence of 450 mg (2.2 mmol) of dicyclohexylcarbodiimide (DCC), to give the compound [12]) in 77% (700 mg) yield. The compound [12] was treated with 4N hydrochloric acid, in dioxane at room temperature, for 4 hr. After this reaction, the pH of the system was adjusted to 7.0, to give the compound [13], having the free amino group, in 61% (240 mg) yield. The compound [13] (400 mg; 1.0 mmol) was allowed to react further with 10 mL dimethylformamide solution at 50°C for 3 days, which gave the final product [15] in 62% (390 mg) yield. In a similar way, the poly(L-glutamic acid) derivative containing adenosine was also prepared.

1D. Isopoly(α-N-adenine-L-lysine).[37] To a solution of N-α-adenine-L-lysine[38] (400 mg, 1.0 mmol) in DMF (1 mL), pentachlorophenyl trichloroacetate (490 mg, 1.2 mmol) was added with stirring at 0°C. Continued stirring for 30 min. under reduced pressure at room temperature gave a clear solution. To this solution, dry Celite-545 (400 mg) was added and the solution was removed under reduced pressure. The residue was heated to 130°C under high vacuum for 1 day to sublimate the pentachlorophenol. After that, the residue was washed with water and the polymer was extracted with 1N hydrochloric acid (30 mL) from Celite. The solution was concentrated and a small amount of methanol was added. The polymer, isopoly(N-α-adenine-HCl-L-lysine), was precipitated from acetone to give a 15% yield (55 mg). In a similar way, isopoly(N-α-uracil-L-lysine) was prepared in 20% yield.

2. Synthesis of Poly(ethyleneimine) Derivatives

2A. Poly(ethyleneimine) Derivatives Containing Thymine Bases.[10]

2Aa. Pentachlorophenyl-3-(thymine-1-yl)propionate, [19]. 3-(thymine-1-yl)propionic acid, [18], (1.9 g, 0.01 mol) was dissolved in 10 mL of dry DMF, and then triethylamine (1.4 mL, 0.01 mol) and pentachlorophenyl trichloroacetate (4.5 g, 0.01 mol) were added in that order. The mixture was stirred for 20 hr. at room temperature. After the reaction, the DMF was distilled off under reduced pressure, and the precipitated product was washed thoroughly with cold water and recrystallized from an acetone benzene mixture to give colorless needles; yield 3.3 g, 73%, m.p. = 236-237°C.

2Ab. Poly(N-[3-thymin-1-yl)propionyl]ethyleneimine), (PEI-T), [20]. A mixture of [19] (4.5 g, 0.01 mol), linear poly(ethyleneimine) (0.43 g, 0.01 mol), and imidazole (0.68 g, 0.01 mol) in 100 mL of DMF was stirred for 3 days at 70°C to afford a clear solution. After distilling the solvent, excess acetone was added to the oily residue which gave rise to the solid polymer. Reprecipitation from DMSO-acetone gave 1.91 g of polymer [9].

2B. Synthesis of Water Soluble Nucleic Acid Analogs.[13]

2Ba. Ethyl-3-(cytosyl-1-yl)-propionate, [**21**]. To a suspension of N^4-acetyl cytosine[14] (27 g; 180 mmol) in ethanol (800 mL), with a catalytic amount of sodium, ethyl acrylate (28 mL; 260 mmol) was added dropwise with stirring. The reaction was carried out at 90°C for 5 hr. to afford a clear solution. After the reaction, the solvent was removed under reduced pressure. The residue was recrystallized from ethanol to give ethyl-3-(cytosyl-1-yl)propionate, [**21**], in 94% yield (35 g); m.p = 199-201°C.

2Bb. 3-(Cytosyl-1-yl)propionic acid, [**22**], (1-β-carboxyethyl cytosine). The ester [**21**] (1.7 g; 6.7 mmol) was hydrolyzed by 1N NaOH aqueous solution at 60°C for 2 hr. After the reaction, the pH of the solution was adjusted to 5.0 by 1N HCl to give a colorless precipitate. The precipitate was collected and was recrystallized from water to give 3-(cytosyl-1-yl)propionic acid, [**22**], in 80% yield (960 mg); m.p = 264-266°C (dec).

2Bc. (±)-α-N-[3-(cytosyl-1-yl)propionyl]amino-γ-butyrolactone, [**24**], (Cyt-Hse-L). To a solution of the 1-β-carboxyethylcytosine (1.3 g; 7.0 mmol) in DMF (40 mL), pentachlorophenyl trichloroacetate (4.5 g; 11 mmol) and catalytic amount of triethylamine were added. The mixture was stirred overnight at 80°C. The precipitate was filtered and washed with diethyl ether to give 2.7 g of pentachlorophenyl-3-(cytosyl-1-yl)propionate, [**23**], in 90% yield. To a solution of (±)-α-amino-γ-butyrolactone hydrobromide (0.91 g; 5.0 mmol) in DMF (20 mL), pentachlorophenyl-3-(cytosyl-1-yl)propionate [**23**] (2.2 g; 5.0 mmol) was added and was stirred for one day at 50°C. After the reaction, the solvent was evaporated under reduced pressure. Acetone (150 mL) was added to the oily residue and was stirred for 4 hr. to give a light brown precipitate. The precipitate was washed with water (50 mL) and dried under reduced pressure. The yield was 1.2 g (88%); m.p. = 236-238°C.

2Bd. (±)-α-N-[3-(hypoxanthyl-9-yl)propionyl]amino-γ-butyrolactone, [**28**], Hy-Hse-L). To a solution of (±)-α-amino-γ-butyrolactone hydrobromide (910 mg; 5.0 mmol) in DMF (20 mL), pentachlorophenyl-3-(hypoxanthyl-9-yl)propionate[15] (2.3 g; 5.0 mmol), triethylamine (0.7 mL; 5.0 mmol) and imidazole (340 mg; 5.0 mmol) were added. The mixture was stirred at 40°C for 20 hr. to afford a clear, light brown solution. After the reaction, the solvent was evaporated under reduced pressure. The residue was washed with acetone and then with water. Recrystallization from ethanol-benzene gave crystals in a 79% yield (1.1 g); m.p. = 218-221°C.

2Be. Poly(ethyleneimine), (PEI). Poly(2-ethyl-2-oxazoline) (PEOX, donated by Dow Chemical Japan Co., MW = 50,000, 40 g) was hydrolyzed in 6N hydrochloric acid (200 mL) at 90°C for 48 hr. After the reaction, the colorless precipitate was filtered and washed with 200 mL of ethanol to give poly(ethyleneimine hydrochloride). Then the polymer was dissolved in water (200 mL) and neutralized with NaOH (25 g, 0.36 mol) at 90°C for 24 hr. The solution was cooled in an ice bath to give the polymer as a precipitate. The polymer was filtered and washed with 3L of water and then dried. The hydrolysis was found to be complete from the NMR and IR spectra.[11]

2Bf. Poly[-N{-2-[3-(cytosine-1-yl)propionyl]amino-4-hydroxy}-butanoyl]ethyleneimine), [**25**], (PEI-Hse-Cyt). To a suspension of Cyt-Hse-L, [**24**], (390 mg; 1.5 mmol) in water (5 mL), 1 N NaOH aqueous solution (1.5 mL; 1.5 mmol) was added. The mixture was stirred at 40°C for 1 hr. to give a clear, light brown solution. After the reaction, the solvent was evaporated under reduced pressure. The resulting residue was washed with acetone (20 mL) and then with diethyl ether (30 mL) to give 430 mg of a powder of the sodium salt of the cytosine derivative (Cyt-Hse-Na) in a 98% yield. To a suspension of (Cyt-Hse-Na) (430 mg;1.5 mmol) in 10 mL of dry DMF, pentachlorophenyl trichloroacetate (TCA-PCP) (910 mg, 2.2 mmol) and catalytic amount of triethylamine were added. The mixture was stirred at 0°C for 2 hr. and then at 50°C until the gas evolution stopped to afford a clear, light brown solution. After the reaction, the solvent was washed with diethyl ether (30 mL) to give a powder of the pentachlorophenyl ester of the cytosine derivative (Cyt-Hse-PCP) (yield 92%, 720 mg).

To a solution of poly(ethyleneimine) (70 mg, 1.5 mmol) in DMF (10 mL), (Cyt-Hse-PCP) (720 mg; 1.5 mmol) and imidazole (90 mg; 1.5 mmol) were added and stirred at 60°C for 48 hr. After the reaction, the solution was concentrated under reduced pressure and was poured into excess acetone. The resulting precipitate was dissolved in a small amount of DMF and reprecipitated twice in excess acetone. The obtained polymer was dissolved in a small amount of water and freeze dried to give a light brown powder (360 mg; 78%; m.p. = 210-215°C).

2Bg. Poly(N-[2-[3-9hypoxanthyl-9-yl)propionyl]amino-4-hydroxy]-butanoyl]-ethyleneimine), **[29]**, (PEI-Hse-Hyp). This polymer was prepared from **[28]** (1.0 g; 3.6 mmol) and poly(ethyleneimine) (160 mg; 3.6 mmol) according to the similar procedure described in the preparation of (PEI-Hse-Cyt), **[25]**, (yield 68%, 820 mg; m.p. = 166-172°C).

3. Hydrolysis of the Polymers

The poly(ethyleneimine) derivatives were hydrolyzed in 6N hydrochloric acid at 80°C for 48 hr., into poly(ethyleneimine hydrochloride) and the carboxyethyl derivatives of the nucleic acid bases. The quantitative calculation was made by using the corresponding carboxyethyl derivatives as the standard sample. The nucleic acid base contents in the polymers are 86% for (PEI-Hse-Cyt), and 92% for (PEI-Hse-Hyp), respectively.

4. Interaction Between the Polymers

Interaction of the polymers was estimated by the hypochromicity values in the UV spectra.[12] The UV spectra were measured with a JASCO UV-660 spectrometer equipped with a temperature controller at 20°C. Poly-(inosinic acid) (sodium salt) ($S^{o}_{20,w}$ = 6-12), poly(cytidylic acid) (sodium salt) ($S^{o}_{20,w}$ = 6-12) were obtained from Yamasa Shoyu Co. Ltd. (PEI-Hse-Cyt), **[5C]**, (PEI-Hse-Hyp), **[5H]**, and the poly(nucleotides) were dissolved in Kolthoff buffer (pH = 7.0) (0.1M KH_2PO_4 + 0.05M $Na_2B_4O_7 \cdot 10H_2O$). These solutions, stored for 2 days at 20°C, were mixed so as to give a polymer mixture of 10^{-4} mol/L total concentration of nucleic acid base units in the aqueous solution.

5. Instrumentation

The UV spectra were measured with a Shimazu UV-180 spectrometer. The NMR spectra were obtained with a JNM-PS-100 spectrometer (JEOL).

The conformation of the polymers prepared was studied by circular dichroism (CD) measurement at different pH values in aqueous solution. A JASCO J-40A spectropolarimeter was used for the CD spectra. The concentration of the polymer solutions were kept in the $2x10^{-3}$-$2x10^{-4}$ mole/L range. The measurement was made with 10- and 1.0-mm path-length quartz cells at ambient temperature.

RESULTS AND DISCUSSION

1. Preparations of Nucleic Acid Analogs

1A. Poly(amino acid) Derivatives. The derivatives of poly(L-lysine) containing the nucleic acid bases, that is adenine, thymine and uracil in the side groups, were first prepared as shown in the following reaction sequence:[7]

```
R-CH2CH2COO-Ph-NO2  +  -NH-CH-CO-   ──→  -NH-CH-CO-
                           |                  |
                          (CH2)4             (CH2)4
                           |                  |
                          NH2·CF3COOH        NH-CO-CH2CH2-R

R = Adenine (1)                          R = Adenine (5)
    Thymine (2)                              Thymine (6)
    Uracil  (3)                              Uracil  (7)
```

Carboxyethyl derivatives of the nucleic acid bases were grafted onto poly(L-lysine) by using the activated ester method[18]. Poly(L-lysine trifluoroacetate), [**4**], was prepared according to the method of Sela[19]. The p-nitrophenyl esters, [1]-[3], were prepared by the reaction of the corresponding carboxyethyl derivatives of the nucleic acid bases with p-nitrophenyl trifluoroacetate, according to the method of Overberger and Inaki,[11] and purified by recrystallization.

The reaction of the p-nitrophenyl esters with the polymer [**4**] was carried out in DMSO solution in the presence of triethylamine at 25°C. The poly(L-lysine) derivatives obtained have different infrared absorption spectra from those of the starting compounds, and have absorptions assigned to the nucleic bases. Poly(ε-N-Ade-L-lysine), [5], was soluble in DMSO and ethylene glycol, and also in water below pH 3, where adenine was in the protonated form. The poly(lysine) containing 53 mol% adenine units was soluble in DMF, while the polymer containing 74 mol% adenine units was insoluble. Poly(ε-N-Thy-L-lysine), [6], and poly(ε-N-Ura-L-lysine), [7], were soluble in DMSO, DMF and 6N hydrochloric acid.

Poly(L-lysine) derivatives containing the nucleosides [8] were next prepared using the active ester method as shown in the reaction sequence below:[8]

The 2',3'-Hydroxy groups of the nucleosides such as uridine, (U), inosine, (I), adenosine, (A), and cytidine, (C), were protected by isopropylidene groups. The products were treated with glutaric anhydride to give nucleoside derivatives containing terminal carboxy groups, and then followed by the reaction with N,N'-disuccinimidocarbonate (DSC) to afford the activated esters. Poly(L-lysine), (PLL), was then prepared by using NCA method, with n-propylamine as the initiator. The degree of polymerization was about 40. The reaction of poly(L-lysine) with the activated ester derivatives was carried out in DMSO solution at 80°C for 7 days. The reaction mixture was reprecipitated in excess acetone. The NMR and UV

HO B HO OH → CH_3CCH_3 (O) → HO B O O (8) → (glutaric anhydride) → $HOCCH_2CH_2CH_2C$-O B O O (9)

(N-O-C)$_2$ / NEt_3 → N-O-$CCH_2CH_2CH_2C$-O B O O (10) → PLL·HBr / NEt_3 → -(NHCHC)- $(CH_2)_4$ NH C-$(CH_2)_3$-CO B O O (11)

spectra of these polymers indicated that they contained 50-70% of the nucleoside units.

In this connection, poly(L-glutamic acid) derivatives containing nucleosides were prepared as shown below:

HO B O O (8) → Boc-N~COOH / DCC → $BocNHCH_2CH_2C$-O B O O (12) → 4NHCl / DOX → $NH_2CH_2CH_2C$O B O O (13)

-(NH-CH-CO)$_n$ $(CH_2)_2$ COONa → TCA-o-PCP → -(NH-CH-CO)$_n$ $(CH_2)_2$ C-O-(C$_6$)Cl_5 (14) → -(NH-CH-CO)$_n$ $(CH_2)_2$ C-$NHCH_2CH_2C$-O B O O (15)

An alternative procedure for obtaining poly(acrylic acid), poly-(methacrylic acid), and poly(L-glutamic acid) derivatives can be attained by the reaction of such polymeric acids with the cyclic derivatives of pyrimidines and their nucleosides.[9] For example, in the case of uracil and uridine, the reaction was performed in an 1:1 unit molar ratio in DMF solution, at 80°C for 7 days under stirring. The polymers were purified by reprecipitation using acetone or methanol. The structure was confirmed by NMR, IR and elemental analysis.

-(CH_2 C(R))- COOH R=H, CH_3

-(HN-CH-CO)- CH_2 CH_2 COOH

(O N X cyclic pyrimidine; O N X CH_2 O O cyclic nucleoside) →

HN X O N C-O-CH_2CH_2 (16)

HN X O N C-O-CH_2 O O (17)

In relation to the case of poly(L-lysine), the isopoly(L-lysine) derivatives appear to be interesting also, because the latter does not form an α-helical structure. For the synthesis of isopoly(L-lysine) with pending nucleic acid bases, suitable monomers bearing the nucleic acid base in the side chain were prepared. The L-lysine derivatives containing the nucleic acid bases at the ε-amino group were prepared by condensation of N-ε-Cbz-L-lysine with the carboxyethyl derivatives of the nucleic acid bases using the activated ester method. After removing the Cbz group, the ε-amino acid derivatives of the nucleic acid derivatives were obtained. The reaction of the amino acid derivatives with pentachlorophenyl trichloroacetate gave the pentachlorophenyl ester trichloroacetic acid salt. This activated ester was heated to 130 to 150°C to give the isopoly(L-lysine) derivatives. The interaction between polymers having complementary bases was studied in these cases by UV and CD spectroscopies. The specific interaction was found to be realized preferably between the neutral forms of the polymers.

1B. Poly(ethyleneimine) Derivatives. The synthesis of the poly(ethyleneimine) derivatives containing nucleic acid bases and amino acids was first studied extensively by Overberger and Inaki.[11] By the reaction of the activated ester [19] with poly(ethyleneimine) in DMSO, in the presence of imidazole as a catalyst, the nucleic acid analogs were prepared.[10] For example, the case of the thymine base is shown in the reaction sequence below:

$$\text{(18)} + CCl_3COO\text{–}C_6Cl_5 \longrightarrow \text{(19)}$$

$$\text{(19)} + \text{–(}CH_2CH_2NH\text{)}_n\text{–} \longrightarrow \text{–(}CH_2CH_2NH\text{)}_{n(1-m)}\text{–(}CH_2CH_2N(COCH_2CH_2\text{-thymine})\text{)}_{mn}\text{–} \quad \text{(20)}$$

More recently, water soluble poly(ethyleneimine) derivatives containing nucleic acid bases have been prepared, which contain homoserine units as a spacer between the poly(ethyleneimine) and the nucleic acid base (Schemes 1 and 2).[13]

1Ba. Homoserine derivatives of nucleic acid bases. Starting with cytosine or adenine, the carboxyethyl derivatives of the nucleic acid bases were prepared by the Michael's type addition reaction of ethyl acrylate. In the case of cytosine, the 4-amino group was first protected by an acetyl group because of low solubility in ethanol, although the acetyl group was removed during the reaction. In the case of hypoxanthine, when the Michael's addition reaction was carried on the hypoxanthine base, the addition occurred at the 7-position of the base. After the carboethoxyethyl derivative of adenine was prepared, the deamination reaction was carried out to afford the carboxyethyl derivatives of hypoxanthine.

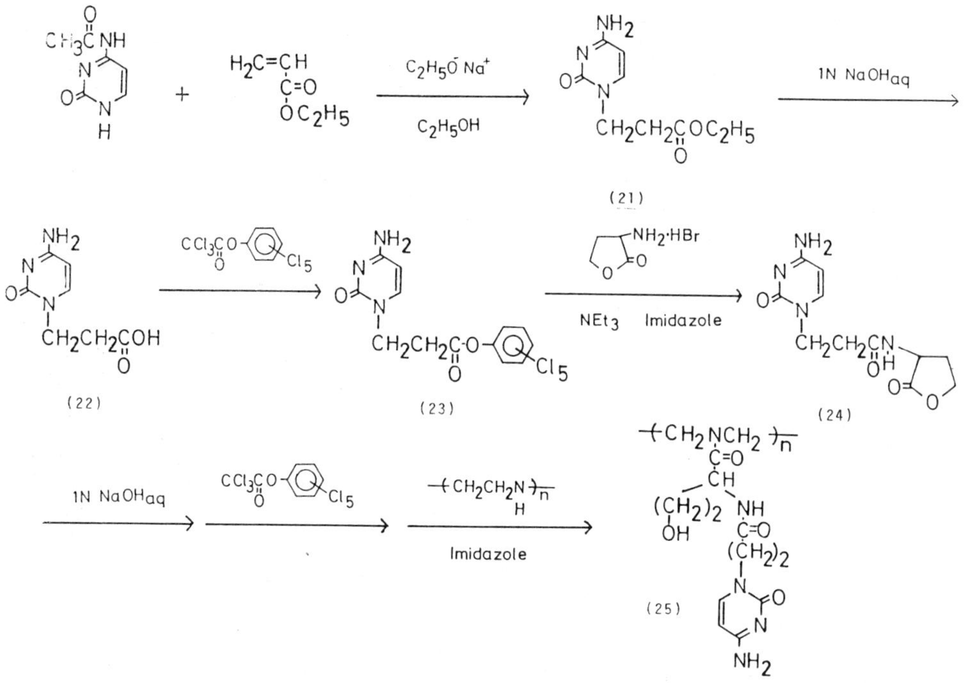

Scheme 1.

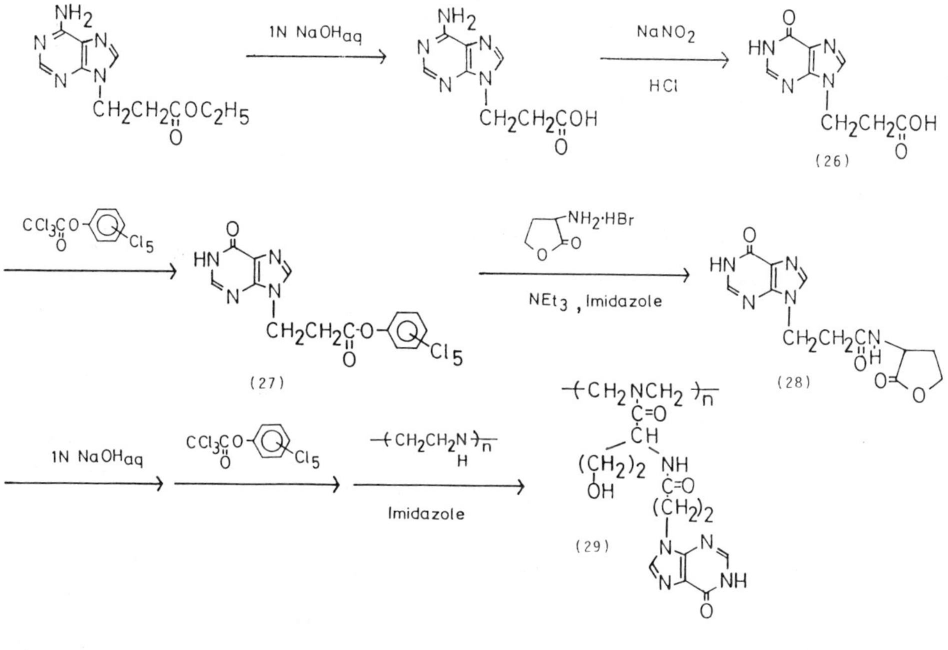

Scheme 2.

The carboxyethyl derivatives of cytosine or hypoxanthine were reacted with (±)-α-amino-γ-butyrolactone hydrobromide to give the corresponding butyrolactone derivatives of cytosine or hypoxanthine. For these reactions, pentachlorophenyl ester derivatives were used with imidazole as a catalyst.[34]

1Bb. Grafting onto poly(ethyleneimine). The grafting of nucleic acid base derivatives having a hydroxyl group onto a hydrophobic polymer backbone was also carried out by the activated ester method. Since the reactivity of the γ-lactone is low, the direct reaction of the lactone derivative with poly(ethyleneimine) hardly occurred. Therefore the lactone derivatives were hydrolyzed to the 3-hydroxybutylic acid ones, followed by condensation with poly(ethyleneimine) using the activated ester method. The grafting reaction was carried out in DMF solution, where a small amount of 4-pyrroridino-pyridine served as an effective catalyst.

The nucleic acid base content of the polymer was determined by UV spectroscopy on their hydrolyzed samples. The quantitative calculation was made by using the corresponding carboxyethyl derivatives as standard samples.

2. Conformation of the Nucleic Acid Analogs

The conformation of the polymers was studied by CD and ORD measurements. Poly(L-lysine) is known to exist in disordered, α-helical and the conformation depends upon the temperature, the pH of the system, and the solvent used. The side chain of the polymer has a significant effect on the backbone conformation. At neutral pH, poly(L-lysine) exists in a random coiled structure, while at pH above 10 the ε-amino group becomes a neutral form and the polymer undergoes transition to a helical structure.

In order to study the effect of base substitutents on the conformation of poly(L-lysine), CD spectra of the copolymers were measured. From the spectra of poly(L-lysine) having adenine units, for example, it was found that the helicity of the polymer increases with rising pH of the system, due to the release of the electrostatic repulsion between positively charged side chains. By adding ethylene glycol to the system, the helicity tends to increase, which shows that ethylene glycol depresses the electrostatic repulsion between the protonated adenine units.

Another important factor for the polymer conformation is the solvent effect. The ORD measurements and analysis clearly showed that the polymer exists in a helical conformation in chloroform solution, while it has a random structure in chloroacetic acid solution. In general, those having a high base content tend to exhibit a helical conformation both in organic solvents and in water.

Like the poly(L-lysine) derivatives containing nucleic acid bases, those containing nucleosides were also found to exist in an α-helical structure at neutral pH regions. The CD spectra of poly(L-lysine) derivatives containing uridine (PPL-Urd, [**11**]) and thymine (PPL-Thy, [**6**]) were studied in detail at different pH values.[36]

3. Specific Interaction of the Nucleic Acid Analogs

3A. Poly(amino acid) Analogs. The synthetic nucleic acid analogs form polymer complexes by the specific interaction between complimentary nucleic acid bases on the side chains, as is the well known case for poly(nucleotides).

The complex formation between the complementary poly(L-lysine) derivatives was first studied by UV spectroscopy in DMSO-ethylene glycol solution. From the mixing curves between adenine and thymine containing poly(L-lysines), it was found that the interaction tended to become weaker, decreasing with the base content of the polymers. That is, the interaction was found to become weaker with increasing the temperature from 25 to 90°C, causing dissociation of the polymer complexes. From this analysis, it was concluded that the polymer complexes are formed by specific base pairing between pendant adenine and thymine or uracil units of poly(L-lysine) derivatives retaining their helical conformation.[20,21]

The helical conformation of thymine-containing poly(L-lysine) was found to be retained also in the system of this polymer and poly(methacryloxyethyladenine), even after forming the polymer complexes with each other. From the 1:1 stoichiometry, the high value of hypochromicity and the helical conformation of thymine containing poly(L-lysine), the polymer complex in question was assumed to exhibit a double helical structure. The polymer complex seems to be held together by the specific base pairing between pendant thymine and adenine bases. A similar experiment was also done on the formation of the polymer complex between thymine containing poly(L-lysine) and poly(methacryloylaminoethyladenine). It was concluded that both the mutual penetration ability and the compatibility of base-base distances of the polymers are sufficient for the formation of the stable polymer complexes by specific base pairing.[22]

The interaction between PPL-Urd, [11], and poly(A) having the complimentary nucleic acid base can be clearly observed in aqueous solution. The rate of such complex formation was found to be slow in aqueous solution. Absorbance of the mixed solution of the both sort of polymers tended to decrease slowly and to became nearly constant after 3 days. The interaction between PLL-Thy, [6], and PLL-Ade, [5], can be observed in a DMSO-ethylene glycol (3/2, v/v) mixture. The poly(ethyleneimine) derivatives having nucleic acid bases also formed polymer complexes with poly(nucleotides) and poly(amino acids), which were studied in detail by UV and CD measurements in aqueous solutions. It was also clarified in this case that the interaction was based upon the specific one between complementary nucleic acid bases, which affected their conformation.

<u>3B. Poly(ethyleneimine) Analogs</u>. In order to clarify the spacer effect in the specific base-base interaction, a series of poly(ethyleneimine) derivatives, and its oligomer models having pendant thymine bases separated by α-alanyl groups as the spacer, were prepared in a similar way as the preparation of the corresponding thymine derivatives without spacers. Spectrophotometric data for the compounds was compared with that of the corresponding thymine derivatives, in order to clarify the intramolecular interaction of thymine bases. The nature of the intramolecular interaction was found to resemble to that of the corresponding compounds without α-alanyl units, while larger UV hypochromicities were found in this series and the increase in intramolecular interaction was suggested to be attributable to the presence of α-alanyl groups. In a similar way, the synthesis and properties of oligomeric models of polyethyleneimine derivatives having spacer-separated adenine bases were also studied.[23,24]

Recently, the interaction of the water soluble poly(ethyleneimine) derivatives of cytosine and hypoxanthine with poly(nucleotides) was studied in neutral aqueous solution. The complex formation of the nucleic acid analogs with poly(nucleotides) in water is very important for the investigation of bioactivity of the polymers.[25] The interaction in water

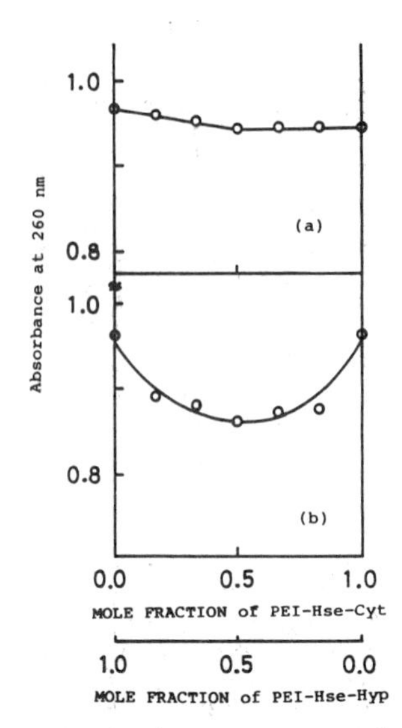

Figure 1. Continuous variation curves of (PEI-Hse-Hyp) and (PEI-Hse-Cyt) in 0.05 M Kolthoff buffer solution (pH = 7.0). Absorbance at 260 nm after (a) 3 hours and (b) 3 days.

was measured between the poly(ethyleneimine) derivatives and the poly(nucleotides) such as poly(inosinic acid) and poly(cytidylic acid).

3Ba. Interaction of (PEI-Hse-Cyt) and (PEI-Hse-Hyp): From the mixing curve between (PEI-Hse-Cyt), [25], and (PEI-Hse-Hyp), [29], in the Kolthoff buffer solution (pH 7.0), the system of (PEI-Hse-Cyt) with (PEI-Hse-Hyp) showed the highest hypochromicity value at the base ratio of about 1:1 (hypoxanthine:cytosine), suggesting the formation of stable 1:1 polymer complex due to complementary nucleic acid base interaction (Figure 1). The hypochromicity value (21%), however, was smaller than those of (PEI-Hse-Hyp)·poly(C) (26%) or poly(I)·poly(C) (33%) systems. In this system, a time dependence of the hypochromicity value was observed. Therefore the conformational change of the synthetic polymers may be important for the formation of a stable polymer complex.[26-28]

3Bb. Interaction of (PEI-Hse-Hyp) with Poly(C): The interaction between (PEI-Hse-Hyp), [29], and Poly(C), which contain the complimentary nucleic acid bases, can be clearly observed in aqueous solution at pH 7.0 (Figure 2). The overall stoichiometry of the complex based on nucleic acid base units was approximately 2:1 (hypoxanthine:cytosine) under the conditions used here. The maximum hypochromicity value (26%) was smaller than that of the poly(I)·poly(C) (33%) system, but higher than the value of another synthetic polymer analog - poly(nucleotide) system.[29,30]

The well known complex, poly(I)·poly(C) was ascertained in the pH 7.0 buffer solution used in this study. The stoichiometry of the complex based on the nucleic acid base units, however, was 2:1 (hypoxanthine: cytosine). The maximum hypochromicity value of this system was 33%. The stoichiometry of poly(I)·poly(C) complex is well known to be 1:1 in tris-HCl buffer solution,[31-33] which was also ascertained in our case. It is also known that the concentration of salt affects the stoichiometry of the poly(I)·poly(C) complex.[31-34] Therefore, the concentration of salt may affect the stoichiometry observed for the (PEI-Hse-Hyp)·poly(C) system in our case (Figure 2).

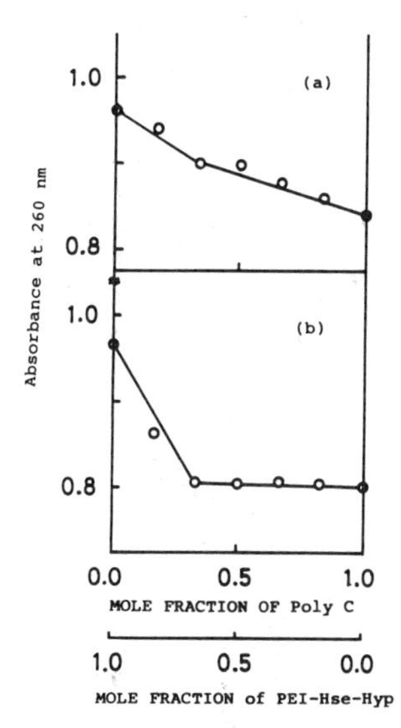

Figure 2. Continuous variation curves of (PEI-Hse-Hyp) and Poly(C) in 0.05 M Kolthoff buffer solution (pH = 7.0). Absorbance at 260 nm after (a) 3 hours and (b) 3 days.

Another factor for the stoichiometry of the complex may be the structure of the polymers. The base-base distance in (PEI-Hse-Hyp) is different from that in poly(C). This fact also can affect the abnormal stoichiometry of the (PEI-Hse-Hyp)·poly(C) complex.

The self association of the nucleic acid bases in the polymer is also an important factor for the formation of the polymer complex. The rate of the complex formation for the poly(ethyleneimine) derivatives was slow compared with that of poly(nucleotide) systems. The absorbance of the mixed solution of the two types of polymers decreased slowly and became constant after 3 days. This fact is caused by the self association of the nucleic acid bases in the poly(ethyleneimine) derivatives, which dissociated slowly to form an intermolecular polymer complex.

3Bc. Interaction of (PEI-Hse-Cyt) and Poly(I): Figure 3 shows the mixing curve between (PEI-Hse-Cyt), [25], and Poly(I) after 3 hr. and 3 days, in an aqueous Kolthoff buffer (pH 7.0) solution. The stoichiometry of the complex is 2:1 (hypoxanthine:cytosine) and the hypochromicity value is *ca.* 7%. The stoichiometry of the (PEI-Hse-Cyt)·poly(I) system is similar to that of the (PEI-Hse-Hyp)·poly(C) system. This stoichiometry of the complex, 2:1 (hypoxanthine:cytosine), may be caused by the difference of the polymer structure between (PEI-Hse-Cyt) and poly(I), and also by the low salt concentration. The hypochromicity value is, however, small compared with the (PEI-Hse-Hyp)·poly(C) system. This may be due to the stability of the self association of hypoxanthine bases in poly(I). Poly(nucleotides) containing purine base form a very stable structure by self association in aqueous solution, while the structure of the poly-(nucleotides) containing pyrimidine bases are rather unstable in aqueous media.[35] This reason is assumed to be caused by the stable structure stacking interaction between the purine bases. Poly(I) forms a stable structure by self association in aqueous solution, which is as stable as poly(G). Therefore, the intermolecular interaction of poly(I) with (PEI-Hse-Cyt) should be considered to be small because of the enhanced intramolecular interaction of poly(I).

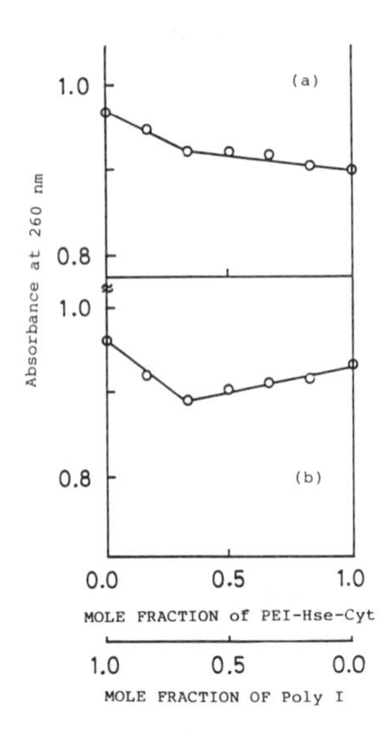

Figure 3. Continuous variation curves of (PEI-Hse-Cyt) and Poly(I) in 0.05 M Kolthoff buffer solution (pH = 7.0). Absorbance at 260 nm after (a) 3 hours and (b) 3 days.

REFERENCES

1. K. Takemoto & Y. Inaki, Adv. Polymer Sci., **41**, 1 (1981).
2. K. Takemoto, Makromol. Che. Suppl., **12**, 293 (1985).
3. Y. Inaki & K. Takemoto. Makromol. Chem. Suppl., **14**, 91 (1985).
4. K. Takemoto in: "*Bioactive Polymeric Systems*", C. G. Gebelein & C. E. Carraher, Eds., Plenum Press, New York, 1985, p. 417.
5. K. Takemoto & Y. Inaki in: "*Funtional Monomers and Polymers*", K. Takemoto, Y. Inaki & R. M. Ottenbrite, Eds., Marcel Dekker, New York, 1987 p. 149.
6. Y. Inaki & K. Takemoto in: "*Current Topics in Polymer Science*", Vol. 1, R. M. Ottenbrite, L. A. Utracki & S. Inoue, Eds., Hanser Publ., 1987, p 79.
7 T. Ishikawa, Y. Inaki & K. Takemoto, Polymer Bull., **1**, 85 (1987).
8. E. Mochizuk, Y. Inaki & K. Takemoto, to be published.
9. E. Mochizuki, Y. Inkai & K. Takemoto, Nucleic Acids Res., Symp. Ser., **16**, 121 (1985).
10. Y. Inaki. Y. Sakuma, Y. Suda & K. Takemoto, J. Polymer Sci., Polymer Chem. Ed., **20**, 1917 (1982).
11. C. G. Overberger & Y. Inaki, J. Polymer Sci., Polymer Chem. Ed., **17**, 1739 (1979).
12. T. Saegusa, H. Ikeda & H. Fujii, Polymer J., **3**, 35 (1972).
13. T. Wada, Y. Inaki & K. Takemoto, Polymer J., **11**, 21 (1988).
14. D. M. Brown, A. R. Todd & S. Varadarajan, J. Chem. Soc., 2384 (1956).
15. Y. L. Kaminskil, I. F. Ivanova & L. S. Gordeeva, Khim. Prikr. Soedin., **1**, 128 (1987).
16. C. G. Overberger & C. C. Chen, J. Polymer Sci., Polymer Chem. Ed., **25**, 373 (1987).
17. G. J. Thomas, Jr. & Y. Kyogoku, J. Amer. Chem. Soc., **89**, 4170 (1987).
18. N. Anand, N. S. R. K. Murthy, F. Naider & M. Goodman, Macromolecules, **4**, 564 (1971).
19. M. Sela, R. Arnon & I. Jacobson, Biopolymers, **1**, 517 (1963).
20. T. Ishikawa, Y. Inaki & K. Takemoto, J. Polymer Sci., Polymer Chem. Ed., **18**, 949 (1980).

21. T. Ishikawa, Y. Inaki & K. Takemoto, J. Polymer Sci., Polymer Chem. Ed., **18**, 1847 (1980).
22. Y, Inaki, T. Ishikawa, S. Sugita & K. Takemoto, J. Polymer Sci., Polymer Lett. Ed., **18**, 725 (1980).
23. Y. Inaki. Y. Sakuma, Y. Suda & K. Takemoto, J. Polymer Sci., Polymer Chem., Ed., **20**, 1917 (1982).
24. Y. Sakuma, Y. Inaki & K. Takemoto, J. Polymer Sci., Polymer Chem., Ed., **20**. 3431 (1982).
25. T. Sato, K. Kojima, T. Ihda, J. Sunamoto & R. M. Ottenbrite, J. Bioacitve and Compatible Polymers., **1**, 448 (1986).
26. S. Fang, Y. Inaki & K. Takemoto, J. Polymer Sci., Polymer Chem., Ed., **22**, 2455 (1984)
27. S. Fang, Y. Inaki & K. Takemoto, J. Polymer Sci., Polymer Chem., Ed., **22**, 3943 (1984)
28. S. Fang, Y. Inaki & K. Takemoto, Polymer J., **17**, 443 (1985).
29. J. Pitha, Adv. Polymer Sci., **50**, 1 (1983).
30. C. G. Overberger, Y Inaki & Y. Nambu, J. Polymer Sci., Polymer Chem., Ed., **17**, 1759 (1979).
31. D. Thiele, W. Guschlbauer & A. Faver, Biopolymers, **8**, 361 (1969). 32. D. Thiele, W. Guschlbauer & A. Faver, Biopolymers, **10**, 143 (1971).
33. D. Thiele, W. Guschlbauer & A. Faver, Biochem. Biophys. Acta., **272**, 22 (1972).
34. K. Fujioka, Y. Baba, A. Kagemoto & R. Fujishiro, Polymer J., **12**, 843 (1980).
35. Z. G. Kudritskaya & V. I Danilov, J. Theor. Biol., **59**, 303 (1976).
36. T. Wada, E. Mochizuki, Y. Inaki & K. Takemoto, Peptide Chem., **41** (1987).
37. Y. Suda, Y. Inaki & K. Takemoto, Polymer J., **16**, 303 (1984).
38. Y. Inaki, M. Omatsu & K. Takemoto, in preparation.

POTENTIAL MEDICAL APPLICATIONS OF NUCLEIC ACID ANALOG POLYMERS

Charles G. Gebelein

Dept. of Chemistry
Youngstown State University
Youngstown, OH 44555

Numerous research groups have prepared polymers which can be considered as synthetic nucleic acid analogs. These have included polymers with a nucleic base on a vinyl backbone, a poly(peptide) backbone or poly(methacrylate) backbones. The synthetic chemistry of these systems is reviewed and some potential medical and non-medical applications of these systems are discussed. The main thrust of this talk will, however, center on the potential application of these pseudo-nucleic acid polymers in cancer therapy. Within this area, the main emphasis will be on poly(methacrylate) polymers with pendant nucleic bases, and related compounds, which could prove useful in chemotherapy. These copolymers show a zero-order release (constant drug level) of the 5-fluorouracil drug unit whereas monolithic dispersions of this drug are released with a declining dosage pattern. Other potential biomedical applications discussed include chromatographic analysis substrates, HIV virus detection, and genetic deficiency disease correction.

INTRODUCTION

Many research groups have prepared polymers which can be classed as synthetic nucleic acid analogs or pseudo-nucleic acids. Unlike the natural DNA or RNA molecules, in which the backbone chain is composed of phosphate and sugar units (ribose or deozyribose), these synthetic polymers and copolymers are systems derived from vinyl, polypeptide and methacrylate backbones, with pendant nucleic base units. In spite of this difference, however, many of these synthetic nucleic acid analogs have shown biological activity and could have some potential biomedical applications. This paper will briefly review some of these systems and will consider some potential biomedical applications of these nucleic acid analog polymers.

The highly specific catalytic activity for enzymes is generally attributed to a lock-in-key mechanism, and many research groups have attempted to duplicate this specific enzymatic activity using synthetic polymers containing the same reactive sites present in the natural enzyme

Biomimetic Polymers
Edited by C. G. Gebelein
Plenum Press, New York, 1990

molecules. These synthetic polymers have been termed enzyme-mimetic polymers, pseudo-enzymes, enzyme-analogs, or synzymes.[1,2] Even though most of these pseudo-enzymes have shown lower catalytic activity than the natural enzymes, they have shown authentic enzyme-like behavior. Potential medical applications of enzyme-mimetic polymers could include enzyme deficiency disease treatment (e.g., phenylketonuria, histidinemia and tyrosinosis) or medical diagnosis. Enzyme-mimetic polymers have been extensively reviewed.[1-3]

1. Vinyl Analogs

Natural nucleic acids have several important biological functions in the human body, including genetic information transfer and the regulation of protein and enzyme synthesis. Like the enzymes, the nucleic acids are also extremely specific in their activities due to the precise arrangement of the individual bases along the macromolecular chain. By analogy with the enzyme-mimetic polymers, several research groups have attempted to develop nucleic acid analogs, pseudo-nucleic acids or nucleo-mimetic polymers in the hope of generating synthetic polymers with biological activity.[4-8] Some of these nucleic acids analogs have exhibited anti-tumor or anti-viral activity. For example, Pitha has prepared poly(9-vinyladenine) and poly(1-vinyluracil), [poly(VU)], which form complexes with the complimentary poly(nucleotides),[9] and the poly(VU) conjugate was shown to be active against murine leukemia virus.[10-13] Some of these vinyl nucleic acid analogs were also template specific inhibitors of *E. coli* RNA polymerase.[14]

2. Synthetic Nucleotides

Many groups have been studying nucleotide polymers such as poly(riboinosinic acid)-poly(ribocytidylic acid), [poly(I·C)]. These are similar to the natural poly(nucleotides) except they consist of a single base distributed along the backbone chain. Some of these synthetic materials have been shown to inhibit tumor growth, including the Ehrlich ascites tumor, fibrosarcoma, lymphoma ascites, leukemia and reticulum cell sarcoma. Research has indicated that effective anticancer activity can occur when the poly(I·C) has a high molecular weight (200,000), is double stranded, contains a 2'-OH group, and has a polymer melting point above 50°C.[15,16] More recently, the stabilized double stranded RNA-like material poly(ICLC) has been shown to be a good interferon inducer which also increases antibody formation in primates.[17] Interferon, in turn, also has been used to treat cancer. The topic of interferon induction by this type polymer has been reviewed.[18]

Another major potential application for nucleic acid analogs is in cancer treatment. Cancer or leukemia treatment is one of the more difficult types of chemotherapy because the drug agents frequently have a low therapeutic index, causing many toxic side effects, such as loss of appetite or hair, nausea, and general patient discomfort. (The therapeutic index is the ratio of the effective dosage level to the toxic level; the higher values are safer to use, but anti-cancer drugs usually have low TI values. While the use of a parental or a controlled release system might alleviate some of these problems, other approaches such as nucleic acid analogs also offer promise. These nucleic acid analogs could function directly as a bioactive polymer (see above), or they could serve as a source of a nucleic base derivative which can act as an anti-cancer drug (see below).

3. Nucleic Acid Analogs With Derivatives of Nucleic Bases

Our research program has focused on polymers containing the antineoplastic agents 5-fluorouracil **[5FU]** and 6-methylthiopurine **[6MTP]**.[19-28] Ballweg, in 1969, appears to be the first to incorporate **[5FU]** into a polymer, via a formaldehyde condensation polymerization method,[29] but this system was not pursued further. More recently, Liu has been published research studies on different types of polymers containing **[5FU]**, which were also prepared by condensation polymerization.[30,31] These studies also reported some positive clinical results.

Butler and coworkers have prepared a 5-fluorouracil monomer by the chemical reaction of **[5FU]** with methyl fumaroyl chloride. When this monomer was cyclocopolymerized, the resulting copolymer released its **[5FU]** very rapidly when placed into an aqueous media.[32] Research has also been done on attaching **[5FU]** onto poly(nucleotides) which resemble the natural DNA materials more closely.[33] The **[5FU]** has also been attached to vinyl polymers containing organosilicon groups.[34] No actual medical or biological data was presented in these studies, however.

Less research has been done with 6-methylthiopurine, but **[6MTP]** has been attached to poly(vinyl alcohol), forming a water soluble polymer system which showed spectral evidence of a charge transfer complex with uracil.[35] Likewise, the 9-vinyl derivative of 6-MTP has been prepared and homopolymerized to form an insoluble, intractable material.[36] Medical or biological activity was not reported in either case, however.

Takemoto and coworkers have made many polymeric derivatives containing the related adenine group in poly(peptide) and vinyl polymer backbone chains.[4,6,8] Some of these analogs appear to have biological activity and several appear to have use in chromatographic applications.

POTENTIAL CANCER APPLICATIONS

1. Preparation and General Features

We have studied the synthesis, polymerization, and copolymerization of monomers containing uracil, 5-fluorouracil or 6-methylthiopurine groups.[19-28] These monomers were prepared by the reaction of acryloyl chloride,[20] vinyl or allyl isocyanates,[19,21-25] or isocyanatoethyl methacrylate **[IEM]**[26-28] with the appropriate purine or pyrimidine. Polymers and copolymers prepared from all of these monomers hydrolyze in an aqueous media to release the pendant bioactive drug.[23,25] In most cases, the drug unit is released too quickly to be of great value for a controlled release system, but the **[5FU]** systems derived from the reaction with **[IEM]**, called **[EMCF]** monomer, release the drug unit fairly slowly and these are promising anti-cancer controlled release systems.

2. Zero-Order Release Systems

The most interesting feature for the **[EMCF]** monomer is that the release rates follow zero order kinetics. This means that there is a constant release of the drug with time. Since cancer normally requires a prolonged treatment with the drug agent, this zero-order kinetic pattern (i.e., a linear relationship between the total amount of drug released and time, or a constant drug level with time) could be very beneficial in cancer therapy.[27,28] In addition to the constant drug level aspect, these copolymers can be tailor-made to release varying amounts of the [**5FU**] into an aqueous environment. This can be accomplished by (1) varying the amount of the [**EMCF**] in the copolymer or (2) by varying the nature of the comonomer involved. These potentially valuable features were achieved without the use of membranes or elaborate polymer sample design. In fact, zero-order kinetics can be obtained with powdered samples as well as pellets, although the release rate will depend upon the sample's total surface area available. Similar results can be obtained with the monomer derived from [**IEM**] and [**6MTP**], although this drug is less used in current anti-cancer therapy than [**5FU**].

The [**5FU**] polymers prepared by Butler, et al., do not seem to show zero-order kinetics. Likewise, a polymeric pro-drug in which [**5FU**] is attached via a carbamoyl bond to an organosilicon moiety, prepared by Ouchi, et al., do not appear to yield a zero-order release.[34] The zero-order characteristics are most probably due to the specific natures of the chemical moieties involved.

3. Potential Use of Nucleic Acid Analogs in Cancer Therapy

In actual cancer treatment, the polymer sample would be implanted at or near the tumor site. The drug could then be released for several months at a constant dose level. The polymer sample would then be removed surgically. Although this surgical removal might be considered as a disadvantage, the overall performance should be a benefit because the drug agent would be implanted directly at or near the tumor site, resulting in smaller total amounts of the drug being used. This could reduce the toxic side effects significantly.

4. Monolithic Systems With 5-Fluorouracil

The behavior of these [**EMCF**] copolymers is significantly different from what occurs in a typical monolithic systems. When [**5FU**] is enclosed in a monolithic matrix the release of the [**5FU**] follows the usual Higuchi kinetic pattern[37] (i.e., a linear relationship between the total amount of drug released and the square root of time, or a continually decreasing amount of the drug released with time).[38] Similar results have been reported by Kiatsu.[39-42] We have also found that annealed monolithic systems can release the [**5FU**] for several months, but the release pattern still follows the Higuchi kinetics. Most monolithic systems in the literature are merely compressed or compacted, rather than annealed. The [**EMCF**] copolymers and the monolithic systems can also be combined to make a system which can give controlled release of the [**5FU**] drug for more than six months.[43,44]

OTHER POSSIBLE MEDICAL USES

1. Chromatographic Substrates

The nucleic acid analogs could also be used as chromatographic substrates for the analysis of natural or synthetic nucleosides, nucleotides and oligonucleotides.[45] Potentially, this type of immobilized nucleic acid analog could be utilized in screening tests for the detection of abnormal nucleic acids or viral materials which cause various diseases. Suitable screening tests could be developed which would permit early detection of genetic deficiency diseases, and facilitate earlier treatment. Thus, if the specific DNA sequence responsible for a particular disease was known, a nucleic acid analog substrate could be synthesized which might complex with this and permit detection. The disease would then be treated by administering the proper drug agents.

2. Possible HIV Virus Detection

In addition to genetic screening procedures, the technique could also be used to detect the HIV virus, providing some unique nucleic base sequence is known to be present in the HIV virus and can be duplicated synthetically. Currently, the presence of the HIV virus is inferred by the presence of antibodies developed toward this virus, but such antibody formation usually requires several months. During this time interval there is no way to detect the HIV virus. If a suitable complementary sequence could be synthesized on a nucleic acid analog, this could be used as a chromatographic substrate to isolate and detect the HIV virus directly.

This earlier detection of the HIV virus, in turn, could make the blood banking procedures safer and also help reduce the incidence of AIDS in the general population.

3. Treatment of Genetic Deficiency Diseases

Additionally, nucleic acid analogs might be useful in treating genetic deficiency diseases, such as cystic fibrosis or sickle cell anemia, if the proper sequencing techniques are developed. An appropriate vinyl nucleic acid analog could be administered to the patient to serve the functions lacking in their own DNA molecules. The vinyl based nucleic acid analogs should be less susceptible to enzymatic hydrolysis in the body and might treat the disease for a longer period of time than a "normal" transplanted nucleic acid.

CONCLUSION

In conclusion, nucleic acid analogs show great potential for biomedical applications in such diverse areas as cancer treatment, disease detection, HIV virus detection, and treatment of genetic deficiency diseases.

REFERENCES

1. J. A. Pavlisko & C. G. Overberger, in: "*Biomedical and Dental Applications of Polymers*," C. G. Gebelein & F. F. Koblitz, Eds., Plenum Publ., New York, 1981, pp. 257-278.
2. Y. Imanishi, J. Polymer Sci., Macromol. Revs., **14**, 1 (1979).
3. Y. Imanishi in: "*Bioactive Polymeric Systems*," C. G. Gebelein & C. E. Carraher,Jr., Eds., Plenum Press, New York, 1985, pp. 435-511.
4. K. Takemoto, J. Polymer Sci., Symposium **55**, 105-125 (1976).

5. J. Pitha, Polymer, **18**, 425 (1977).
6. K. Takemoto, in: "*Polymeric Drugs*," L. G. Donaruma & O. Vogl, Eds., Academic Press, New York, 1978, pp. 103-129.
7. J. Pitha, Proc. IUPAC, Macromol. Symp. IUPAC, Oxford, U.K., **1982**, p. 380.
8. K. Takemoto in: "*Bioactive Polymeric Systems*," C. G. Gebelein & C. E. Carraher, Jr., Eds., Plenum Publ. New York, 1985, pp. 417-433.
9. L. Noronha-Blob & J. Pitha, Biochim. Biophys. Acta, **519**, 285-290 (1978).
10. J. Pitha, S. H. Wilson & P. M. Pitha, Biochem. Biophys. Res. Comm., **81**, 217-223 (1978).
11. V. E. Vengris, P. M. Pitha, L. Sensenbrenner & J. Pitha, Molecular Pharmacology, **14**, 271-277 (1978).
12. J. Pitha, M. Akashi & M. Draminski, in: "*Biomedical Polymers. Polymeric Materials and Pharmaceuticals for Biomedical Use*," E. P. Goldberg & A. Nakajima, Eds., Academic Press, New York 1980, pp. 271-297.
13. J. Pitha, in: "*Biomedical and Dental Applications of Polymers*," C. G. Gebelein & F. F. Koblitz, Eds., Plenum Publ., New York, 1981, pp. 203-213.
14. H. J. Chou, J. P. Froehlich & J. Pitha, Nucleic Acids Res., **5**, 691-696 (1978).
15. N. Stebbing, Arch. Virol., **68**, 291 (1981).
16. E. DeClerq, Texas Reports on Biology and Medicine, **35**, 29 (1977).
17. H. B. Levy & C. Bever, Jr., in: "*Applied Bioactive Polymeric Materials*," C. G. Gebelein, C. E. Carraher, Jr. and V. R. Foster, Eds., Plenum Press, New York, 1988, pp. 205-221.
18. H. B. Levy & T. Quinn in: "*Bioactive Polymeric Systems*," C. G. Gebelein & C. E. Carraher, Jr., Eds., Plenum Press, New York, 1985, pp. 387-415.
19. C. G. Gebelein & R. M. Morgan, Polymer Preprints, **18** (1), 811 (1977).
20. C. G. Gebelein, R. Glowacky, Polymer Prepr., **18** (1), 806 (1977).
21. C. G. Gebelein, R. M. Morgan & R. Glowacky, Polymer Preprints, **18** (2), 513 (1977).
22. C. G. Gebelein & M. W. Baig, Polymer Preprints, **19** (1), 543 (1978).
23. W. Baig & C. G. Gebelein in: "*Controlled Release of Bioactive Materials, 8th Internat. Symp.*," S. Z. Mansdorf & T. J. Roseman, Eds., Controlled Release Society, 1981, p. 170.
24. C. G. Gebelein, Org. Coatings & Plastics Chem., **42**, 422 (1980).
25. C. G. Gebelein, W. Baig, R. M. Morgan & R. Glowacky, Trans. Soc. Biomaterials, **5**, 95 (1982).
26. C. G. Gebelein, Proc. Polym. Mat. Sci. Eng., **51**, 127 (1984).
27. R. R. Hartsough & C. G. Gebelein, Proc. Polym. Mat. Sci. Eng., **51**, 131 (1984).
28. R. R. Hartsough & C. G. Gebelein in: "*Polymeric Materials in Medication*," C. G. Gebelein & C. E. Carraher, Jr., Eds., Plenum Press, New York, 1986, pp. 115-124.
29. H. Ballweg, D. Schmael & E. von Wedelstaedt, Arzneim. Forsch., **19**, 1296 (1969).
30. Z.-H. Liu, "*27th International Symposium on Macromolecules*," Strasbourg, France, 1981, p. 1335.
31. Z.-H. Liu, J. Wuhan U. (PRC), Nat. Sci. Ed., **3**, 114 (1983).
32. P. P. Umrigar, S. Ohashi & G. B. Butler, J. Polym. Sci., Chem. Ed., **17**, 351 (1979).
33. J. L. Alderfer, R. E. Loomis & T. J. Zielinski, Biochemistry, **21**, 2738 (1982).
34. T. Ouchi, K. Hagita, M. Kwashima, T. Inoi & T. Tashiro, J. Controlled Release, **8**, 141-150 (1988).
35. T. Seita, M. Kinoshita & M. Imoto, J. Macromol. Sci.-Chem., **A7**, 1297 (1973).
36. S. Hoffman, W. Witkowski & H. Schubert, Z. Chem., **14**, 14 (1974).
37. T. Higuchi, J. Pharm. Sci.,**59**, 353 (1961).

38. C. G. Gebelein, M. Chapman & T. Mirza in: "*Applied Bioactive Polymeric Materials,*" C. G. Gebelein, C. E. Carraher, Jr. and V. R. Foster, Eds., Plenum Press, New York, 1988, pp. 151-163.
39. M. Yoshida, M. Kumakura & I. Kaetsu, Polymer, **19**, 1375 (1978).
40. M. Yoshida, M. Kumakura & I. Kaetsu, Polymer, **11**, 775 (1979).
41. I. Kaetsu, M. Yoshida, M. Kumakura, A. Yamada & Y. Sakurai, Biomaterials, **1**, 17 (1980).
42. I. Kaetsu, M. Yoshida & A. Yamada, J. Biomed. Mater. Res., **14**, 185 (1980).
43. C. G. Gebelein, M. Davison, T. Gober & M. Chapman, Polym. Mater. Sci. Eng., **59**, 798-802 (1988).
44. C. G. Gebelein, M. Chapman, M. K. Davison & T. G. Gober in: "*Progress in Biomedical Polymers,*" C. G. Gebelein & R. L. Dunn, Eds., Plenum Press, New York, 1990, pp. 321-333.
45. Y. Inaki, S. Nagae, T. Miyamoto, Y. Sugiura & K. Takemoto in: "*Applied Bioactive Polymeric Materials,*" C. G. Gebelein, C. E. Carraher, Jr. and V. R. Foster, Eds., Plenum Press, New York, 1988, pp. 185-204.

GROWTH RATE INCREASE IN NORMAL WISTAR RATS CATALYZED BY INSULIN

Paul W. Wang

Laboratory of Chemical Biology
Institute of Biomedical Engineering
Faculty of Medicine
University of Toronto
Toronto, Ontario, Canada M5S 1A8

Suboptimal growth and development are major factors limiting more efficient output of consumer food products derived from agriculturally important animals. New approaches are being developed through advances in biotechnology, and analogs to pituitary growth hormones made by rDNA have been shown to enhance growth by daily injections at a dose of 2 mg/100kg body weight. Pancreatic insulin is also an anabolic hormone which can be readily formulated in the form of sustained release implants. In a study with 4 groups of 10 normal Wistar rats each, implants made of 12% bovine insulin in palmitic acid were used to maintain hypoglycemia at 3.1 ± 0.7 mM/L in one group, and another group was injected daily with a 12 U dose which induced hypoglycemia for ~8 hr. In a 14-day period, the controls gained about 10.6 ± 8.5 g, the insulin injected group showed an increase of 21.3 ± 12.6 g in body weight, while the animals with implant grew 29.3 ± 11 g with about 10% higher food consumption. Since the hypoglycemia caused no apparent ill effects, the results suggest a feasible alternative to the pituitary hormones in catalyzing growth rate of animals in agriculture.

INTRODUCTION

It has been known for some time that pituitary extracts containing growth hormones (abbrev. as GH) can stimulate growth and increase lactation in animals.[1] With the advance of biotechnology, biosynthetic somatotropin is now available at a reasonable cost. Thus, the use of exogenous GH to increase the quantity and quality of dairy and meat products has become economically feasible.

One of the side effects reported for exogenous GH therapy is hypoglycemia.[2] In the complex series of events which converts food intake to body growth, there may be alternative steps which can enhance these events, aside from an increased level of GH. As insulin is also an anabolic hormone, and the hypoglycemic symptoms observed with increased GH

Biomimetic Polymers
Edited by C. G. Gebelein
Plenum Press, New York, 1990

suggests hyperinsulinemia, the maintenance of an increased level of insulin in a normal animal should also lead to increase in the growth rate. Insulin promotes growth by more efficient carbohydrate utilization from increased intakes, which provides more energy for protein and fat syntheses in the animal body.

This report describes the use of a compressed admixture of palmitic acid and insulin as a sustained release implant to test the aforementioned hypothesis in Wistar rats. The results show that enhancement of growth can indeed be effected.

EXPERIMENTAL

Materials

Palmitic acid and bovine insulin (Zn salt, 24 U/mg) were purchased from Sigma Chemicals, St. Louis, MO. The Betadine[R] solution was prepared by Purdue Fredrick, Toronto, Ontario. The Glucometer[R] and Dextrostix[R] are products of Miles Laboratories, Etobicoke, Ontario. The mechanical shaker used for mixing the pellet components was the Wig-L-Bug Amalgamator model 3110B made by Crescent Dental, Lyons, IL. The 13-mm Specac Pellet Die was obtained from Analytical Accessories, Dorval, Quebec. The model C hydraulic Laboratory Press was purchased from Fred S. Carver, Menomonee Falls, WI. Young (8 to 10 wk old) prepubertal female Wistar rats (body weight: 235-270 g) were bred and supplied by Jackson Laboratories, Bar Harbor, ME. Due to hypophysectomy, the prepubertal female hypox rats were smaller and weighed about 120 to 140 g.

Preparation of Pellet Discs

Appropriate amounts of palmitic acid and insulin were weighed separately and then transferred into an 1.5 mL round bottomed polyethylene vial with a screw cap. A clean stainless steel bearing ball was put atop the powder, and the contents of the capped vial were shaken vigorously for 15 s on the Wig-L-Bug shaker. After the bearing ball was removed, the well-mixed powder was transferred to the well of the pellet die which was then placed at the center of the platform of the hydraulic press. A moderate initial compression was applied to seal the die chamber, which was then evacuated for 2.5 min, before a final compression of 5 metric tons was applied for another 3 min. After the compression and vacuum were released, the die was detached from its base, and the resulting pellet was pushed out of the well by tapping the protruding plunger stem of the die. A standard size pellet disc thus made weighed about 200 mg, was 13 mm in diameter, and about 1.5 mm thick.

Implantation of Pellet Disc Pieces

Under ether anesthesia, the hair on the abdominal skin of a Wistar rat was closely shaven, and the skin was swabbed with the Betadine[R] solution. A midline incision of 7 mm in length was made about 2 cm below the sternum with a pair of small scissors. A subcutaneous pocket at least 3 cm from the cut was made towards the flank of the animal by blunt dissection. An 1/8-size piece of the pellet disc (about 25 mg) was inserted towards the end of the pocket, and the skin wound closed with a small

Michel clip. The whole procedure required only about 3 min to complete, and the animal recovered from the anesthesia within 1 min.

Blood Glucose Content

By needle puncture of the tail vein, a drop of blood was smeared directly onto the tip of the Dextrostix^{R} reagent strip. After 1 min., the tip was washed with water, and the blue color developed was read on the reflectance photometer which gave the result in mM glucose/L of blood .

Growth Determination

Each individually caged animal was given sufficient supply of a measured amount of food, which was freely available with water. Their body weight was recorded daily at a set time between 3 and 3:30 P.M. The temperature of the room was maintained at 21°C throughout, but the lighting was on a 12-hr on and 12-hr off cycle daily. In addition, the tail length of the rats was measured weekly by restraining its torso in a plastic rectangular container with an opening at one end. The tail was pulled gently and extended to its full length through the opening. The measurement was taken to the nearest mm over a graph paper.

Food Consumption

Standard rat chow in the amount of 50 g was given to each animal daily. On the following day, any food which was not eaten, including food fragments that had fallen and were collected in a plastic tray beneath the cage, were weighed and subtracted from the initial amount of the food given the previous day. This amount was recorded as the total food consumption for one day.

RESULTS

Maintenance of Hypoglycemia

In less than 1 hr after the insertion of the small insulin implant, tests showed that all the 10 Wistar rats in the study group showed an hypoglycemia blood glucose level of 1.8 to 4.1 mM/L. The 4 animals with blood glucose near the lower limit soon after implantation were given a 10% sugar supplement dissolved in the drinking water for a few days. Usually, on the third day, their blood glucose level rose slightly to over 3.0 mM/L and the sugar supplement was no longer necessary. Thereafter the blood glucose was monitored daily until the end of the study period on day 14. The average daily blood glucose values for 8 rats and the standard errors of the mean were calculated. The results obtained demonstrated that it was possible to maintain a relatively steady low blood glucose level in the normal rats with the implant.

In an exploratory run, 5 hypophysectomized (hypox) rats which would not grow after the neurosurgery were also used. An 1/8-size piece of a 10% insulin pellet was implanted subcutaneously into each rat in order to observe whether growth would resume. The hypox rats are also endogenously hypoglycemic which made this study more difficult, since convulsions often occurred due to the additional insulin released from the implant. In addition, the health status of hypox rats is usually poor, which is caused by the deprivation of other hormones when a section of the pituitary gland was removed. Therefore, morbidity and mortality among these rats are frequent and only 2 animals survived. Before implantation, they had a blood glucose level of 3.5 ± 1.0 mM/L, which would be lowered to 2.0 ± 0.2 mM/L with the implant.

In another group of 10 normal Wistar rats, each was given an 1 mL injection of 12 U insulin suspended in saline before 10 A.M. daily, including weekends. Blood samples taken after 4 hr showed that the injected insulin exerted its action quickly, and the hypoglycemia in the range of 1.8 to 4.1 mM glucose/L blood persisted for 4 to 8 hr. The next morning, however, when the blood glucose level was tested before injection, the normoglycemic state was always found to have been restored. Therefore, this group was hypoglycemic for only part of the 24-hr day.

The first control group of 10 normal rats was given a saline injection daily, and the other 10 were given a placebo implant made of palmitic acid without any insulin. Their blood glucose in the 2-wk period displayed a remarkable consistency of 5.30 ± 0.48 mM/L for saline injected controls, and 5.60 ± 0.82 mM/L for placebo implanted controls.

Food Intake

In a previous study,[3] the food intake of 5 normal Wistar rats was followed on the hourly basis during an 8-hr period. The total overnight consumption in a period of 16-hr duration was also recorded. The hourly meals were found to be remarkably consistent at 2.3 ± 0.7 g during the day. The overnight feeding had a larger range of fluctuation at 21.5 ± 4.1 g. The consistency of the food intake observed in this group, which was not included in the present project, helped to locate the occasional morbidity of the confined animals in the larger groups of the present project when respiratory, gastrointestinal or other unknown ailments might develop. The ailments often affect their appetite and invariably the body weight which must be taken into consideration in the interpretation of the overall results.

In the control group of 10 rats given saline injection and the other 10 implanted with the placebos, only 1 animal experienced a weight loss for 3 consecutive days and was excluded. There were also 2 rats in the implant test group that developed diarrhea for 1 to 5 days. Another 2 test animals out of the 10 in the daily insulin injected group became sick and lethargic in part of the 14-day period. For calculating the mean, only 8 healthy animals of the implant group were available. This restriction compelled the random selection of an equal number of healthy animals in each of the other groups. The mean food consumption for the 14-day period and the deviations indicated that the control group ate less feed throughout the whole period of the study. Some overlaps were apparent between the other 2 test groups, but the food intake was about 12.7% higher overall for the implant group, and the insulin injected group showed an intermediate level of consumption of 5.1% higher than its own controls. The deviations from the mean are comparable for the 2 groups and their controls.

Tail Length

Stimulated growth of an animal as observed in this study may be attributed to the increase in the mass of soft tissue as well as bone structure. Classical assays for the linear bone structure growth involve the measurement of the increase of thickness of the proximal epiphyseal cartilage of the tibia,[4] which requires the sacrifice of the test animal. Subsequently, an alternative method, which can provide an indication of continuing growth, employs the measurement of tail length.[5] When the controls and the 2 test groups were measured twice weekly, the results indicate that there was little, if any, increase in their tail lengths. It was possible that the 14-day period was too short. After the study, 3 animals in each group were randomly selected and kept until they reached the adult body weight of 350-400 g which took about 8 wk. The tail length was then measured once again, and the linear rate of growth was found to be only 0.37 ± 0.29 cm to 1.5 ± 0.78 cm for the 8-wk period. This magnitude of change did not seem to be a sufficiently sensitive growth index for a short term study.

Body Weight Gain

Exponential growth of a young animal is composed of increases in soft and osseous tissue mass. Therefore, measurement of weight gain is an indirect indication which may be affected by food intake, waste discharge, morbidity, and perhaps, the psychological well-being of the confined animal. For these reasons, growth as measured by body weight gain was assessed through the weighing of all animals in this study within 30 min at the set time of 3 - 3:30 P.M. Unnecessary and excessive handlings were avoided, in addition to noise control and adherence to the lighting schedule in the animal quarters.

Although 10 animals were used at the beginning of the study, events as stated previously led to the exclusion of some. A total of 8 animals in each of the 4 groups were finally taken into consideration for comparison of growth enhancement by the exogenous insulin. The data obtained show that the implant group gained body weight quickly even in the first week of the study, and by the end of the 14-day period, the apparent increase was almost double its control group. A doubling in weight was also noted for the injection group at the end of the study. Since this group was hypoglycemic for part of the 24-hr day, it is quite unexpected that the effect of the anabolic polypeptide hormone was enough to sustain the body weight gain like the implant group which was exposed to a lower dose hourly for the entire 24-hr period. However, it is noted that the deviations from the mean for both the insulin implant and insulin injected groups are slightly higher than the controls. Statistical analysis indicated that the enhanced growth observed for the implant group is significantly higher ($P > 0.1$).

As already noted by others,[13] the surviving hypox rats with the insulin implant were found to gain body weight steadily, which confirmed the anabolic effect of insulin. However, the weight gain without the endogenous GH may be due mostly to the overgrowth of fatty tissue.

Health Status

The young female Wistar rats used in this study were randomly selected from litters born in the same colony at the breeder. Upon arrival, boarding was arranged at the same quarters to avoid environmental variations, and after 1 wk of observation, they were divided into the 4 groups of 10 animals, while their body weights were matched as closely as possible between corresponding members of each group. As a result of the preconditioning and matching, there were no visible signs of ailment noted, and before the experiment, the average group body weight of each group was 235.6 g for insulin implanted, 238.3 g for placebo implanted, 238.3 g for insulin injected and 266.3 g for saline injected.

Since the minor surgery for implantation requires ether anesthesia, which may result in some degree of physical stress, the control group of 10 animals at the start was subjected to the same surgery as the insulin implanted group by implanting placebo pellets to include the additional burden. Indeed, the data obtained shows that the control group as a whole actually experienced a slight weight loss for the first 4 days of the study. Since the 8 animals finally taken for the statistical analysis from each group were all healthy, and if the physical stress affected the 2 test groups as well, then the apparent weight gain observed in the test groups may have been the difference after slight amount of similar weight loss occurred due to surgery or injections. The ailment, if any, of one animal in the control group which lost a total of 32 g in 3 days was not diagnosed. The 2 animals in the implant group and the 2 in the injection group that showed weight loss were due to hypoglycemia-caused diarrhea which was an occasional consequence known to occur in such conditions. Otherwise, the general health of all the remaining animals in the 4 groups was excellent, and it is therefore reasonable to attribute the enhanced growth observed to the exogenous insulin administered.

DISCUSSION

Improvements in the product output from farm animals have been achieved steadily over the years through better health care methods, antibiotics, breeding, nutrition, and economic management. New approaches are being developed through advances in biotechnology. The most promising outcome is the availability of the otherwise very scarce GH. Although the rDNA-produced somatotropin contains an extra methionine unit, several studies[1,6] have not shown any ill-effects resulting from the daily injections over several months. In pigs, these studies have demonstrated that the enhanced growth is accompanied by further improvement in the red meat quality and reduction in fat content.[7] When given to lactating cows, the biosynthetic GH also has the effect of increasing the milk production by about 27% with only a slight increase in feed consumption.[1,6] However, other studies in rats have shown that excess polypeptide hormone may result in abnormal hepatic development.[5] As well, undesirable symptoms have been observed clinically in growth enhancement induced by somatotropin.[2]

Insulin is also an anabolic hormone and its highly potent action on glycemia has been thoroughly studied for many years in the treatment of diabetes mellitus. Further, insulin has only 31 amino acids with rather limited variations in its structure among different species. Recent studies[8,9] in this laboratory have shown that this hormone can be readily formulated into a sustained delivery preparation by compression of an admixure with palmitic acid for subcutaneous implantation. The sustained action reduces hyperglycemia in diabetic Wistar rats[3,8,9] and rabbits[3] for as long as 4 months.

Sustained delivery implants have not yet been reported for the sparingly soluble somatotropin which also requires special effort in preparing the injectable formulation.[10] Since the simple implant for sustained delivery of insulin is available, this study was undertaken to compare the difference in growth enhancement resulting from 1 insertion of an implant and daily injection.

Before the start of this study, special care was taken in planning of the experiments, because the method used to measure the enhanced growth rate was by means of the body weight gain of the young normal female rats. It is surprising to find that the daily injections which made the rats hypoglycemic for only part of the day can result in the same extent of growth increase as the implant. However, it should be noted that the implant functions for 24 hr each day at a lower hourly dose. In contrast during the 4-8-hr period, the anabolic effect was maintained by 12 U of insulin from a single injection. Also perhaps, even if hypoglycemia was continuously maintained by an implant, by repeated injections or by pump infusion,[11] there must be a maximum limit to its anabolic effect, and therefore the growth enhancement would not have been substantially different from what has been observed between the 2 test groups. But the cost factor for repeated injections or pump infusion would have been much higher. As well, a difference of almost 20 g at the end of the 14-day period between the test groups in this study and the controls is higher than the results obtained in another study[12] where somatotropin was given to Wistar rats by daily injections. Therefore, the sustained delivery of insulin can, in fact, exceed the effect of GH given by injection. It is not known, however, if the higher weight gain observed in the present study represents similar structural growth induced by the exogenous somatotropin.

The model study, as presented herein, clearly demonstrates that insulin can be safely used to enhance growth without causing chronic ill effects. The high purity, low cost, and high potency of insulin in comparison to the other GH, as well as the promising results observed in this study, suggest the potential use of this polypeptide hormone as an alternative to improving the suboptimal growth in farm animals raised for the purpose eventually to become consumer products.

ACKNOWLEDGMENT

I thank the Medical Research Council of Canada for support. Additional technical assistance was provided by D. Shelton, and M. Lee.

REFERENCES

1. P. J. Eppard, D. E. Bauman & S. N. McCutcheon, J. Dairy Sci., **68**, 1109 (1985).
2. Product Information Booklet on Humatrope - Somatotropin (rDNA origin) for Injection, Eli Lilly and Co., Indianapolis, IN, 1987.
3. P. Y. Wang, to be published.
4. F. S. Greenspan, C. H. Li, M. E. Simpson & H. M. Evans, Endocrinology, **45**, 455 (1949).
5. M. D. Groesbeck, A. F. Parlow & W. H. Daughaday, Endocrinology, **120**, 1963 (1987).
6. T. J. Fronk, C. J. Peel, D. E. Bauman & R. C. Gorewit, J. Animal Sci., **57**, 699 (1983).
7. M. N. Sillence private communications.
8. P. Y. Wang, Diabetes, **36**, 1068 (1987).
9. P. Y. Wang, ASATO Transactions, **33**, 319 (1987).
10. M. J. Hageman, Abstract of Paper, 194th ACS Meeting, MBTD 12, (1987).

11. K. Prestele, M. Franetzki & H. Kresse, Diabetes Care, 3, 362 (1980).
12. M. N. Sillence & P. Y. Wang, to be published.
13. J. M. Salter, I. W. F. Davidson & C. H. Best, Can. J. Biochem. Physiol., **35**, 913 (1957).

CONTRIBUTORS

David E. Albert 115
Anatrace, Inc.
1280 Dussel Dr.
Maumee, OH 43537

A. J. Aleyamma 191
Biosurface Technology Division
Sree Chita Tirunal Institute for Medical Sciences and Technology
Biomedical Technology Wing
Poojapura, Trivandrum-695012, India

Brent A. Burdick 15
Meiogenics
9160 Red Branch Road
Columbia, MD 21045

Charles E. Carraher, Jr. 71
Florida Atlantic University
Department of Chemistry
Boca Raton, FL 33431

Gustavo Cei 39
Department of Polymer Science
University of Southern Mississippi
Hattiesburg, Mississippi 39406-0076

Charles G. Gebelein 269
Dept. of Chemistry
Youngstown State University
Youngstown, OH 44555.

Norman Herron 81
E. I. duPont de Nemours and Company
Central Research and Development Department
P. O. Box 80328
Wilmington, DE 19880-0328

Donald Hilvert 95
Department of Molecular Biology
Research Institute of Scripps Clinic
10666 North Torrey Pines Road
La Jolla, CA 92037

Yukio Imanishi 203
Department of Polymer Chemistry
Kyoto University
Yoshida Honmachi, Sakyo-ku
Kyoto 606, Japan

Yoshiaki Inaki 253
Faculty of Engineering
Osaka University
Suita, Osaka 565, Japan

Melvin H. Keyes 115
Anatrace, Inc.
1280 Dussel Dr.
Maumee, OH 43537

Shunsaku Kimura 203
Department of Polymer Chemistry
Kyoto University
Yoshida Honmachi, Sakyo-ku
Kyoto 606, Japan

Pawel Klosinski 203, 223
Center of Molecular and Macromolecular Studies
Polish Academy of Sciences
90-363 Lodz, Sienkiewicza 112, Poland

Robert J. Linhardt 135
Division of Medicinal and Natural Products Chemistry
College of Pharmacy
University of Iowa
Iowa City, Iowa 52242 USA

Duraikkannu Loganathan 135
Division of Medicinal and Natural Products Chemistry
College of Pharmacy
University of Iowa
Iowa City, Iowa 52242 USA

Isabel Lopez 71
Wright State University
Department of Chemistry
Dayton, OH 45435

Lon J. Mathias 39
Department of Polymer Science
University of Southern Mississippi
Hattiesburg, Mississippi 39406-0076

Eiko Mochizuki 253
Faculty of Engineering
Osaka University
Suita, Osaka 565, Japan

Stanislaw Penczek 203, 223
Center of Molecular and Macromolecular Studies
Polish Academy of Sciences
90-363 Lodz, Sienkiewicza 112, Poland

James R. Schaeffer 15
Life Sciences Research Laboratories
Eastman Kodak Company
Rochester, New York 14650

C. P. Sharma 191
Biosurface Technology Division
Sree Chita Tirunal Institute for Medical Sciences and Technology
Biomedical Technology Wing
Poojapura, Trivandrum-695012, India

Kiichi Takemoto 253
Faculty of Engineering
Osaka University
Suita, Osaka 565, Japan

Ching-Leou C. Teng 175
College of Pharmacy
The University of Michigan
Ann Arbor, Michigan 48109-1065

Louis G. Tissinger 71
Florida Atlantic University
Department of Chemistry
Boca Raton, FL 33431

Rajeev A. Vaidya 39
Department of Polymer Science
University of Southern Mississippi
Hattiesburg, Mississippi 39406-0076

Takehiko Wada 253
Faculty of Engineering
Osaka University
Suita, Osaka 565, Japan

Paul W. Wang 277
Laboratory of Chemical Biology
Institute of Biomedical Engineering
Faculty of Medicine
University of Toronto
Toronto, Ontario, Canada, M5S 1A8

Melanie Williams 71
Florida Atlantic University
Department of Chemistry
Boca Raton, FL 33431

G. Wulff 1
Institute of Organic Chemistry & Macromolecular Chemistry
University of Düsseldorf, Universitätsstr. 1
D-400 Düsseldorf, F.R.G.

Victor C. Yang 175
College of Pharmacy
The University of Michigan
Ann Arbor, Michigan 48109-1065

INDEX